Hagios Charalambos

A Minoan Burial Cave in Crete

II. The Pottery

Hagios Charalambos

A Minoan Burial Cave in Crete

II. The Pottery

by

Louise C. Langford-Verstegen

with contributions by

Philip P. Betancourt, Costis Davaras, Susan C. Ferrence, and Eleni Nodarou

edited by

Philip P. Betancourt, Costis Davaras, and Eleni Stravopodi

Published by
INSTAP Academic Press
Philadelphia, Pennsylvania
2015

Design and Production
INSTAP Academic Press, Philadelphia, PA

Hagios Charalambos
A Minoan Burial Cave in Crete

I. Excavation and Portable Objects (2014), Philip P. Betancourt, ISBN 978-1-931534-80-2 (print), ISBN 978-1-623033-93-4 (ebook)
II. The Pottery (2015), Louise C. Langford-Verstegen, ISBN 978-1-931534-83-3 (print), ISBN 978-1-623034-02-3 (ebook)

Library of Congress Cataloging-in-Publication Data

Betancourt, Philip P., 1936–
  The Hagios Charalambos cave I : the excavation and the portable artifacts / by Philip P. Betancourt ; with contributions by Costis Davaras, Heidi M.C. Dierckx, Susan C. Ferrence, Panagiotis Karkanas, Tanya J. McCullough, James D. Muhly, Natalia Poulou-Papadimitriou, David S. Reese, Antonia Stamos, Eleni Stravopodi, Maria Tsiboukaki, Louise L. Langford-Verstegen, and Gayla M. Weng ; edited by Philip P. Betancourt, Costis Davaras, and Eleni Stravopodi.
  pages cm. — (Prehistory monographs ; 47)
  Includes bibliographical references and index.
  ISBN 978-1-931534-80-2
  1. Crete (Greece)—Antiquities. 2. Hagios Charalambos (Greece)—Antiquities. 3. Caves—Greece—Hagios Charalambos. 4. Ossuaries—Greece—Hagios Charalambos. 5. Human remains (Archaeology)—Greece—Hagios Charalambos. 6. Grave goods—Greece—Hagios Charalambos. 7. Excavations (Archaeology)—Greece—Hagios Charalambos. 8. Minoans—Greece—Hagios Charalambos—Antiquities. 9. Bronze age—Greece—Hagios Charalambos. I. Davaras, Costis. II. Dierckx, Heidi. III. Title.
  DF221.C8B5635 2014
  939'.18—dc23
  2014022231

This book is dedicated to my mother
Maris Clymer Langford

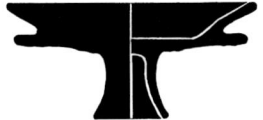

Table of Contents

List of Tables in the Text

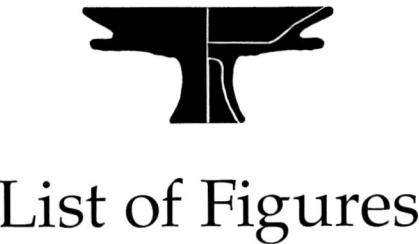

List of Figures

List of Plates

voids with remnants of burned organic material. Fabric Group 5: red fabric with quartz and clay pellets (e). Fabric Group 6: red fabric with quartz (f).

Plate 6.   Fabric Group 7 (Mirabello Fabric): granodiorite fabric Group 7A (a), Group 7B (b). Fabric Group 8: very fine fabric with rare nonplastics (c). Fabric 9: fine calcareous with pellets (d). Fabric 10: fine calcareous with biotite mica (e). Fabric 11: fine fabric with rounded sandstones (f).

Plate 7.   Fabric 12: very fine fabric with pellets (a); note layer of slip on the upper part of field. Fabric 13: very fine fabric with serpentinite (b).

Overview

The study of finds from the cave at Hagios Charalambos in the Lasithi Plain provides an illustration of the practice of secondary burial in Early and Middle Bronze Age Crete. The cavern adds to our understanding of Early and Middle Minoan Lasithi and illuminates more clearly the function of the cave at Trapeza, which has close parallels for most classes of objects found at Hagios Charalambos. A total of 18,065 sherds and vases come from the excavations, and they were all studied either statistically or in detail.

A majority of the pottery from the site is made locally, but a selection of imports ranges in date from EM I or earlier to MM IIB. Imports can be recognized from the Gulf of Mirabello area, the Mesara, Malia, the Pediada, and Knossos. The pottery shows a shift in the use of imports during the site's history. Objects from the Mesara decline after MM I, and objects from Malia increase in MM IIB. This change may reflect a shift in economic and/or political dominance and influence in Lasithi.

Typical of pottery associated with burials, the types of vessels were mostly used for the pouring and drinking of liquids. In addition, the assemblage has small-sized vessels that most likely contained precious oils or liquids and several other classes including several small jars and spouted containers, a few jars in large sizes, and a fine cylindrical pyxis. The assemblage also has offering tables made from the local clay fabric of the Lasithi region. The tables were designed to be carried by a short stem, and they could have held a liquid or solid offering.

The local clay sources used to make the distinctive fabric of the plain's pottery are represented in the majority of the ceramics found at the cave. Petrographic analysis provides a thorough description, and the fabric is referred to as the Lasithi Red Fabric Group.

Petrographic analysis of samples of imported wares confirms the origin of manufacture for some of the vases and allows for conclusions regarding the implications of certain imports.

Overall, the pottery here shows that the people who deposited their dead in the secondary burial cave at Hagios Charalambos were in contact with ceramic production centers in East Crete, the Mesara, Knossos, the Pediada, and Malia. This range of influences speaks not only of trade relations and political spheres of influence but also of tastes in pottery production and consumption.

Acknowledgments

Many people offered invaluable help and encouragement toward the completion of this study. Two seasons of excavation were conducted in July and August 2002 and in July 2003 by a team of archaeologists from the Tyler School of Art of Temple University with a permit issued by the Greek Ministry of Culture, under the auspices of the American School of Classical Studies at Athens. The excavation was directed jointly by Philip P. Betancourt and Costis Davaras in 2002, with Eleni Stravopodi joining the team as co-director in 2003. The Institute for Aegean Prehistory (INSTAP), Temple University, the University of Pennsylvania, and private donors funded the excavation. Additionally, several study seasons were conducted at the INSTAP Study Center for East Crete (SCEC). Many thanks go to Philip Betancourt and Costis Davaras for allowing me to participate as trench supervisor during the two seasons of excavation and also for allowing me to study the pottery.

I am grateful to the various libraries and institutions that have aided my research: in Greece, the INSTAP-SCEC; in Rome, the German Archaeological Institute; and in Florence, the Berenson Library at Villa I Tatti. I was greatly aided by the administration at Temple University in Philadelphia. My research was generously supported by Temple University travel grants.

A most crucial debt goes to those conservators and scientists who shared unpublished results of their work and all generously provided analyses and discussed their significance. These individuals include Sylvie Müller-Celka, Yiannis Papadatos, Kostas Christakis, Giorgos Rethemiotakis, David Wilson, Nicoletta Momigliano, Donald Haggis, and Evi Sikla. Alekos Nikakis contributed, most importantly, as excavation specialist during excavation seasons and in the conservation of objects now on display in the Hagios Nikolaos Museum, Crete. Stephania Chlouveraki, Chief Conservator at the INSTAP-SCEC, contributed many hours of expertise. Eleni Nodarou, Chief Petrographer at the INSTAP-SCEC, contributed invaluable scientific expertise and interpretation. Thanks go to Thomas M. Brogan, Director of the INSTAP-SCEC, and Chronis Papanikolopoulos, Chief Photographer at the INSTAP-SCEC. Special thanks goes to Registrar Mary Betancourt for cataloging pottery, and to my colleagues, Gayla Weng, Susan Ferrence, and Tanya McCullough, for their assistance in pottery study.

Finally, to my dissertation committee members, Philip Betancourt, Jane Evans, and George Myer, I owe a great debt of thanks. I am especially indebted to Philip Betancourt who welcomed me to the field of Aegean Prehistory and has guided my research ever since. For the countless doors he opened, and the cordial environment he created at Temple University, I cannot thank him enough.

List of Abbreviations

B	base	L	leg
C	complete	LM	Late Minoan
cm	centimeter(s)	LN	Late Neolithic
d.	diameter	m	meter(s)
EM	Early Minoan	max. dim.	maximum dimension(s)
FN	Final Neolithic	mm	millimeter(s)
H	handle	μm	micrometers
h.	height	MM	Middle Minoan
HCH	Hagios Charalambos excavation number	MNI	minimum number of individuals
		MNV	minimum number of vases
HNM	Hagios Nikolaos Archaeological Museum number	NC	nearly complete
		pres.	preserved
IGME	Institute of Geology and Mineral Exploration, Athens, Greece	R	rim
		rest.	restored
		S	spout
INSTAP-SCEC	INSTAP Study Center for East Crete, Pacheia Ammos, Crete, Greece		

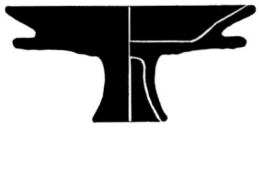

1

Minoan Caves:
Points to Consider Regarding Function

The caves of Crete are the subject of major works by Faure (1964) and Tyree (1974). Adding to these studies, closer analytical treatments have focused on common elements among the caves to gain insight into the functions of the cave sites with respect to Minoan religion (Rutkowski 1986; Marinatos 1993; Nowicki 1996). The beginning of the Middle Bronze Age marks a high point in the popularity of natural sanctuaries, with religious practices taking place at both peak sanctuaries and caves. Peak sanctuaries have been regarded as a correlate of the greater consolidated powers of the first palaces (Peatfield 1987, 89). Thus, the associated practices are of a socially complex nature. In his article dealing with the character of the peak sanctuary and palace relationship, however, Peatfield (1987, 90) argued that the cult practices at peak sanctuaries originated in Early Minoan (EM) religious practices conducted at cemeteries:

> Though a domestic shrine was discovered at Phournou Koriphi, [...] the main cult places appear to have been the tombs. The Mesara tholoi, with their complexes of rooms, outside paved areas, finds indicative of shrines [...], and the continuity of their offerings, provide the clearest evidence of extended ritual use beyond immediate funerary rites [...]. Scholars [...] have suggested that these tombs were used by kinship groups. Within any culture, funerary rituals perform an important function in reaffirming group solidarity when faced with the crisis of death.

This situation supports that all the pieces were in place for a highly developed culture of cave burial and ritual practice.

Unlike peak sanctuaries, caves like Hagios Charalambos have been less extensively studied than other types of tombs. They are fewer in number, and often they have been victims of looting. Consequently, they are harder to define as sacred spaces. While caves used for burial are sometimes used later for sacred or cult practices, the implicit model is that these two functions are unrelated and chronologically distinct (Rutkowski 1986, 65). Though we do not see a distinct shift in function from secondary burial to a site for cult practice, it is possible that some kind of intangible ritual took place upon the occasion of the secondary deposit. I propose that we, at least, consider the act of making a secondary deposit a ritual act in itself. This chapter, therefore, will illustrate the cave's function as a burial deposit, nominate the existence of *ritual* (as conceived more narrowly than cult activity), and consider parallels to other artifact assemblages. Lastly, it will review the implications of this evidence in light of the other caves used for various purposes in the Lasithi Plain.

Defining and Distinguishing Ritual and Cult

According to the standard definitions, Hagios Charalambos does not qualify as a sacred cave (Tyree 1974; Rutkowski 1986; Nowicki 1996). Tyree distinguished sacred caves by the presence of architectural elements including paved areas, partition walls, and low walls surrounding stalagmites. In addition to architecture, her other criterion is the presence of cult implements. These implements include models of weapons, figurines, votive double axes, lamps, ritual vessels, and small pots. Larger cult implements include large votive double axes, storage vessels, stone offering tables, manmade altars, and horns of consecration (Tyree 1974, 6; for the vessels for feasting, see Tyree, Kanta, and Robinson 2008). Rutkowski and Nowicki add to Tyree's corpus but merely implicitly continue the use of her criteria for their assignment of particular caves to the categories that they address (certainly, probably, and possibly used for cult purposes).

The objects excavated from the cave that were once associated with the primary burials are quite typical of the objects found in other burial contexts from this time period. For example, there are parallels to objects found in tholoi in the Mesara such as anthropomorphic and zoomorphic figurines (Betancourt and Ferrence 2014a) and seals (Betancourt and Ferrence 2014b; Muhly 2014c). Also, the assemblage has models of items, such as weapons (Muhly 2014a, 55) and jewelry (Muhly 2014b). Despite the inability to distinguish items that were exclusively deposited at the time of secondary interment in the ossuary, we should not be mistaken in thinking that there were no activities taking place there with religious significance. A definition can be given for religious ritual that affirms the existence of practices for the care of the dead that can remain analytically separate from our traditional understanding of cult.

In order to discuss ritual as it relates to the cave and secondary deposit of burials at Hagios Charalambos, we must establish the bases for recognizing and defining ritual practice. However, first, it is useful to reflect on the physical context of burial in which these rituals would arise. Undoubtedly the tombs of Prepalatial Crete differ in size and shape according to region, but most are monumental in character. One exception is the cemetery at Pseira (Betancourt 2003; Betancourt and Davaras, eds., 2003). The graves there are of several different styles (Betancourt 2003, 3). Though not monumental, they are situated in an area that is removed from the town but within sight of it (Betancourt and Davaras, eds., 2003). There seems to be a concern with visibility. This is true also of monumental structures, such as the tholoi of the Mesara, which were constructed with a regard to visibility and permanence (Xanthoudides 1924; Branigan 1970b). Their large size, though not necessarily their shape, is explained by the fact that they were designed for communal burial. Nevertheless, they are imposing structures designed specifically to house the remains of the dead, and this fact reflects a unique consideration for the dead. The importance of the deceased and the importance of honoring the deceased are evidenced also in the richness of the objects that were associated with the remains. Therefore, the belief system that was manifested in the communal burials with their associated rich objects may have resonated further when the secondary disposal was made.

There are myriad definitions of both ritual and cult and many studies that seek to explain the function of each (Durkheim 1912; Goody 1961; Goodman 1988; Lawson and McCauley 1993; Leach 2002). Social scientists have tried to categorize religious rituals in a number of ways. A successful and popular attempt was made by French ethnographer and folklorist Arnold van Gennep (1961). Quite simply, he proposed that the multitude of worldwide rituals can be grouped into three types: those of separation, of transition, and of incorporation. He pointed out that rituals accompany people throughout their lives and mark situations such as birth, puberty, marriage, and death. Ritual facilitates the passage from one social condition to the next. Discussions of the functions of ritual—such as the meaning of ritual, ritual as social communication, ritual as power, or ritual as belief—are beyond the scope of this study. Rather, it is important to consider that the secondary deposits that took place at Hagios Charalambos were, in van Gennep's terms, a ritual action marking the condition of death (1961, 165). Though ritual and cult practice are not mutually exclusive, in the case of the Hagios Charalambos cave there may have been only ritual.

In the Lasithi Plain, while the cave site of Hagios Charalambos served a ritual function, the neighboring Psychro Cave (Hogarth 1899–1900) served a cultic/religious function. The Psychro Cave is at an altitude of 1,025 m and is situated above the modern village of Psychro on the southern edge of the plain (Figs. 1, 2). An extensive terrace is located east of the cave opening, and the cave consists of two parts, a large upper vestibule and a large lower chamber. In the northwestern area of the upper chamber, four layers were reported by Hogarth. Finds of animal bones and pottery indicate that this area was used for habitation in the Late Neolithic and Early Minoan I periods. Psychro was used for burials in the Early Minoan II and III periods, and it has been closely associated with the legendary Dictaean Cave. Rutkowski and Nowicki mention that the burial use at Psychro was similar to that at Trapeza (Rutkowski 1986; Nowicki 1996). They add that, "The nearest settlement to the Dictaean Cave has been identified as Katsoucheiroi, and it was perhaps people from there that had designated part of the Upper Grotto for their cemetery, as was the case of Trapeza and settlement on Kastello near Tzermiado" (Rutkowski and Nowicki 1996, 18). Here, a reassessment shows that the burials at Trapeza are not similar to Psychro, but that they reflect an altogether different practice of secondary burial, much like that at Hagios Charalambos. At Psychro, the cave was used for a different class of function as indicated by the abundance of cult objects and equipment including bronze figurines, double axes, gems, blades, razors, tweezers, and offering receptacles. The continuity of cult usage is attested by finds from the Geometric, Roman, and Byzantine periods.

The difficult task is inferring ritual or cult activity from archaeological finds. In a large study dealing with Minoan religion, Marinatos (1993, 14) states that the Prepalatial communal tombs provide us with evidence for the latter. This has bearing on Hagios Charalambos as the cave was an ossuary for the remains interred initially in communal burials. Unfortunately, the original burial site or sites are not known. Perhaps there are undiscovered tholoi in the Lasithi area that housed the primary burials. Marinatos corroborates the ideas of Tyree (1974) and Rutkowski (1986). In recognizing the involvement of a respective community with regard to burial, she states (Marinatos 1993, 14):

> The large size of the communal tombs indicates that they were deliberately constructed for conspicuous display. In addition, many had paved areas for gatherings and altars. Cult implements and offerings were found *outside* the tombs, thus suggesting periodic visitation to the cemeteries *by sizable groups of people*. Evidence of *collective* secondary burial implies treatment of the dead. It suggests the involvement of not one but several families.

If we are to regard the Hagios Charalambos Cave as a collective deposit from communal primary burials suggesting the involvement of several families, then according to Marinatos' model, Hagios Charalambos, being a communal deposit regardless of the lack of cult equipment, is a reflection of ritual practice. It is the result of the efforts of a community to redispose of their dead. It is a group effort, that group being larger than the immediate families as the shear bulk of material suggests. If we are to accept Marinatos' model, then the mere existence of the cave as a place of secondary deposit is evidence for ritual and cult activity. Again, a closer examination of the nuances that distinguish ritual and cult is a larger problem not to be undertaken here.

Discussion

The closest parallel for the Hagios Charalambos Cave is the Trapeza Cave (Pendlebury, Pendlebury, and Money-Coutts 1935–1936), which lies at an altitude of 850 m on the northeastern outskirts of the village of Tzermiado on the eastern side of the Lasithi Plain (Fig. 2). The cave is a single corridor with extensions containing many stalagmites and stalactites. The overall stratigraphy of the cave was mixed and deposited over an intact area that had traces of Late Neolithic occupation. Human bones and objects including pottery, figurines, weapons, jewelry, and sealstones were found. At both Trapeza and Hagios Charalambos, the skeletons and broken artifacts were mixed and scattered among the different rooms or areas of the respective caves. The finds of the Trapeza Cave provide the closest parallels to those excavated at Hagios Charalambos. The objects themselves are paralleled as well as the classes of objects found together in both contexts.

In light of the excavations at Hagios Charalambos, it is suggested that the cave at Trapeza was used as a secondary deposit much in the same way that the cave at Hagios Charalambos was. Pendlebury's conclusion (Pendlebury, Pendlebury, and Money-Coutts 1935–1936, 23) that there were EM II and III primary burials on top of a Final Neolithic (FN) level, therefore, is probably wrong. Tyree points out that Pendlebury initially felt that Trapeza was abandoned in MM I, but he then changed his mind to say that the cave was used later (Pendlebury 1939, 126, notes MM II material). Tyree suggested that it was used for cult practice in MM I later to be replaced in function by the Psychro Cave (Tyree 1974, 11). The "MM I" finds were a Twelfth Dynasty scarab, a clay seated figurine, three bronze dagger blades, a bronze knife, and potsherds. It

should be noted that Rutkowski and Nowicki seem to be in agreement with Pendlebury in that they put the Trapeza cave within their category of "caves possibly used for cult purpose," but again, they do not explain their criteria (Rutkowski and Nowicki 1996, 68–69; see also Jones 1999, 4). As at Hagios Charalambos, the objects were from mixed burials, and it is not possible to discern subsequently deposited objects. As at Hagios Charalambos again, however, the act of secondary deposit speaks of ritual practice at Tzermiado.

The two caves seemed to have served the same purpose during the same time periods. Notably, they were in relatively close proximity to each other. Their proximity strongly suggests that those utilizing one cave must have known of the other. They were used concurrently, so this is not a case of one being filled, and then the other coming into use. Rutkowski posits that the cave at Hagios Charalambos may have served many sites that occupied the plain. He states, "The number of bodies may indicate that it was a very special site used by the majority of the plateau rather than by a small local community in the vicinity of it" (Nowicki 1996, 34). Perhaps we may want to consider the Trapeza cave as well, as servicing many settlements rather than a local community in its vicinity, but the statistics do not support this conclusion because one burial per year does not suggest more than a single settlement (see discussion in this volume). Nonetheless, we then have two sites (caves that were repositories of secondary burials) where ritual manifested and resonated with a significant population of the Lasithi region in MM IB/II, a critical period of political transition that brings with it changes in belief systems.

2

Methodology

History of Excavation and Research

Lasithi is a high plain in East-Central Crete (Figs. 1, 2). In 1976, during the construction of a segment of road that roughly follows the plain's outer perimeter, dynamite blasted open an entrance into the Hagios Charalambos Cave, a natural cavern that had once facilitated drainage of the plain (Karkanas 2014). In order to guard against destruction and looting, a rescue excavation was conducted at the cave (Davaras 1976, 1982, 1983). Costis Davaras removed much of the contents of the cave in order to store them safely. Finds were placed in the Hagios Nikolaos Museum. A study season on the pottery from Davaras' 1976 excavation was conducted during the summer of 2001 at the INSTAP Study Center for East Crete (INSTAP-SCEC).

Two seasons of excavation followed in 2002 and 2003, directed jointly by Costis Davaras, Philip P. Betancourt, and Eleni Stravopodi (in 2003). An additional study season was conducted in January of 2004 at the Hagios Nikolaos Museum. On that occasion, vessels were removed from storage and were cataloged, drawn, and photographed. During the summers of 2004 and 2005, study seasons were conducted at the INSTAP-SCEC. The author had the opportunity to take part in all of these aspects of the Hagios Charalambos project beginning in 2001, working as a survey technician, trench supervisor, and pottery specialist.

Methodology

This study of the pottery from Hagios Charalambos has several objectives that contribute different types of historical information. First, thorough descriptions of distinct pottery classes are provided. Adding to this, the pottery assemblage as a whole is discussed. The assemblage represents objects, both manufactured locally and imported, that were chosen to accompany burials. The importance of these objects, in terms of function and as symbols of external relations, is investigated. The imported pottery represents influences on the local community, and because of this situation it provides evidence for political and economic boundaries. With a more diachronic perspective, the implications of a chronological continuum represented in the assemblage is discussed. Questions regarding a perceived change in the function of the cave with regard to practices of habitation, secondary burial, and ritual use are addressed also.

The pottery from the cave is a very heterogeneous assemblage. It includes examples of vessels whose characteristics are well known, vessels whose class has not been published before, and vases whose cultural and technological affiliation is unclear because of poor preservation or other causes. Most of the pottery survives as small fragments. All of the pottery was kept and studied, and the results of the study are presented in this volume. The statistics are provided in Appendix B. Specific examples of the classes presented in Appendix B were cataloged to illustrate the various classes present in the assemblage. All complete or nearly complete vessels were cataloged, and examples of sherds were added as well to provide a more complete range of what was present.

The nomenclature follows Betancourt (2008a). The following terms are used in this volume:

Class. Any group of pottery vessels.

Fabric. The materials used for a vessel, including both the original ingredients and aspects derived from the forming, the firing, and the post-burial conditions.

Style. A class of vessel defined by the surface treatment and decoration.

Ware. A class of vessel defined both by the style and by the fabric. The fabric must be defined by ceramic petrography.

With this system, a class may exist both as a Style (with the fabric undefined) or as one or more Wares, manufactured from specific fabrics. Examples of the main classes that occur in the text are defined using this system for the nomenclature. The most important ones include the following:

Barbotine Style. A pottery with rough, raised clay added to the surface (Ch. 9) was described in detail by Foster (1982). When the class is manufactured from Cretan South Coast Fabric (as defined by several studies including Nodarou 2013, 157–158), it is called Barbotine Ware. This is an important class in the Mesara, but it has not been recognized from Hagios Charalambos. The pottery from the cave has not had its fabric defined, and it is called the Barbotine Style.

Diagonal Line Style. A class of vessels with similar diagonal line decoration whose fabric has not been defined is called the Diagonal Line Style (Ch. 8).

East Cretan White-on-Dark Ware. Two classes of White-on-Dark Ware are recognized and defined at present, East Cretan White-on-Dark Ware made of Mirabello Fabric (Betancourt 1984; for the fabric, see Nodarou 2013, 160, with bibliography) and Lasithi White-on-Dark Ware manufactured from the Lasithi Red Fabric Group as defined in this volume (App. A, fabric 2).

Hagios Onouphrios Style. Pottery with red linear decoration (Ch. 4) has been described many times (Branigan 1970a, 23–27; Blackman and Branigan 1982, 29–32; Betancourt 2008a, 47–53). It has a technological distinction from ceramics with dark brown or black linear decoration in that the red decoration is fired in an oxidizing atmosphere, not a reducing one. If it is made of Cretan South Coast

Fabric, it can be called Hagios Onouphrios Ware. If the fabric is different or it is undefined, it is called Hagios Onouphrios Style.

Kamares Ware. Polychrome vessels manufactured from fine fabrics or Cretan South Coast Fabric are called Kamares Ware. The polychrome pottery from the cave is called Kamares Ware (Ch. 13).

Koumasa Style. The EM pottery decorated with dark linear decoration (Ch. 4) was called Hagios Onouphrios II by Branigan (1970a, 28–29). Here it is called Koumasa Ware if it is made of Cretan South Coast Fabric and Koumasa Style if its fabric is undefined (Betancourt 2008a, 47–48).

Light-on-Dark Class. Many styles of light-on-dark pottery were manufactured in Middle Minoan (MM) Crete. (Ch. 7)

Monochrome Style. Vases whose surfaces are a single color are called the Monochrome Style.

Vasiliki Ware. Mottled pottery from EM IIB manufactured from Mirabello Fabric (Ch. 6) is called Vasiliki Ware (Betancourt 1979). It was apparently manufactured in several related workshops on the coast of the Gulf of Mirabello (not at Vasiliki).

The pottery was organized and studied using several different types of criteria. It was sorted first by date, and then by fabric and class, and finally by shape. The fabrics were sorted in a preliminary way using features (including color, texture, coarseness, and color of visible inclusions) that were visible by macroscopic examination. The division by class considered technology, ornament, shape, and style. As the final step in the organization, the pottery was sorted based on completeness (whole vases, vases that were largely restorable, and sherds), and the sherds were sorted by their position on the vase. The vases and sherds were then counted, and the resulting total picture is presented in statistical form in Appendix B.

Divisions by fabrics were confirmed and sometimes modified based on ceramic petrography. This information is presented in Appendix A. The methodology uses both the traditional stylistic examination of shapes, decoration, surface treatment, and manufacturing technology along with the more recently developed techniques of ceramic petrography. Petrographic anlaysis provides valuable information on inclusions, microscopic structure, and other elements of the clay fabric (Peterson 2009). The conclusions reflect the multivariate analysis. For example, one of the workshop groups is represented by a range of shapes that was produced locally using the local clay (Ch. 17). Decoration in white is added over the dark undercoat in imitation of the East Cretan White-on-Dark Ware from the northeastern coast of the island. The local Lasithi production uses a more limited repertoire of decorative motifs than is used in the main region of this ware, which extends along the northern coast of eastern Crete from Malia to Palaikastro. Petrographic analysis of samples conducted by Eleni Nodarou at the INSTAP-SCEC identifies the group as being local in production (App. A). Analysis on pottery of local manufacture provides a thorough description of the Lasithi fabric, and it will be valuable to future ceramic studies regarding pottery from the Lasithi region. Analysis of imported pottery at the site confirms identification as such. Petrographic analysis provides data suitable for further studies regarding all of the wares that were analyzed.

Previous Work on the Ceramics of the Lasithi Region

In addition to the work that accompanied the excavations in the Trapeza Cave in the 1930s (Pendlebury, Pendlebury, and Money-Coutts 1935–1936; Pendlebury 1936–1937), several other studies have examined the pottery from Lasithi and its vicinity. Before the excavations at Trapeza, Spyridon Marinatos had uncovered a tholos tomb with many vases at Krasi, located just north of the highland plain (Marinatos 1929; Galli 2014). More recently, Gerald Cadogan discussed the pottery from Lasithi in comparison with the ceramics from Malia and Pyrgos-Myrtos (Cadogan 1990, 1995). In addition, distinctive fabrics of a "Lasithi group" were discussed in a study regarding EM III–MM IA pottery from Knossos (Momigliano 2000, 76; Rethemiotakis and Christakis 2004; Macdonald and Knappett 2007, 39, 156; see also Momigliano 1991, 261–264; MacGillivray 1998, 88–89). Regional fabric groups of the Pediada region have been examined in a study regarding pottery from Galatas and Kastelli (Rethemiotakis and Christakis 2004, 169–175). The pottery from Hagios Charalambos fits well with the conclusions of these earlier studies, which identify a series of production centers making pottery for use in the Pediada and nearby Lasithi using the local phyllite-rich clays that fire to a red color. The present study provides much more detail on some of the local and regional pottery productions and allows them to be characterized more completely. It also

demonstrates that several different production centers were operating in the large region that supplied pottery to the communities of the Lasithi and Pediada regions, that these workshops sometimes specialized in specific classes of vessels, and that they developed and changed their output through time.

Several preliminary reports have been published by those who worked at Hagios Charalambos on the site's pottery and the conclusions that it can provide for chronology, style, and other classes of information (Davaras 1976; 1982; 1983; 1986; 1988; Betancourt 2002; 2003; 2005; 2007; 2008b, 2011a; Ferrence 2007; 2008; Betancourt and Muhly 2008; Betancourt et al. 2008; 2014b; Dierckx 2008; Hickman 2008; Karkanas 2008; Langford-Verstegen 2008; McGeorge 2008; Muhly 2008). The first volume in the final report was published in 2014 (Betancourt 2014a).

3

Coarse Fabrics, Neolithic to EM I

Incised Coarse Dark Burnished Class, Late Neolithic II

Though no intact Neolithic level was detected at Hagios Charalambos, sherds of a Neolithic open vessel with punctation marks were found. The shape is open with an everted rim. The interior and exterior are burnished. The diameter of the vessel suggests that it is a bowl, and its use may not have been restricted to drinking. Knossian parallels suggest a date in Late Neolithic (LN) II (Tomkins 2007, 31, fig. 1.8:24–26). As such, it stands apart from the predominant drinking vessels from subsequent periods in the assemblage. The coarse carbonate and rarer phyllitic fabric was analyzed petrographically. It was probably from the Pediada region.

Parallels for this shape, and much of the ceramic assemblage and the other finds, come from the Trapeza Cave (Pendlebury, Pendlebury, and Money-Coutts 1935–1936).

Carinated Bowl with Everted Rim

1 (HCH 160; Fig. 3). Carinated bowl with everted rim, body sherds. D. of rim 22 cm. A coarse fabric (reddish brown, 5YR 4/3), with carbonate temper; burnished on interior and exterior. Incised punctation marks on the exterior. *Comments*: from Rooms 1–3 (1982 season). Analyzed by ceramic petrography, sample no. HCH 04/25 (Hagios Charalambos Fabric 3). *Parallels*: Pendlebury, Pendlebury, and Money-Coutts 1935–1936, 27, fig. 6:T16 (Trapeza Cave). *Date*: LN II. *Bibl.*: Betancourt et al. 2008b, 166, fig. 28:a; 2014, 29, fig. 17:a; Langford-Verstegen 2008, 556, fig. 10:15.

Monochrome Style, Dark Burnished Fabrics

A few of the open vessels from FN–EM I are made in a fabric that is tempered with angular fragments of white calcium carbonate. One example is also tempered with chaff. The vessels have a rounded profile, and they are burnished on the interior and exterior. The shape suggests that they are drinking vessels.

Final Neolithic–EM I fabrics tempered with phyllite are more numerous, and the shapes have a range of sizes, albeit a small one. The vessels include shapes with rim diameters from 10 to 12 cm, an open vessel with straight profile and lug handle that has a 16 cm diameter, an open vessel with an open spout, and a jar with a strap handle. Petrographic analysis has been used to describe the calcite-tempered fabric and the fabric tempered with phyllite. The use of phyllitic inclusions is a distinctive characteristic of the regional pottery of Lasithi and the adjacent Pediada.

MONOCHROME STYLE, DARK BURNISHED CLASS WITH CALCITE

Open Vessel

Open vessels designed for beverages are already important in the assemblage from its early periods. Most of these earliest pieces are not well preserved, and many of the original shapes cannot be restored. This class is made from a clay recipe that often includes short stems of vegetal matter (presumably chaff) as well as fragments of white calcium carbonate, which is either calcite or marble.

The clay recipe that includes angular fragments of white calcium carbonate was used over a wide geographic region in Crete (and also in the Cyclades; for references, see Betancourt 2008a, 28–29). Significant differences in regional shapes suggest that the technology was used in many workshops.

2 (HCH 03-215; Fig. 3). Open vessel, body sherd. Max. dim. 3.9 cm. A coarse fabric (red, 2.5YR 5/6), tempered with calcite and chaff; burnished on interior and exterior. Rounded profile. Monochrome. *Comments*: from Room 5, area 5, level 2 (unit HCH 02-3-5-2). Analyzed by ceramic petrography, sample no. HCH 04/28 (Hagios Charalambos Fabric Group 1). *Date*: FN–EM I.

Closed Vessel with Handle

3 (HCH 02-120; Fig. 3). Closed vessel, probably a jar, body sherd with scar from handle. Max. dim. 5.7 cm. A coarse fabric (varies from black at the core to red, 2.5YR 5/6, at the surface), carbonate tempered; burnished on exterior. Very thick handle. Monochrome. *Comments*: from the Room 4/5 Entrance, level 4 (unit HCH 02-2/3 Ent-4). Analyzed by ceramic petrography, sample no. HCH 04/19 (Hagios Charalambos Fabric Group 3). *Date*: FN–EM I.

MONOCHROME STYLE, DARK BURNISHED CLASS WITH PHYLLITE

Open Vessels

Vessels tempered with phyllite (either added by the potters or occurring as a natural part of the clay deposits) are important in the assemblage from the cave from its earliest periods. Some of these sherds come from outside the cave, and they are part of a Final Neolithic deposit that is not associated with the cave's later funerary use. This class of fabric continues to be used in EM I, and the date of specific sherds cannot be determined on ceramic evidence alone.

4 (HCH 03-164; Fig. 3). Open vessel, body sherd. Max. dim. 2.3 cm. A coarse fabric, (red, 2.5YR 5/6), with phyllite; burnished on interior and exterior. Monochrome. *Comments*: from the cave exterior, Trench 11 (unit HCH 03-11-2). Also tempered with chaff(?) as well as phyllite. *Date*: FN–EM I.

5 (HCH 05-449; Fig. 3). Cup, complete profile. H. 7.4; d. of rim 11.5; d. of base 7.7 cm. A coarse fabric (reddish brown, 2.5YR 4/4), with phyllite; burnished on interior and exterior. Rounded profile. Monochrome. *Comments*: from Rooms 1–4 (1983 season). *Date*: FN–EM I.

6 (HCH 05-448; Fig. 3). Cup, base sherd. D. of base 5.7 cm. A coarse fabric (light reddish brown, 5YR 6/4); highly burnished on interior and exterior. Monochrome. *Comments*: from Rooms 1–4 (1983 season). *Date*: FN–EM I.

7 (HCH 03-169; Fig. 3). Open vessel, body sherd. Max. dim. 3.1 cm. A coarse fabric (red, 2.5YR 5/6), with phyllite; burnished on interior and exterior. Monochrome. *Comments*: from the cave exterior, Trench 12 (unit HCH 03-12-1). *Date*: FN–EM I.

8 (HCH 03-189; Fig. 3). Open vessel, body sherd. Max. dim. 2.9 cm. A coarse fabric (reddish brown, 2.5YR 2.5–3/4), with phyllite; burnished on exterior. Monochrome. *Comments*: from the cave exterior, Trench 12 (unit HCH 03-12-8). *Date*: FN–EM I.

9 (HCH 49; Fig. 3). Cup, complete profile. H. 3.6; d. of rim 10; d. of base 6 cm. A coarse fabric (red, 2.5YR 5/6), with phyllite; burnished on interior and exterior. Rounded profile and flat base. Monochrome. *Comments*: from Rooms 1–4 (1983 season). *Date*: FN–EM I.

10 (HCH 23; Fig. 3). Cup, rim sherd. D. of rim 12 cm. A coarse fabric (from black to red to reddish brown, 5YR 3/4), with phyllite; burnished on interior and exterior. Rounded profile. Monochrome. *Comments*: from Rooms 1–4 (1983 season). *Date*: FN–EM I.

11 (HCH 05-441; Fig. 3). Cup, profile almost complete. D. of base 10 cm. A coarse fabric (varies from black to red, 2.5YR 5/6), with phyllite; burnished on interior and exterior. Rounded profile. Monochrome. *Comments*: from Rooms 1–4 (1983 season). Analyzed by ceramic petrography, sample no. HCH 04/26 (Hagios Charalambos Fabric Group 2a, Lasithi Red Fabric Group). *Date*: FN–EM I.

12 (HCH 05-442; Fig. 3). Cup, rim sherd. D. of rim 12 cm. A coarse fabric (varies from black to red, 2.5YR 5/6), with phyllite; burnished on interior and exterior. Rounded profile. Monochrome. *Comments*: from Rooms 1–4 (1983 season). Analyzed by ceramic petrography, sample no. HCH 04/29 (Hagios Charalambos Fabric Group 2a, Lasithi Red Fabric Group). *Date*: FN–EM I.

13 (HNM 13,789; Fig. 3). Open vessel (cup?), base sherds. D. of base 7.5 cm. A coarse black fabric, with phyllite; burnished on interior and exterior. Rounded profile and flat base. Monochrome. *Comments*: from Rooms 1–4 (1976–1983 seasons). *Date*: FN–EM I.

14 (HCH 04-271; Fig. 3). Open vessel (bowl?), base sherd. D. of base 12–14 cm. A coarse black fabric (varies between very dark gray, 10YR 3/1, and black, 10YR 2/1), with phyllite; burnished on interior and exterior. Flat pronounced base. Monochrome. *Comments*: from Room 4 (1983 season). *Date*: FN–EM I.

15 (HCH 05-440; Fig. 3). Open vessel (cup or bowl?), base and body sherds. D. of base 6 cm. A coarse fabric (varies from black to red, 2.5YR 5/8), with fine-grained phyllite; burnished on interior and exterior. Shallow open vessel with barely flattened base. Monochrome. *Comments*: from Rooms 1–4 (1983 season). Analyzed by ceramic petrography, sample no. HCH 04/34 (Hagios Charalambos Fabric Group 6). *Date*: FN–EM I.

Open Vessel with Lug Handle

16 (HCH 196; Fig. 3). Open vessel (cup?) with lug handle, rim sherd with handle. D. of rim 16 cm. A coarse fabric (varies from black to red, 2.5YR 4/8), with phyllite; burnished on interior and exterior. Straight profile with lug handle. Monochrome. *Comments*: from Room 4, upper levels (1983 season). *Date*: FN–EM I.

Jar with Strap Handle

17 (HCH 04-287; Fig. 3). Jar with strap handle, handle sherd. Max. dim. 4.5 cm. A coarse fabric (red, 2.5YR 5/6), with phyllite; burnished on exterior. Monochrome. *Comments*: from Room 4/5 Entrance, level 4 (unit HCH 02-2/3 Ent-4). *Date*: FN–EM I.

Monochrome Style, Red Fabric with Phyllite, Burnished

Jar

18 (HCH 05-450; Fig. 4). Jar, rim sherd. D. of rim 11.7 cm. A coarse fabric (red, between 2.5YR 4/6 and 2.5YR 5/6), with phyllite; burnished on exterior. Monochrome. *Comments*: from Rooms 1–4 (1983 season). *Date*: FN–EM I.

Scored Style in Phyllite Fabrics

The style (Scored Style) is a class of ceramics from Early Minoan (EM) I that is characterized by parallel scratches on the exterior (for discussion and bibliography, see Betancourt 1985, 31; 2008a, 64; Sakellarakis and Sapouna-Sakellaraki 1997, 377). Most examples have come from the western and central parts of Crete. It is represented in the assemblage by very few sherds. Fragments of open vessels with both a round and straight profile have a surface that is scored horizontally. The finish suggests use of a comb or brushlike instrument to achieve the parallel scratches. The rarity of this ware may suggest that it was imported from elsewhere on Crete. Petrographic analysis (sample no. HCH 04/30) was used to describe this coarse, phyllitic fabric, a metamorphic fabric with epidote.

Open Vessel, EM I

19 (HCH 05-437; Fig. 4). Open vessel, body sherd. Max. dim. 4.5 cm. A coarse fabric (red, 2.5YR 5/8), with phyllite; burnished(?) interior; horizontally scored exterior. Slightly rounded profile. *Comments*: from Room 4, upper levels (1983 season). Interior of fabric is black. Scored Style. *Parallels*: Wilson and Day 2000, pl. 8:164–175 (Knossos). *Date*: EM I.

20 (HCH 05-438; Fig. 4). Open vessel, body sherd. Max. dim. 2.8 cm. A coarse fabric (red, 2.5YR 4/6), with phyllite; burnished interior; horizontally scored exterior. Straight profile. *Comments*: from Room 4, upper levels (1983 season). Scored Style. Analyzed by ceramic petrography, sample no. HCH 04/30 (a loner containing metamorphic rocks with epidote). *Date*: FN–EM I.

Monochrome Style, Dark Burnished Fabrics with Phyllite

These open vessels, both with a straight profile, have diameters ranging in size from 10 to 16 cm. The open, conical shape suggests that they are drinking vessels. The clay fabrics here are harder and thinner than those dated to the Final Neolithic period discussed above. Interior and exterior surfaces are well burnished. Dark colors are usual.

Open Vessels

21 (HCH 04-407; Fig. 4). Cup or bowl, rim sherd. D. of rim 10 cm. A coarse fabric (dark reddish brown, 5YR 3/2), with phyllite; burnished on interior and exterior. Straight profile. Monochrome. *Comments*: from cave exterior, Trench 11, level 1 (unit HCH 03-11-1). *Date*: EM I.

22 (HCH 04-408; Fig. 4). Cup or bowl or small chalice, rim sherd. D. of rim 16 cm. A coarse fabric (dark reddish brown, 5YR 3/2), with phyllite; burnished on interior and exterior. Straight profile. Monochrome. *Comments*: from cave exterior, Trench 14, level 2 (unit HCH 03-14-2). *Date*: EM I.

Dark Burnished Fabrics without Phyllite, Red to Black Fabric

Chalices (large open vessels supported on conical bases) are the most common shapes of this class. The fabric is hard and thin with burnishing on the surface. Because the fabric does not contain any phyllite (which is a distinctive characteristic of the local Lasithi-Pediada production), the vessels are imports from elsewhere on Crete. For discussion of the shape, see Haggis 1997. Parallels come from many sites: Wilson 1985, 297–299; Hood 1990, 369 and fig. 1:1, 2; Momigliano 1990, 477–479 (Knossos); Blackman and Branigan 1982, 27–29 (Hagia Kyriaki); Todaro 2001, 17, fig. 7 (Hagia Triada); Haggis 1996, 664–668 (Kalo Chorio); Xanthoudides 1918, 156–159 (Pyrgos Cave); Davaras and Betancourt 2004, tomb 1 no. HN 2484, tomb 7 no. HN 2842-1, tomb 53 no. HN 2773, tomb 85 no. 3972, and others (Hagia Photia).

Chalice, Pyrgos Style

The Pyrgos Style is a class of Minoan pottery with red to dark surfaces that are decorated with burnished lines (for discussion and the range of shapes, see Betancourt 2008a, 56–63). The chalice is the most common shape in this fabric at Hagios Charalambos. The Pyrgos Style is well known from EM I contexts (for a list of sites in Crete with the Pyrgos Style, see Betancourt 2008a, 59). Various fabrics are present at Hagios Charalambos.

23 (HNM 13,856; Fig. 4). Chalice, body sherd from waist (join between bowl and base). D. of waist 11.3; max. dim. 8.1 cm. An almost black fabric (very dark gray, 2.5YR N3/0), tempered with fragments of carbonate rock; burnished on interior and exterior. Bowl on a tall conical base. Two incised grooves at the waist. *Comments*: from Rooms 1–4 (1976–1983 seasons). Pyrgos Style. For the shape, see Alexiou and Warren 2004, fig. 6:15 (Lebena); Karantzali 2008, 258, fig. 25.15:MH 7489 (Pyrgos Cave). Analyzed by ceramic petrography, sample no. HCH 04/31 (Hagios Charalambos Fabric 4). *Date*: EM I.

24 (HCH 03-202; Fig. 4). Chalice, body sherd. Max. dim. 3.8 cm. A medium-coarse fabric (red, 10R 5/6), with calcite and chaff. Open vessel with straight profile.

Decorated with incised horizontal lines on exterior. *Comments*: from cave exterior, Trench 12 (unit HCH 03-12-8). Pyrgos Style. Analyzed by ceramic petrography, sample no. HCH 04/27 (Hagios Charalambos Fabric Group 1). *Date*: EM I.

25 (HCH 86; Fig. 4). Chalice, body sherd from the bowl. Max. dim. 3 cm. Straight profile. A medium-coarse fabric (reddish brown, 5YR 5/6); burnished on interior and exterior. Bowl above conical base. Decoration of burnished horizontal striations on exterior. *Comments*: from Rooms 1–4 (1983 season). Pyrgos Style. *Date*: EM I.

26 (HCH 04-275; Fig. 4). Chalice, body sherd from the bowl. Max. dim. 2.7 cm. Straight profile. A medium-coarse fabric (dusky red, 2.5YR 3/2); burnished on interior and exterior. Bowl above conical base. Decoration of burnished zigzag pattern on exterior. *Comments*: from Room 4 (1983 season). Pyrgos Style. *Date*: EM I.

27 (HCH 04-273; Fig. 4). Chalice, rim sherd. D. of rim 16–18 cm. Straight profile. A medium-coarse fabric (between very dark gray, 2.5YR N3, and dusky red, 2.5YR 3/2); burnished on interior and exterior. Bowl above conical base. Traces of pattern burnish(?) on exterior. *Comments*: from Room 4 (1983 season). Pyrgos Style. *Date*: EM I. *Bibl.*: Betancourt et al. 2008b, 166, fig. 28:b.

Dark Burnished Fabrics with Phyllite or Calcite, Red to Brown Fabric

Like the chalices discussed above, these open shapes were intended for drinking. The shape is an open vessel supported on a conical base. Most of the fabrics do not contain any calcite inclusions, and they do contain inclusions of phyllite, suggesting a different production from the calcite-tempered vases. They are substantially redder than those in the preceding group. The date is EM I (Wilson 2007, 51–54).

Open Vessels

28 (HCH 04-371; Fig. 4). Open vessel (chalice?), body sherd. Max. dim. 4.8 cm. A medium-coarse fabric (brown, 7.5YR 5/4, with the exterior surface dark, between very dark gray, 2.5YR N3, and black, 2.5YR N2.5); burnished on exterior. Straight profile. Decoration of burnished horizontal striations on exterior. *Comments*: from Room 4 (1983 season). Pyrgos Style. *Date*: EM I.

29 (HCH 04-377; Fig. 4). Open vessel (chalice?), body sherd. Max. dim. 5.8 cm. A medium-coarse fabric (yellowish red, 5YR 4/6); burnished on interior(?) and exterior. Straight profile. *Comments*: from Room 4, upper levels (1983 season). *Date*: EM I.

Chalice

30 (HCH 04-274; Fig. 4). Chalice, body sherd. Max. dim. 4.9 cm. A medium-coarse fabric (reddish brown, 2.5YR 4/4), with phyllite; burnished on interior and exterior. Rounded profile. *Comments*: from Room 4, upper levels (1983 season). *Date*: EM I.

31 (HCH 04-355; Fig. 4). Chalice, body sherd from waist (join between bowl and base). D. of waist 14 cm. Straight profile. A medium-coarse fabric (dark reddish brown, 5YR 3/4, with the surface dark reddish brown, 5YR 3/2), with fine-grained phyllite; burnished on interior and exterior. Bowl above conical base. Burnished horizontal lines on the interior and vertical lines on the exterior. *Comments*: from Room 4, upper levels (1983 season). Analyzed by ceramic petrography, sample no. HCH 04/32 (Hagios Charalambos Fabric Group 6). *Date*: EM I.

32 (HCH 04-272; Fig. 4). Chalice, body sherd. Max. dim. 6.1 cm. A medium-coarse fabric (varies from black to red to reddish brown, 2.5YR 5/4), with phyllite; burnished on interior and exterior. Rounded profile. Burnished horizontal lines on exterior. *Comments*: from Room 4, upper levels (1983 season). For the shape of the bowl on a tall base, see Alexiou and Warren 2004, fig. 6:15 (Lebena). *Date*: EM I.

33 (HCH 05-443). Chalice, sherd from base of bowl. Max. dim. 5.6 cm. A medium-coarse fabric (dark reddish gray, 5YR 4/2), with carbonate rock; burnished. Black surface (5Y 2.5/1). Sherd shows where base joined lower part of bowl. *Comments*: from Rooms 1–4 (1983 season). Analyzed by ceramic petrography, sample no. HCH 04/33 (Hagios Charalambos Fabric Group 4). *Date*: EM I.

Goblet

34 (HCH 149; Fig. 5). Goblet, body sherd from waist (join between bowl and stem). D. of waist 3.5 cm. A coarse fabric (reddish brown, between 2.5YR 4/4, and 2.5YR 5/4), with phyllite; burnished interior(?) and exterior. Bowl supported on a stem. Two incised grooves at waist. *Comments*: from Rooms 1–3 (1982 season). *Date*: EM I.

Monochrome Style in Soft Reddish-Brown to Black Fabric

Only a single vessel made in this pale brown to black fabric was recognized from the cave. Too little of the vessel survives to identify the shape.

Open Vessel with Lug Handle

35 (HCH 123; Fig. 5). Open vessel (bowl or small chalice?), rim sherd with vertically pierced lug handle. D. of rim 16 cm. A coarse fabric (dark yellowish brown, 10YR 5/6); burnished on interior and exterior. Rounded profile, straight rim. Either a bowl or a small chalice. Burnished. *Comments*: from Rooms 1–3 (1982 season). *Date*: FN–EM I.

Miniature Vessel with Pierced Lug Handle

The soft fabric of this miniature vessel suggests an early date. Too little of the vessel survives to identify the shape.

36 (HCH 04-373; Fig. 5). Miniature open vessel, rim sherd with vertically pierced lug handle. D. of rim 4 cm. A coarse fabric (reddish brown, 5YR 5/8). Straight profile. Dark slip on exterior. *Comments*: from Room 4, upper levels (1983 season). *Date*: EM I.

Monochrome Style in Calcite-Tempered Fabric with Variegated Color

Only one example of this vessel has been identified. The diameter of the rim may suggest that the jug was of average size, intended for pouring, and not a miniature.

Large Jug with Pointed Spout

37 (HCH 04-404; Fig. 5). Jug, rim sherds with spout and handle scar. D. of rim 7 cm, not including spout. A coarse fabric (red, 10R 5/6), with calcite and chaff. Irregularly colored surface ranging from light reddish brown (5YR 6/3) to dark reddish gray (5YR 4/2) to very dark gray (5YR 3/1); burnished on interior of rim and exterior. Rounded profile, pointed spout, outturned rim. Incised band at neck. *Comments*: from Room 4, upper levels (1983 season). Analyzed by ceramic petrography, sample no. HCH 04/22 (Hagios Charalambos Fabric Group 1). *Date*: EM I.

4

Coarse and Fine Fabrics, EM I–IIA

Hagios Onouphrios Style

A few examples of the Hagios Onouphrios Style (called Hagios Onouphrios Ware when made of Cretan South Coast Fabric) were found in the ceramic assemblage. This class is made of fabrics with a pale color decorated with red linear ornament. It is a hallmark of EM I (Branigan 1970a, 23–27; Blackman and Branigan 1982, 29–32; Wilson 2007; Betancourt 2008a, 47–53). Along with the Pyrgos Style chalices and the two examples of the Scored Style, discussed above, EM I is represented by several vessels discussed in this chapter.

An open vessel with straight profile and a jug, for pouring, are the most diagnostic examples. The jug has the typical painted linear decoration of the Hagios Onouphrios Style. Petrographic analysis was used to provide a description of this finely textured calcareous fabric with rounded sandstones (Hagios Charalambos Fabric Group 11).

Open Vessel

38 (HCH 61; Fig. 5). Open vessel, body sherd. Max. dim. 6.6 cm. A fine pale fabric (reddish yellow, 5YR 5/6); burnished on interior and exterior. Straight profile. Burnished, no paint survives. *Comments*: from Room 4, upper levels (1983 season). Probably Hagios Onouphrios Style. The sherd is assigned to this date on the basis of the fabric. *Date*: EM I–IIA.

Jugs

39 (HCH 04-278; Fig. 5). Jug with wide mouth, complete profile. H. (excluding spout) 9.3; d. of rim 8; d. of base 6.1 cm. A coarse, pale fabric (pink, 7.5YR 7/4); slipped and burnished on exterior. Rounded profile, straight rim, flat base. Decoration of two thin horizontal bands at rim, four thin diagonal bands from rim to mid body, three horizontal bands at base. *Comments*: from Room 4, upper levels (1983 season). Hagios Onouphrios Style. Analyzed by ceramic petrography, sample no.

04/36 (Hagios Charalambos Fabric Group 11), containing sandstone and phyllite fragments. *Date*: EM IIA. *Bibl.*: Langford-Verstegen 2008, 554, fig. 9:4; Betancourt et al. 2014, 29, fig. 17:c.

40 (HCH 04-321; Fig. 5). Jug, handle sherd. Max. dim. 4.8 cm. A fine, pale fabric (very pale brown, 10YR 8.4); burnished on exterior. Rounded profile. No paint survives. *Comments*: from Room 4, upper levels (1983 season). Probably Hagios Onouphrios Style. *Date*: EM I.

41 (HCH 75; Fig. 5). Jug, base sherd. D. of base 6 cm. A fine, pale fabric (pink, 5YR 8/4); slipped and burnished on exterior. Rounded profile. Decoration of thin diagonal lines, some cross-hatched. *Comments*: from Room 4, upper levels (1983 season). Hagios Onouphrios Style. Analyzed by ceramic petrography, sample no. HCH 04/35 (Hagios Charalambos Fabric Group 11). *Date*: EM I.

42 (HCH 06-457; Fig. 5). Jug, sherd from base of neck. D. of neck 2.8 cm. A fine, pale fabric (reddish yellow, 5YR 6/6); slipped and burnished on exterior. Rounded profile. Decoration of diagonal lines. *Comments*: from Rooms 3–4 (1983 season). Hagios Onouphrios Style. *Date*: EM I.

43 (HCH 02-138; Fig. 5). Jug(?), body sherd. Max. dim. 3.7 cm. A fine, pale fabric (light red, 2.5YR 6/6); slipped (pink, 7.5YR 7/4) and burnished on exterior. Red lines. *Comments*: from the Room 4/5 Entrance (unit HCH 02-2/3-Surface). Hagios Onouphrios Style. *Date*: EM I.

44 (HCH 02-140; Fig. 5). Jug, body sherd with handle. Max. dim. 5.2 cm. A fine, pale fabric (reddish yellow, 5YR 7/6); slipped and burnished on exterior. Traces of red slip, probably lines. *Comments*: from the Room 4/5 Entrance, level 3 (unit HCH 02-2/3-3). Possibly Hagios Onouphrios Style. *Date*: EM I.

Koumasa Style Pottery with Pale Fabrics, EM IIA

The Koumasa Style is a loosely related series of styles using black linear ornament on pale pottery (for the style, see Branigan 1970a, 28–29, where it is called Hagios Onouphrios II; see also Blackman and Branigan 1982, 29–32; Betancourt 1983, 41–43). Several shapes in this style are present here. They include cups, bowls, and spouted vessels. The surfaces are burnished and decorated in the typical, linear dark-on-light decoration.

Cup

45 (HCH 04-303; Fig. 5). Cup with horizontal handle, complete profile. H. 5.6; d. of rim 10.5; d. of base 7 cm. A pale fabric (pink, 5YR 7/4); burnished. Straight profile, horizontal handle, straight rim, straight base. Decoration of dark band on interior of rim and top of rim. Added white: short diagonal lines on top of rim. *Comments*: from Room 4, upper levels (1983 season). Koumasa Style. *Date*: EM IIA. *Bibl.*: Betancourt et al. 2008b, 166, fig. 28:c.

Wide-Mouthed Vessel

46 (HCH 04-353; Fig. 6). Wide-mouthed vessel, rim, handle, and base sherds.. H. 9; d. of rim 12; d. of base 6.9 cm. A fine, pale fabric (reddish yellow, 5YR 7/6); burnished. Rounded profile, vertical handle with eliptical section. Decoration of dark band on interior of rim and bands and dark cross-hatching on exterior of rim.

Comments: from Room 4, upper levels (1983 season). Koumasa Style. *Date*: EM IIA.

Wide-Mouthed Vessel with Spout

47 (HCH 04-286; Fig. 6). Wide-mouthed vessel, spout sherd. Max. dim. 8.9; length of spout 4 cm. A fine, pale fabric (pink, 7.5YR 7/4); burnished on interior and exterior. Rounded profile, open spout, straight rim. Decoration of two thin, dark vertical bands on spout, horizontal bands and cross hatching at rim. *Comments*: from Room 4, upper levels (1983 season). Koumasa Style. For the shape, see Alexiou and Warren 2004, fig. 20:41 (Lebena). *Date*: EM IIA.

48 (HCH 74; Fig. 6). Wide-mouthed vessel with bridged spout, rim sherd with spout. D. of rim 12 cm. A fine, pale fabric (pink, 7.5YR 8/4). Dark band of cross-hatched lines at rim, two dark bands partly around the spout, band at base of spout, and tip of spout painted. *Comments*: from Room 4, upper levels (1983 season). Koumasa Style. For the shape, see Alexiou and Warren 2004, figs. 20:46, 47 (Lebena). *Date*: EM IIA.

49 (HCH 04-285, Fig. 6). Wide-mouthed vessel with bridged spout, rim sherd with spout. D. of rim 16 cm. A fine, pale fabric (between pink, 5YR 7/4, and light reddish brown, 5YR 6/4); burnished. Rounded profile, bridged spout, straight rim. Decoration of dark horizontal band on interior and exterior of rim, vertical band on spout. *Comments*: from Room 4, upper levels (1983 season). Koumasa Style. For the shape, see Alexiou and Warren 2004, fig. 20:41 (Lebena). *Date*: EM IIA.

50 (HCH 04-279; Fig. 6). Wide-mouthed vessel with unknown spout, rim sherd. D. of rim 22–26 cm. A fine, pale fabric (pink, 5YR 8/3); burnished. Rounded profile. Decoration of dark band on interior of rim, diagonally cross-hatched area on exterior near rim. *Comments*: from Room 4, upper levels (1983 season). Koumasa Style. *Date*: EM IIA.

Monochrome Style, Pale Burnished, EM IIA

The Monochrome Style consists of vessels that are either unpainted, covered with a uniform coat of slip, or wiped and burnished over the entire surface (for discussion of the style, see Betancourt 2008a, 65–71). The style is very common in Crete from EM I to MM II. If the surface is in poor condition or missing, as with many of the vases from this cave, it can be difficult or impossible to know if decoration was originally present.

The bowl in this class was probably imported for its own sake, but among the other vessels are containers that were more likely used in Lasithi for their contents. This group includes a pyxis, a group of rather irregular jugs in small sizes, and a group of small cylindrical jugs. All of these vessels have the low-fired, pale, burnished fabrics typical of EM IIA. The cylindrical jugs are made from an especially finely textured, pale fabric, and their surfaces are very heavily burnished. Perhaps they held a special oil or perfume. An unpublished parallel for the cylindrical jug is known from Myrsini, near the northeastern coast of Crete.

Bowl with Horizontal Handle

51 (HCH 04-356; Fig. 6). Bowl, lower part. D. of base 7.8 cm. A fine, pale fabric (pink, 5YR 7/4); burnished. Conical container, horizontal handle placed low on the body. *Comments*: from Room 4, upper levels (1983 season). Dated by the quality of the soft, heavily burnished pale fabric. *Parallels*: the shape is related to a bowl from EM I with a rounded base (Hood 1990, fig. 1:8, 9 [Knossos]), but the base is flat in this example. *Date*: EM IIA.

Cylindrical Jug

52 (HCH 02-41; Fig. 6; Pl. 1). Cylindrical jug, partial profile with handle scar. H. 4.9; d. of opening 1.8; d. of base 6.8 cm. A fine, pale fabric (yellowish red, 5YR 5/6); slipped or self-slipped and burnished. White slip on exterior. Straight profile and small neck. *Comments*: from the Room 4/5 Entrance (unit HCH 02-2/3 Ent). *Parallels*: Myrsini (unpublished). *Date*: EM IIA.

53 (HCH 02-146; Fig. 6). Cylindrical jug, body sherd with handle. Max. dim. 4.5 cm. A fine, pale fabric (reddish yellow, 5YR 7/6); slipped or self-slipped and burnished. Straight profile and small neck. *Comments*: from Room 3, cleaning (unit HCH 02-2-1). *Parallels*: see **52**. *Date*: EM IIA.

54 (HCH 03-225; Fig. 6). Cylindrical jug, body sherd with handle. Max. dim. 4.4 cm. A fine, pale fabric (light red, 2.5YR 6/6); slipped or self-slipped and burnished. Straight profile and small neck. *Comments*: from Room 5, area 3, level 5 (unit HCH 02-5-3-5). *Parallels*: see **52**. *Date*: EM IIA.

Conical, Irregular Jug

The handmade jug with a wide base and an irregular but generally conical shape can be dated to EM IIA (for the date, see Soles 1992, pl. 4:G I-4, from Gournia).

55 (HNM 13,842; Fig. 6; Pl. 1). Conical jug, lower part. Rest. h. ca. 7.5; d. of base 9.5 cm. A medium-coarse fabric (red, 2.5YR 5/6). Conical shape, wide base, and irregular walls and surface. *Comments*: from Room 5, area 3, level 5 (unit HCH 02-5-3-5). *Parallels*: Marinatos 1929, 112, fig. 9, lower row second from right (Krasi); Soles 1992, pl. 4:G I-4 (Gournia, Tomb I). *Date*: EM IIA.

56 (HCH 171; Fig. 6). Conical jug, lower part. Rest. h. ca. 6; d. of base 6.8 cm. A medium-coarse fabric (yellowish red, 5YR 5/8). Conical shape, wide base, and irregular walls and surface. *Comments*: from Room 4, upper levels (1983 season). *Parallels*: see **55**. *Date*: EM IIA.

57 (HCH 156; Fig. 6). Conical jug, three-quarters complete. Rest. h. ca. 10–12; pres. h. 8.0 cm. A medium-coarse fabric (between reddish yellow, 5YR 7/6, and pink, 5YR 7/4). Conical shape, wide base, and irregular walls and surface. *Comments*: from Room 4, upper levels (1983 season). *Parallels*: see **55**. *Date*: EM IIA.

Pyxis

58 (HCH 82; Fig. 6). Pyxis, body sherd. Rest. h. ca. 10–12; d. of rim ca. 7 cm. An atypical fabric (light red, 2.5YR 6/8). Vertical neck, rounded body, and lugs on upper shoulder. *Comments*: from Room 4, upper levels (1983 season). *Parallels*: for the shape, see Alexiou and Warren 2004, fig. 27:318 (Lebena). *Date*: EM IIA?

5

Dark Burnished Offering Tables in Monochrome Style with Coarse Phyllite Fabric, EM I–IIA

A notable addition to the Hagios Charalambos ceramic assemblage (which, as a whole, is largely made up of both locally made and imported vessels for pouring and drinking) is a local or regional production consisting of open vessels that may have been used as offering tables. Unique and usually restricted to burial contexts, the shape begins in EM I with a shallow form that has outturned sides (Hood 1990, fig. 2:18), but the main development (with more vertical sides) is in EM IIA. The main period is dated by evidence that comes from Tholos Tomb Gamma at Archanes (Papadatos 2005, 14, no. P 78) and from Tomb I at Gournia (Soles 1992, 13–14, nos. G I-15, G I-16). The date is reinforced by the EM IIA stratigraphy of a dark, burnished, coarse fabric in the West Court House at Knossos (Wilson 1985, 353).

The shape is a shallow, circular receptacle with a flat bottom and low, slightly flared to almost straight walls (range of height is 2.2–3.5 cm), attached to a short, cylindrical stem. Plastic decoration is sometimes applied to the surface of the underside of the vessel in the form of raised concentric rings, bosses, and (in one example with no parallel), a radiating motif.

The function of these objects, sometimes considered to be lids or stands, seems more reasonable as an offering table. Although considered lids by Xanthoudides (1924, 10), Banti (1930–1931, 166–167), and Pendlebury (1939, 65), both Xanthoudides and Pendlebury acknowledged the absence of vases that would have been covered by such lids. While we may consider that as lids they covered an object or vessel made from a perishable material, the fact that no matching vase exists holds true for all extant examples. Evans and Seager considered the examples from Mochlos to be fruitstands (Evans 1921–1935, I, 75; Seager 1912, 18, 20, 71). Soles argued further that the hollow pedestals and wear on the interior also suggest that they were used as stands (Soles 1992, 26). When we consider fruitstands from subsequent periods like, for example, the elaborate MM II Kamares Ware stands (see Betancourt 1985, pl. 10:D), it seems that the early stands are reasonable predecessors on structural elements alone. The interpretation as a container is corroborated by an example from EM I consisting of a plate on a stand, which could never be used as a cover (Boyd 1905, 180, fig. 1:Gie).

What speaks loudly in the argument for function is the consistency of contexts for most examples. Published contexts are communal primary burials or secondary ossuaries. It is appropriate that objects designed for offerings were present there. Further, from Lebena (Alexiou and Warren 2004, 31), the authors believe an example was found in situ, resting on the pedestal base. With regard to shape and consistency of find spots (in burial contexts), it therefore seems reasonable that we may regard these objects as handheld, portable offering tables. They may have held an offering and may have been carried in addition to being placed down on the flat surface of the cylindrical stem. As such, the tables are evidence for ritual at the occasion of primary burial.

The offering tables are made of a coarse, red to brown fabric in the Monochrome Style. The surface is black, and the exterior and interior are highly burnished. Fabric samples were analyzed petrographically, and the presence of phyllite and sandstone inclusions indicates that the fabric is typical of the Lasithi production (in App. A, the fabric is the Lasithi Red Fabric Group 2). Consequently, the distribution of parallels is important evidence for export out of the Lasithi region. Moreover, the distribution of these specialized items is extensive, as parallels are known from elsewhere in Lasithi, the northern coast of Crete, the Mesara, and the southeastern coast of Crete. Examples are known from the following sites: Sphoungaras (Hawes et al. 1908, fig. 37:3), Trapeza (Pendlebury, Pendlebury, and Money-Coutts 1935–1936, 53), Malia (Pelon 1970, fig. 3, pl. 24), Mochlos (Seager 1912, 18, 71), Gournia (Soles 1992, figs. 5, 10, pls. 5, 11), Myrsini (unpublished), Koumasa (Xanthoudides 1924, pls. 18, 20), Platanos (Gerontakou 2003, 309, no. 16), Lebena (Alexiou and Warren 2004, pls. 7b, 58b, 58c), Archanes (Papadatos 2005, 14, no. P78), Knossos (Hood 1990, 373, fig. 2:18; Wilson 2007, 55, fig. 2.4:7), and Pyrgos (Xanthoudides 1918, 152, fig. 12).

The typology of the offering tables shows 12 variations of plastic decoration that are added to the basic shape (Table 5.1). The plastic decoration is found on the underside of the vessel body. The descriptive terms "inner and outer" refer to the placement of decoration with regard to where it occurs in proximity to the outer rim. Concentric ridges occur in pairs or singly. Bosses occur, three in number, in sets of three or four.

Type B

59 (HNM 12,569; Fig. 7). Offering table, complete. H. 10.6; d. of rim 21.5; d. of base 6.4 cm. A coarse fabric (red, 2.5YR 4/6), with phyllite; black burnished surface on interior and exterior. Shallow, circular receptacle with flat bottom and low, flared walls (h. 3.5), attached to a short, cylindrical stem (handle). Plastic decoration of one ridge on underside of bowl. *Comments*: from Rooms 1–4 (1976–1983 seasons). Type B. *Date*: EM IIA.

60 (HCH 04-391; Fig. 7). Offering table, side of bowl. D. of rim 19.7 cm . A coarse fabric (red, 2.5YR between 4/6 and 5/6), with phyllite; black burnished surface on interior and exterior. Shallow, circular receptacle with flat bottom and low, flared walls (h. 2.8) attached to a short, cylindrical stem (handle). Plastic decoration of one ridge on underside of bowl. *Comments*: from Room 1 (1982 season). Surface is mottled black to red. Type B. Analyzed by ceramic petrography, sample no. HCH 04/38, Hagios Charalambos Fabric Group 2a (Lasithi Red Fabric Group). *Date*: EM IIA.

Type D

61 (HNM 12,570; Fig. 7). Offering table, complete. H. 10.2; d. of rim 21.4; d. of base (handle) 6.2 cm. A coarse fabric (red, 2.5YR 4/6), with phyllite; black burnished surface on interior and exterior. Shallow, circular receptacle with flat bottom and low, flared walls (h. 3.5), attached to a short, cylindrical stem (handle). Plastic decoration of two inner ridges and two outer ridges on underside of bowl. *Comments*: from Rooms 1–4 (1976–1983 seasons). Type D. *Parallels*: Pendlebury, Pendlebury, and Money-Coutts 1935–1936, pl. 9:404 (Trapeza Cave). *Date*: EM IIA. *Bibl*: Langford-Verstegen 2008, 556, fig. 10:16.

Type E

62 (HCH 04-389; Fig. 7). Offering table, rim sherd. D. of rim 18 cm. A coarse fabric (red, 2.5YR 3/4), with phyllite; black burnished surface on interior and exterior. Shallow, circular receptacle with flat bottom and low, flared walls (h. 2.2), attached to a short, cylindrical stem (handle). Plastic decoration of inner ridges and two outer ridges on the underside of the bowl, and an unknown number of sets of three bosses. *Comments*: from Room 3 (1983 season). Surface is mottled black to red. Type E. *Date*: EM IIA.

Type	Description	Examples
A	No plastic decoration	Gournia: G I-15, G II-8 (Soles 1992, figs. 5, 10, pls. 5, 11)
		Myrsini: HNM 11,980, unpublished
		Trapeza: 402, 411 (Pendlebury, Pendlebury, and Money-Coutts 1935–1936, 53)
B	One inner ridge	Hagios Charalambos: **59**, **60**
C	One outer ridge	Trapeza: 406, 407, 408, 409 (Pendlebury, Pendlebury, and Money-Coutts 1935–1936, 53)
D	Two inner ridges, two outer ridges	Trapeza: 404 (Pendlebury, Pendlebury, and Money-Coutts 1935–1936, 53) Hagios Charalambos: **61**
E	Three inner ridges, two outer ridges, three sets of three bosses	Hagios Charalambos: **62**, **63**, **64**
F	Three inner ridges, three outer ridges	Trapeza: 403, 405 (Pendlebury, Pendlebury, and Money-Coutts 1935–1936, 53)
G	Three inner ridges, two outer ridges	Trapeza: 401 (Pendlebury, Pendlebury, and Money-Coutts 1935–1936, 53)
H	Three inner ridges, two outer ridges, four sets of bosses	Gournia: G I-16 (Soles 1992, figs. 5, 10, pls. 5, 11)
I	One inner ridge, four sets of three "rays" joined at one end extending outward	Hagios Charalambos: **65**
J	Bowl shape with no plastic decoration	Lebena: Tomb I 38, Tomb II 222, 223 (Alexiou and Warren 2004, pls. 7, 58B, 58C)
K	Flared walls and incised decoration of "rays"	Mochlos: Seager 1912, 71
L	One inner ridge, one outer ridge, three single bosses; tall straight walls	Mochlos: Seager 1912, 71 Koumasa: unpublished

Table 5.1. Typology of EM IIA offering tables.

63 (HCH 04-390; Fig. 7). Offering table, rim sherd. D. of rim 18 cm. A coarse fabric (red, 2.5YR 5/4), with phyllite; black burnished surface on interior and exterior. Shallow, circular receptacle with flat bottom and low, flared walls (h. 2.5) attached to a short, cylindrical stem (handle). Plastic decoration of inner ridges and two outer ridges on the underside of the bowl, and sets of three bosses. *Comments*: from Room 1 (1982 season). Surface is mottled black to red. Type E. Analyzed by ceramic petrography, sample no. HCH 04/37, Hagios Charalambos Fabric Group 2a (Lasithi Red Fabric Group). *Date*: EM IIA.

64 (HNM 12,568; Fig. 8; Pl. 1). Offering table, complete. H. 10.3; d. of rim 18.5; d. of base (handle) 6.3 cm. A coarse fabric (red, 2.5YR 4/6), with phyllite; black burnished surface on interior and exterior. Shallow, circular receptacle with flat bottom and low, flared walls (h. 2.8) attached to a short, cylindrical stem (handle). Plastic decoration of three inner ridges and two outer ridges on the bottom of the bowl, and three sets of three bosses. *Comments*: from Rooms 1–4 (1976–1983 seasons). Type E. *Date*: EM IIA.

Type I

65 (HCH 1; Fig. 8). Offering table, complete. H. 9.2; d. of rim 22; d. of base (handle) 6.8 cm. A coarse fabric (reddish brown, 5YR 4/4), with phyllite; black burnished surface on interior and exterior. Shallow, circular receptacle with flat bottom and low, flared walls (h. 3) attached to a short, cylindrical stem (handle). Plastic decoration of one inner ridge and four sets of three "rays" joined to the inner ridge at one end and "radiating" outward. *Comments*: mended from sherds from the Room 4/5 Entrance (unit HCH 02-3/4 Ent) and Rooms 1–3 (1982 season). Surface is mottled black to red. Type I. *Date*: EM IIA.

Offering Table with Unknown Plastic Decoration

66 (HCH 24; Fig. 8). Offering table, rim sherd. Max. dim. 2.6 cm. A coarse fabric (red, 2.5YR 3/6), with phyllite; black burnished surface on interior and exterior. Shallow, circular receptacle with flat bottom and low, almost straight walls attached to a short, cylindrical stem (handle). Unknown plastic decoration. *Comments*: from Rooms 1–4 (1983 season). *Date*: EM IIA.

67 (HCH 126; Fig. 8). Offering table, base (handle) sherd. H. 6 cm. A coarse fabric (ranges from black to red, 2.5YR 4/4), with phyllite; black burnished surface (not high burnish) on interior and exterior. Shallow, circular receptacle with flat bottom and low, almost straight walls, attached to a short, cylindrical stem (handle). Unknown plastic decoration. *Comments*: from Rooms 1–4 (1983 season). Surface is mottled black to red. *Date*: EM IIA.

68 (HNM 13,885, formerly HCH 02-76; Fig. 8). Offering table, base (handle) sherd. H. 5.7 cm. A coarse fabric (ranges from black to red, 2.5YR 4/8), with phyllite; black burnished surface (not high burnish) on interior and exterior. Shallow, circular receptacle with flat bottom and low, almost straight walls, attached to a short, cylindrical stem (handle). Unknown plastic decoration. *Comments*: from Room 5, area 5, level 3 (unit HCH 02-5-5-3). *Date*: EM IIA.

69 (HCH 05-409; Fig. 8). Offering table, rim sherd. D. of body ca. 14; max. dim. 4.4 cm. A coarse fabric (light reddish brown, 2.5YR 6/4) with phyllite; black burnished surface (not high burnish) on interior and exterior. Shallow, circular receptacle with flat bottom and low, almost straight walls, attached to a short cylindrical stem (handle). Unknown plastic decoration. *Comments*: from Room 4 (1976–1983 season). Analyzed by ceramic petrography, sample no. HCH 04/1, Hagios Charalambos Fabric 2A (Lasithi Red Fabric Group). *Date*: EM IIA.

70 (HCH 05-439; Fig. 8). Offering table, handle and body sherds. Max. dim. 6.8 cm. A coarse fabric (reddish brown, 5YR 4/3), with phyllite; black burnished surface on interior and exterior. Shallow, circular receptacle with flat bottom and low, almost straight walls, attached to a short cylindrical stem (handle). Unknown plastic decoration. *Comments*: from Rooms 1–4 (1983 season). Surface mottled black to brown. Analyzed by ceramic petrography, sample no. HCH 04/39, Hagios Charalambos Fabric Group 2a (Lasithi Red Fabric Group). *Date*: EM IIA.

Small Unburnished Offering Table with Unknown Plastic Decoration

71 (HCH 05-435; Fig. 8). Offering table, rim sherds. D. of rim 16 cm. A coarse fabric (between light red, 2.5YR 6/6, and red, 2.5YR 5/6), with phyllite; black burnished surface (not high burnish) on interior and exterior. Shallow, circular receptacle with flat bottom and low, almost straight walls, attached to a short, cylindrical stem (handle). Unknown plastic decoration. *Comments*: from Rooms 1–4 (1983 season). Surface mottled black to red. *Date*: EM IIA.

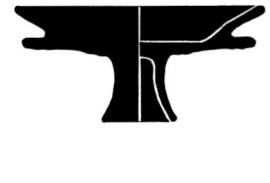

6

Vasiliki Ware and Its Imitations, EM IIB

Several examples of Vasiliki Ware imported from the area of the Gulf of Mirabello, as well as imitations, are present in the assemblage (for discussion of this pottery, see Betancourt 1979). The examples of true Vasiliki Ware include a spouted goblet and jugs in a range of sizes. The burnished and mottled surface (ranging from black to red), that of true Vasiliki Ware, survives on most examples. Typical details, such as a pronounced base, a groove above the base, and "eyes" or bosses on the rim, are present.

The presence of Vasiliki Ware, the hallmark of EM IIB, contributes nicely to our chronological continuum linking the earlier EM IIA Koumasa Style with the later EM III–MM I pottery with light-on-dark decoration. This demonstrates continuity of occupation for the community represented by this deposit. The presence of Vasiliki Ware and its imitations, both local to the Lasithi region and from elsewhere, is further evidence for early trade relationships with production centers on the north coast of Crete.

Vasiliki Ware

Vasiliki Ware often contains small grains of granodiorite, a rock that is only present in the region of the Gulf of Mirabello that extends from Gournia on the east to Priniatikos Pyrgos and Kalo Chorio on the west (Fig. 1; Dierckx and Tsikouras 2007). Its surface is slipped, burnished, and fired to create a mottled surface with colors ranging from red to brown to black.

Spouted Goblet

Spouted goblets are not as common as the unspouted variety. They consist of bowls with small spouts but no handles. They are supported on small bases.

72 (HCH 62; Fig. 9). Spouted goblet, spout sherd. Max. dim. 4.4 cm. A pale fabric (between pink, 5YR 7/3, and light reddish brown, 5YR 6/3); burnished. Dark slip (traces). Slightly rounded profile. *Comments*: from Rooms 1–4 (1983 season). Vasiliki Ware. For discussion of the shape, see Betancourt 1979, 34; for parallels of the shape, see Hall 1912, 47, fig. 21:a, f (Sphoungaras); Betancourt 1985, fig. 27:C (Vasiliki). *Date*: EM IIB.

Jugs, Small Size

Vasiliki Ware jugs have piriform bodies and low or tall vertical spouts. A vertical handle is opposite the spout, and a small groove is sometimes present above the base. A few jugs have small knobs at the sides of the spout. Two small jugs (**73**, **74**), four medium jugs (**75–78**), and one large jug (**79**) were recovered from the cave.

73 (HNM 12,422; Fig. 9; Pl. 1). Small jug, complete profile. H. 7; d. of base 3.6 cm. A pale fabric (reddish yellow, 5YR 7/8); burnished. Rounded profile, raised spout. Slipped and burnished, mottled black to red. *Comments*: from Room 4 (1983 season). Vasiliki Ware. *Parallels*: for the shape, see Betancourt 1985, fig. 27:M (Vasiliki). *Date*: EM IIB.

74 (HNM 12,460; Fig. 9; Pl. 1). Small jug, complete profile. H. 7; d. of base 4.6 cm. A pale fabric (pink, 7.5YR 7/6); burnished. Rounded profile, raised spout, bosses on either side of neck, handle with circular section. Dark slip. Surface mottled black to red. *Comments*: from Rooms 1–4 (1983 season). Vasiliki Ware. *Parallels*: for the shape, see Betancourt 1985, fig. 27:M (Vasiliki). *Date*: EM IIB. *Bibl.*: Langford-Verstegen 2008, 554–555, fig. 9:5.

Jugs, Medium Size

75 (HCH 131; Fig. 9). Medium jug, almost complete profile. Spout opening 2.5; h. 16.5; d. of base 5.8–6.2 cm. A pale fabric (pink, 5YR 7/4); burnished. Rounded profile, raised spout, handle with circular section. Dark slip, light red (2.5YR 6/8). Surface mottled red to black. *Comments*: from Rooms 1–4 (1983 season). Vasiliki Ware. *Date*: EM IIB.

76 (HCH 107; Fig. 9). Medium jug, base sherd. D. of base 4.2 cm. A pale fabric (pink, 5YR 7/4); burnished. Rounded profile, pronounced base. Dark slip. Surface mottled (varies from light red, 2.5YR 6/8, to red, 2.5YR 5/6). *Comments*: from Rooms 1–4 (1983 season). Vasiliki Ware. *Date*: EM IIB.

77 (HCH 144; Fig. 9). Medium jug, base sherd. D. of base 4.1 cm. A pale fabric (pink, 5YR 7/4); burnished. Rounded profile, conical shape at base, incised horizontal groove at base. Surface mottled (light reddish brown, 2.5YR 6/6, to light red, 2.5YR 6/8). *Comments*: from Room 1–3 (1982 season). Vasiliki Ware. *Date*: EM IIB.

78 (HCH 102; Fig. 9). Medium jug, body sherd with spout and handle. Max. dim. 6.8; d. of opening 1.4 cm. A pale fabric (pink, 5YR 7/4); burnished. Rounded profile, small neck and small raised spout, handle with circular section. Surface mottled black to red. *Comments*: from Rooms 1–4 (1983 season). Vasiliki Ware. *Date*: EM IIB.

Jug, Large Size

79 (HCH 04-276+277; Fig. 9). Large jug, spout, handle, and body sherds. Rest. h. 27; d. of opening 6.8 cm. A fine fabric (between pink, 5YR 7/4, and light reddish brown, 5YR 6/4); burnished. Piriform shape, upraised spout, handle with circular section. Dark slip. Surface mottled black to brown to red. *Comments*: from Room 4 (1983 season). Vasiliki Ware. Analyzed by ceramic petrography, sample no. HCH 04/2 (Hagios Charalambos Fabric Group 7b), confirming that this sherd was an import from the region of the Gulf of Mirabello. *Parallels*: for the shape, see Betancourt 1985, fig. 27:J (Vasiliki). *Date*: EM IIB. *Bibl.*: Betancourt et al. 2014, 29, fig. 17:d.

Closed Vessels, Probably Jugs

80 (HCH 158; Fig. 9). Jug(?), body sherd with edge of handle. D. of opening 1.8; max. dim. 7.8 cm. A pale fabric (pink, 5YR 7/4); burnished. Rounded profile, rounded carination at shoulder, handle with circular section. Dark slip (light red, 2.5YR 6/8). Surface mottled black to red to brown. *Comments*: from Rooms 1–4 (1983 seasons). Vasiliki Ware. *Parallels*: for the shape, see Betancourt 1985, fig. 27:I (Sphoungaras). *Date*: EM IIB.

81 (HCH 159; Fig. 9). Jug(?), body sherd. Max. dim. 3.9 cm. A pale fabric (reddish yellow, 5YR 7/6); burnished. Rounded profile, rounded carination at shoulder. Surface mottled black to brown to red. *Comments*: from Rooms 1–3 (1982 seasons). Vasiliki Ware. *Parallels*: for the probable shape, see Betancourt 1985, fig. 27:I (Sphoungaras). *Date*: EM IIB.

82 (HCH 103; Fig. 9). Jug(?), body sherd. Max. dim. 5.6 cm. A pale fabric (reddish yellow, 5YR 7/4); burnished. Rounded carination at shoulder. Dark slip. Surface mottled black to brown. *Comments*: from Rooms 1–4 (1983 season). Vasiliki Ware. *Parallels*: for the probable shape, see Betancourt 1985, fig. 27:I (Sphoungaras). *Date*: EM IIB.

Vasiliki Style Imitations in Pale Fabrics

Imitations of Vasiliki Ware made from pale fabrics of unknown origin copy the typical shapes of the mottled pottery. They include both goblets and jugs. On most examples the surface condition is poor, so it is difficult to discern how the imitation of the mottled surface of true Vasiliki Ware is handled. Nevertheless, some examples have a dark slip. Details such as a raised spout, pronounced base, a groove above the base, and "eyes" or bosses on the rim are present.

Goblet (Egg Cup)

83 (HNM 13,843; Fig. 9; Pl. 1). Goblet, rim sherd. D. of rim 6.2 cm. A pale fabric (reddish yellow, 7.5YR 7/6). Rounded bowl on flared, hollow stem/base. Surface missing. *Comments*: from Room 1 (1976–1983 seasons). Vasiliki Style. *Parallels*: for the shape, see the list of parallels given by Betancourt 1979, fig. 11:1–6, and add Betancourt and Davaras, eds., 2003, 8, 77, nos. 1.11, 9.35–9.41 (Pseira). *Date*: EM IIB.

84 (HCH 141; Fig. 9). Goblet, base sherd. D. of base 5 cm. A pale fabric (between pink, 5YR 7/4, and reddish yellow, 5YR 7/6, with a light gray core); burnished(?). Rounded bowl on flared, hollow stem/base. Dark slip varies from black to red (2.5YR 5/8). *Comments*: from Room 4 (1983 season). Vasiliki Style. *Parallels*: for the shape, see Betancourt 1985, fig. 27:H (Vasiliki). *Date*: EM IIB.

85 (HCH 142; Fig. 9). Goblet, base sherd. D. of base 4 cm. A pale fabric (between pink, 5YR 7/4, and reddish yellow, 5YR 7/6, with a light gray core); burnished(?). Rounded bowl on flared, hollow stem/base. Dark slip from black to red (2.5YR 5/8). *Comments*: from Room 4, upper levels (1983 season). Vasiliki Style. *Parallels*: for the shape, see Betancourt 1985, fig. 27:H (Vasiliki). *Date*: EM IIB.

86 (HCH 04-305; Fig. 9). Goblet, base sherd. D. of base ca. 6 cm. A coarse pale fabric (reddish yellow, 5YR 6/6), burnished(?). Rounded bowl on flared, hollow stem/base. Surface in poor condition. *Comments*: from Rooms 1–3 (1976–1983 seasons, unit HCH-Dump). Vasiliki Style. *Parallels*: for the shape, see Betancourt 1985, fig. 27:H (Vasiliki). *Date*: EM IIB.

Jug, Small Size

87 (HCH 129; Fig. 9). Small jug, almost complete rounded profile. D. of base 3.7 cm. A pale fabric (pink, 5YR 7/4); burnished. Slipped and burnished, mottled black to red (2.5YR 6/6). *Comments*: from Room 4, upper levels (1983 season). Vasiliki Style. *Parallels*: for the shape, see Betancourt 1985, fig. 27:M (Vasiliki). *Date*: EM IIB.

Jug, Carinated

88 (HNM 12,417; Fig. 10). Carinated jug, almost complete. H. 10; d. of base 4.5 cm. A pale fabric (reddish yellow, 5YR 7/6); burnished(?). Rounded profile, rounded carination at shoulder, small neck, short spout, handle with circular section. Dark slip not well preserved. *Comments*: from Room 4 (1983 season). Vasiliki Style. *Parallels*: for the shape, see Hall 1912, 47, fig. 21:c and Betancourt 1985, fig. 27:I (Sphoungaras). *Date*: EM IIB. *Bibl.*: Betancourt et al. 2008b, 166, fig. 28:j.

89 (HNM 12,425; Fig. 10; Pl. 1). Carinated jug, almost complete. H. 9.3; d. of base 4.2 cm. A pale fabric (reddish yellow, 5YR 7/6); burnished(?). Rounded profile, rounded carination at shoulder, small neck, short spout, handle with circular section, pronounced base. Surface mostly missing. *Comments*: from Room 1 (1976–1983 seasons). Vasiliki Style. *Parallels*: for the shape, see Betancourt 1985, fig. 27:H (Vasiliki). *Date*: EM IIB.

Jug, Globular Body with Incised Dashes

90 (HNM 12,418; Fig. 10; Pl. 1). Globular jug, almost complete. H. 12; d. of base 6 cm. A pale fabric (light reddish brown, 2.5YR 6/4); probably burnished. Globular profile, small neck, handle with circular section, incised groove above base. Decoration of double lines of incised dots at neck and mid body. *Comments*: from Rooms 1–4 (1976–1983 seasons). Vasiliki Style. *Parallels*: for a Vasiliki Ware jug with incised dots, see Hall 1914, fig. 46, upper right (Priniatikos Pyrgos). *Date*: EM IIB.

Vasiliki Style Imitations in the Lasithi Red Fabric Group

The pottery workshops that made ceramics somewhere in the Lasithi region also produced copies of the Vasiliki Style. The vases, recognizable because they use the Lasithi Red Fabric Group, include the spouted conical bowl, the goblet, and both small and larger jugs. Like the originals and the copies in pale fabrics, include both drinking and pouring vessels.

Spouted Conical Bowl

91 (HNM 12,449; Fig. 10). Spouted conical bowl, complete. H. 5.3; d. of rim uneven; d. of base 8 cm. Lasithi Red Fabric Group, light red (2.5YR 6/8); possibly burnished. Conical shape, open spout at rim opposite a small knob, groove above the base. *Comments*: from Room 4, upper levels (1983 season). Vasiliki Style. This bowl copies most aspects of the East Cretan bowls, including the groove above the base, but the proportions are not the same (this bowl is more shallow). *Parallels*: for the genuine Vasiliki Ware spouted conical bowl, see the list published by Betancourt 1979, 34–35, and add Betancourt and Davaras, eds., 2003, 65, 78, nos. 7.11, 9.49 (Pseira). *Date*: EM IIB.

Goblets

92 (HNM 13,841; Fig. 10; Pl. 1). Goblet, almost complete. H. 7.5; d. of rim 8.8; d. of base 4.5 cm. Lasithi Red Fabric Group (red, 2.5YR 5/6). Rounded bowl on flared stem/base. *Comments*: from Rooms 1–4 (1976–1983 seasons). Vasiliki Style. *Parallels*: for the shape, see Betancourt 1985, fig. 27:H (Vasiliki). *Date*: EM IIB.

93 (HCH 13; Fig. 10). Goblet, base sherd. Max. dim. 3.9 cm. Lasithi Red Fabric Group (light red, 2.5YR 6/8); burnished(?). Rounded bowl on flared, hollow stem/base. Surface in poor condition. *Comments*: from Room 4, upper levels (1983 season). Vasiliki Style. *Parallels*: for the shape, see Betancourt 1985, fig. 27:H (Vasiliki). *Date*: EM IIB.

94 (HCH 190; Fig. 10). Goblet, base sherd. Max. dim. 5.7 cm. Lasithi Red Fabric Group (reddish brown, 2.5YR 5/4); burnished(?). Rounded bowl on flared, hollow stem/base. Dark slip on exterior. *Comments*: from Room 4, upper levels (1983 season). Vasiliki Style. *Parallels*: for the shape, see Betancourt 1985, fig. 27:H (Vasiliki). *Date*: EM IIB.

Jugs, Medium Size

95 (HCH 29; Fig. 10). Medium jug, upper part with upper part of handle. Max. dim. 7.1 cm. Lasithi Red Fabric Group (2.5YR 4/6). Low spout, bosses on side of spout, handle with circular section attached at rim. Surface in poor condition. *Comments*: from Room 4, upper levels (1983 season). Vasiliki Style. *Date*: EM IIB.

96 (HCH 02-75; Fig. 10; Pl. 1). Medium jug, upper part with handle. Max. dim. 13.8 cm. Lasithi Red Fabric Group (light red, 2.5YR 6/8 to 5/8). Low spout, bosses on side of spout, handle with circular section attached at rim. Surface in poor condition. *Comments*: from Room 5, area 5, level 3 (unit HCH 02-3-5-3). Vasiliki Style. *Date*: EM IIB.

97 (HCH 02-80; Fig. 10). Medium jug, almost complete. Rest. h. ca. 14; d. of base 4.7 cm. Lasithi Red Fabric Group (light red, 2.5YR 6/8). Rounded body, raised spout, bosses on side of spout, handle with circular section attached at rim (as shown by scar). Surface mostly missing but probably once slipped and mottled. *Comments*: from Room 1 (1976 season). Vasiliki Style. *Date*: EM IIB.

98 (HNM 12,412; Fig. 10). Medium jug, almost complete. Rest. h. ca. 11–12; pres. h. 9.5; d. of base 4.4 cm. Lasithi Red Fabric Group (light red, 2.5YR 6/8). Raised spout, bosses on side of neck, handle with circular section attached below rim and on shoulder. Surface in poor condition. *Comments*: from Room 4, upper levels (1983 season). Vasiliki Style. *Date*: EM IIB.

99 (HNM 13,806; Fig. 10). Medium jug, almost complete. Rest. h. 13.7; d. of base 4.7 cm. Lasithi Red Fabric Group (light red, 2.5YR 6/8). Raised spout, bosses on side of spout, groove just above base. Surface missing. *Comments*: from Room 1 (1976–1983 seasons). Vasiliki Style. *Date*: EM IIB.

Jug, Large Size

A jug is considered large (**100**) if it is more than 20 cm in height.

100 (HCH 02-81; Fig. 10). Large jug, almost complete. Rest. h. ca. 22–24; d. of base 7.2 cm. Lasithi Red Fabric Group (light red, 2.5YR 6/8); burnished(?). Piriform body, raised spout, bosses on side of spout, handle with circular section attached below rim. Dark slip (mottled) on exterior. *Comments*: from Room 5, area 5, level 3 (unit HCH 02-3-5-3). Vasiliki Style. *Date*: EM IIB.

7

Vessels in Pale Fabrics
with Mostly Light-on-Dark Decoration,
EM III–MM II

Light-on-dark decoration characterizes much of the EM III and later pottery production of Middle Bronze Age Crete. White-on-Dark Ware, a class from the beginning of the tradition, was especially popular in EM III and MM IA. It originated in East Crete (Betancourt 1984). The assemblage of related vases from the cave includes: cups; tumblers; and jugs in small, medium, and large sizes. Over a dark slip, white decoration is added in the form of bands, dots, and circles. These vases are all imports into Lasithi from other parts of Crete, and often the white decoration survives only in traces. The fabric is pale, and the exact origins of the production are unknown. This class, however, nicely represents EM III and early MM in the advancing chronological continuum, and we continue to see many vessels designed for pouring and drinking.

Shallow Conical Bowl

101 (HCH 14; Fig. 11). Shallow bowl, almost complete. D. of rim ca. 20 cm. A pale fabric (pink, 7.5YR 7/4). Shallow conical profile with flat outturned rim. Dark slip on interior of rim and on exterior. Added white band inside rim and chevrons on top of rim. *Comments*: from Room 4, upper levels (1983 season). *Parallels*: Hall 1912, 46, fig. 20 (Sphoungaras); Betancourt 2006, 91, fig. 5.7:89 (Chrysokamino). *Date*: EM III–MM I.

Cups

102 (HCH 73; Fig. 11). Cup, rim sherd with handle. D. of rim 8–10 cm. A pale fabric (reddish yellow, 5YR 7/6). Semiglobular or straight-sided cup, flat handle (no dark slip on underside). Dark slip on exterior and interior of rim. Added white bands on handle. *Comments*: from Room 4, upper levels (1983 season). Analyzed by ceramic petrography, sample HCH 04/40 (Hagios Charalambos Fabric Group 7A, containing granodiorite). The fabric (Mirabello Fabric) demonstrates that this cup is an import from the region of the Gulf of Mirabello. The flat handle means that the cup is no earlier than MM IB. *Date*: MM IB–II.

103 (HNM 12,445; Fig. 11). Semiglobular cup without handle, complete. H. 6.0; d. of rim 10.3; d. of base 3.3 cm. A pale fabric (pink, 5YR 4–7/4). Dark slip on interior and exterior. Added white missing. *Comments*: from Room 4, upper levels (1983 season). *Parallels*: Pelon 1970, pl. 15:5 (Malia). *Date*: MM IA.

104 (HCH 05-445; Fig. 11). Rounded cup, almost complete. H. 5.7; d. of rim 7.8 cm. A pale fabric (reddish yellow, 5YR 5/6). Pronounced base, handle not preserved. Dark slip (red, 2.5YR 4/8) on interior and exterior. Added white not preserved. *Comments*: from Rooms 1–4 (1976–1983 seasons). *Date*: MM I–IIB.

105 (HCH 69; Fig. 11). Cup, base sherd. Rest. h. 8–10; d. of base 4.2 cm. A pale fabric (light red, 2.5YR 6/6). Rounded or semiglobular shape. Dark slip on interior and exterior. Added white bands on lower body, part of a frieze on body, not fully preserved. *Comments*: from Rooms 1–4 (1983 seasons). *Date*: MM IB–IIB.

106 (HCH 05-431; Fig. 11). Cup, base sherd. Rest. h. ca. 10; d. of base 8 cm. A fine, pale fabric (reddish yellow, 5YR 6–7/6). Straight-sided profile. Dark slip (red) on interior and exterior. Added white missing. *Comments*: from Room 4, upper levels (1983 season). *Date*: EM III–MM IA

107 (HCH 182; Fig. 11). Cup, complete profile (handle missing). H. 5.7; d. of rim 10; d. of base 4.0 cm. A pale fabric (reddish yellow, 5YR 7/6). Conical profile, straight rim, straight walls. Dark slip on interior and exterior. No added white preserved. *Comments*: from Rooms 4, upper levels (1983 season). *Date*: MM I–IIB.

108 (HCH 139; Fig. 11). Cup, rim sherd. D. of rim 14.0 cm. A pale fabric (pink, 5YR 8/4). Rounded or semiglobular shape. Dark slip on interior and exterior of rim. Added white bands on exterior. *Comments*: from Room 4, upper levels (1983 season). The form of this cup is uncertain because of the small amount of the vessel that is preserved, and it may have been a two-handled cup. *Date*: MM I–II.

Tumblers

109 (HCH 05-436; Fig. 11). Tumbler, base sherd. D. of base 4 cm. A pale fabric (reddish yellow, 5YR 7/6, with the core light red, 2.5YR 6/6). Flat base, straight sides. Dark slip on interior and exterior (light red, 2.5YR 6/6). Added white band at base, vertical bands, dots. *Parallels*: MacGillivray 1998, pl. 46:144, 145 (Knossos). *Comments*: from Room 4, upper levels (1983 season). *Date*: EM III–MM I.

110 (HNM 13,823; Fig. 11). Tumbler, almost complete. H. 7.2; d. of rim 6; d. of base 3.5 cm. A fine fabric (pink, 5YR 7/4). Tall cup, slightly pronounced base. Dark slip on interior of rim and on exterior. Traces of added white. *Comments*: from Rooms 1–4 (1976–1983 seasons). *Parallels*: among many others, see Zois 1969, pl. 29 (Gournes); Alexiou and Warren 2004, fig. 44:118 (Lebena). *Date*: EM III–MM I.

Small Jugs

111 (HNM 12,409; Fig. 11). Jug, almost complete. H. 9.5; d. of base 3.5 cm. A pale fabric (pink, 7.5YR 8/4). Piriform shape, small spout, handle with circular section. Dark slip on exterior. Added white bands at base of neck and on lower body, elliptical design on body. *Comments*: from Rooms 1–4 (1976–1983 seasons). Poorly preserved. *Date*: EM III–MM I.

112 (HNM 13,808; Fig. 11). Jug, almost complete. H. 8.7; d. of base 2.9 cm. A pale fabric (very pale brown, 10YR 7/4). Piriform shape, small spout, pronounced base, handle with circular section. Dark slip on exterior. Added white bands at base of neck and lower body, circles on shoulder bounded by bands. *Comments*: from Rooms 1–4 (1976–1983 seasons). *Date*: EM III–MM I.

113 (HNM 12,426; Fig. 11; Pl. 1). Jug, almost complete. H. 11.4; d. of rim 2.6; d. of base 4.3 cm. A pale fabric (pink, 7.5YR 7/4). Piriform shape, small spout, handle with circular section. Dark slip on exterior. Added white traces. *Comments*: from Room 1 (1976–1982 seasons). *Date*: EM III–MM I. *Bibl.*: Betancourt et al. 2008b, 166, fig. 28:k.

114 (HNM 12,458; Fig. 11). Jug, complete profile. H. 9.8; d. of base 3.7 cm. A pale fabric (reddish yellow, 5YR 7/6). Rounded profile, small spout, handle with circular section. Dark slip on exterior (red, 2.5YR 5/8). Added white not preserved. *Comments*: from Rooms 1–4 (1976–1983 seasons). Burned. *Date*: EM III–MM I.

115 (HNM 12,436; Fig. 11). Jug, almost complete. H. 8.4; d. of base 3.5 cm. A pale fabric (pink, 5YR 7/4). Rounded profile, small spout, handle with circular section. Dark slip on exterior (light red, 2.5YR 6/8). Added white not preserved. *Comments*: from Room 4, upper levels (1983 season). Burned. *Date*: EM III–MM I.

116 (HNM 12,401; Fig. 11). Jug, complete. H. 7.1; d. of base 4 cm. A fine pale fabric (reddish yellow, 5YR 7/8). Squat globular shape, small spout, handle with circular section. Traces of dark slip on exterior. Added white missing. *Comments*: from Room 4, upper levels (1983 season). *Date*: MM I–IIB.

117 (HCH 38; Fig. 11). Jug(?), base sherd. D. of base 5.1 cm. A pale fabric (pink, 5YR 7/4). Dark bands on lower body. *Comments*: from Room 4, upper levels (1983 season). *Date*: MM I–II.

118 (HNM 12,402; Fig. 12). Jug, almost complete. H. 6.9; d. of base 4.3 cm. A fine pale fabric (reddish yellow, 5YR 7/6). Squat globular form, small spout, handle with circular section. Traces of dark slip on exterior. Added white missing. *Comments*: from Room 4, upper levels (1983 season). *Date*: MM I–IIB.

119 (HNM 13,825; Fig. 12). Jug, almost complete. H. 8.7; d. of base 2.9 cm. A fine pale fabric (reddish yellow, 5YR 7/6). Squat globular form, small spout, handle with circular section. Dark slip (red, 5YR 5/6) on exterior. Added white traces. *Comments*: from Rooms 1–4, upper levels (1976–1983 seasons). *Date*: MM I–IIB.

120 (HNM 12,410; Fig. 12; Pl. 1). Jug, complete. H. 7; d. of base 3.5 cm. A fine pale fabric (very pale brown, 10YR 8/3). Globular form, small spout, handle with circular section. Traces of dark slip on exterior. Added white missing. *Comments*: from Room 4, upper levels (1983 season). *Date*: MM I–IIB.

121 (HNM 13,838; Fig. 12; Pl. 1). Jug, base with handle. Pres. h. 4.5; d. of base 5.4 cm. A pale fabric (reddish brown, 2.5YR 5/4). Convex (slightly carinated) profile, handle with elliptical section. Dark slip on exterior. Added white traces. *Comments*: from Rooms 1–4 (1976–1983 seasons). *Date*: EM III–MM I.

122 (HCH 56; Fig. 12). Jug, three-quarters complete. Rest. h. ca. 9–12 cm. A fine pale fabric (reddish yellow, 5YR 7/6, to pink, 5YR 7/4). Piriform shape, small spout, handle with circular section. Dark slip on exterior. Added white missing. *Comments*: from Rooms 1–4 (1976–1983 seasons). *Date*: MM I–IIB.

123 (HCH 167; Fig. 12). Jug, neck sherd with handle and upper shoulder. Max. dim. 4.1 cm. A pale fabric (light red, 2.5YR 6/8). Small spout, handle with circular section. Dark slip on exterior. Added white horizontal bands and frieze of quirks on neck. *Comments*: from Rooms 1–4 (1983 season). *Date*: MM I–II.

Medium to Large Jugs

A medium-sized jug has a height above 15 cm, and a large jug measures over 20 cm tall.

124 (HCH 183; Fig. 12). Jug, upper part with spout and edge of handle. Max. dim. 10.2; d. of opening 3.4 cm (no tip of spout surviving). A pale fabric (between light reddish brown, 5YR 6/4, and reddish yellow, 5YR 7/6). Medium-sized jug with rounded profile, boss on either side of neck, handle with elliptical section, double wall construction for spout. Dark slip on exterior and interior of rim. Added white bands on spout and at base of neck, group of diagonal lines on shoulder (possibly part of a zigzag motif). *Comments*: from Rooms 1–4 (1983 season). *Date*: EM III–MM I.

125 (HCH 119; Fig. 12). Jug, neck sherd. Max. dim. 4.7 cm. A pale fabric (light red, 2.5YR 6/6). Bumps on interior of sherd. Dark slip on exterior. Added white horizontal bands at neck. *Comments*: from Rooms 1–4 (1983 season). *Date*: EM III–MM I.

126 (HCH 116; Fig. 12). Jug, body sherd from shoulder. Max. dim. 5 cm. A pale, fine fabric (pink, 5YR 7/4). Medium-sized or large jug. Dark slip on exterior. Added white vertical lines on shoulder. *Comments*: from Rooms 1–4 (1983 season). *Date*: EM III–MM I.

127 (HCH 85; Fig. 12). Jug, neck sherd. D. of base of neck 5.5 cm. A pale fabric (between pink, 5YR 7/4, and reddish yellow, 5YR 7/6). Raised spout, tall neck. Dark slip on exterior. Added white horizontal bands on neck. *Comments*: from Room 4 (1983 season). *Date*: MM I–II.

128 (HCH 31; Fig. 12). Jug, neck and shoulder sherd. D. at base of neck 6 cm. A pale fabric (light red, 2.5YR 6/8). Dark slip on exterior. Added white horizontal bands and zigzag band at upper shoulder. *Comments*: from Room 4 (1983 season). *Date*: MM I–II.

129 (HCH 157; Fig. 12). Jug, upper part with spout and handle. Max. dim. 14.3; d. of opening 6.5 cm. A pale fabric (between pink, 5YR 7/4, and reddish yellow, 5YR 7/6). Large jug with rounded profile, small spout, handle with circular section. Dark slip on exterior and interior of rim. Added white traces. *Comments*: from Room 4, upper levels (1983 season). *Date*: MM I.

130 (HCH 03-185; Fig. 12). Jug, upper part with spout and body sherds. Max. dim. 10.6 cm. A pale fabric (light yellowish brown, 10YR 6/4). Rounded profile, small spout. Surface missing. *Comments*: from Room 7, washed in from Room 5 in post-burial times (unit HCH 03-7-1). *Date*: EM III–MM I.

131 (HCH 04-372; Fig. 13). Jug, spout sherd. Rest. h. of vessel ca. 20–22 cm. A pale fabric (light red, 2.5YR 6/6–8). Low spout. Dark slip on exterior. Added white bands on rim, neck, and upper shoulder. *Comments*: from Room 7, washed in from Room 5 in post-burial times (unit HCH 03-7-1). *Date*: MM I–II.

Closed Vessel

132 (HCH 05-410; Fig. 13). Closed vessel, body sherd. Max. dim. 4.4 cm. A fine pale fabric (pinkish gray, 7.5YR 6/2, to brown, 7.5YR 5/2), with many clay pellets. Rounded profile. Dark slip. *Comments*: from Rooms 1–4 (1976–1983 seasons). Analyzed by ceramic petrography, sample no. HCH 04/24 (Hagios Charalambos Fabric 9). *Date*: EM II–MM II.

Jar with Pierced Vertical Handles

133 (HCH 194; Fig. 13). Jar, two rim sherds with vertical handles. D. of rim 8.5 cm. A pale fabric (reddish yellow, 7/5YR 7/6). Straight profile, pierced vertical handles rising from the rim. Dark slip on exterior (traces). Added white missing. *Comments*: from Room 4, upper levels (1983 season). *Date*: MM I–II.

Miscellaneous Pouring Vessels

134 (HCH 04-348; Fig. 13). Jug without spout, almost complete, handle missing. D. of rim 8.5 cm. A pale fabric (reddish yellow, 7/5YR 7/6). Vertical neck or rim, one vertical handle (missing). Dark slip on exterior. Added white bands and friezes of horizontal chevrons. *Comments*: from Room 4, upper levels (1983 season). *Parallels*: for the motif see Zois 1969, pls. 28, 29 (Tylissos). *Date*: MM I–II.

135 (HCH 153; Fig. 13). Wide-mouthed jug, rim sherd with spout and handle. D. of rim ca. 14 cm. A pale fabric (reddish brown, 5YR 4/4). Wide mouth, straight vertical rim, open spout at the rim, one vertical handle with circular section opposite the spout. Dark slip on exterior. Added white bands and frieze of hatched triangles. *Comments*: from Room 4, upper levels (1983 season). *Date*: EM III–MM II.

136 (HCH 04-351; Fig. 13). Spouted bowl, complete profile. H. 8.3; d. of rim ca. 16; d. of base 6.5 cm. A pale fabric (reddish yellow, 5YR 7/6). Straight, slightly inturned rim, two horizontal handles with circular sections below rim, open spout at the rim, lug at the rim opposite the spout. Poorly preserved dark slip on interior and exterior. Added white traces. *Comments*: from Room 4, upper levels (1983 season). *Parallels*: for the shape, see Alexiou and Warren 2004, fig. 45:6 (Lebena). *Date*: EM III–MM I.

137 (HNM 6,828; Fig. 13; Pl. 1). Teapot, half complete. H. 7.6; d. of rim 8.0; d. of base 5.0 cm. A pale fabric (reddish yellow, 7.5YR 7/6). Globular form with straight, slightly inturned rim, two horizontal handles with circular sections on shoulder, teapot-type spout below rim. Dark slip on exterior and interior of rim. Added white sets of three curved diagonal lines extending from base to rim. *Comments*: from Rooms 1–2 (1976 season). *Parallels*: for the ornament, see Zois 1969, pl. 21 (Gournes); Sakellarakis and Sapouna-Sakellaraki 1997, 407, fig. 362 lower left (Archanes); Betancourt and Davaras, eds., 2003, fig. 32:4.53 (Pseira); for both shape and decoration, see Alexiou and Warren 2004, fig. 43:68 (Lebena). *Date*: MM IB. *Bibl.*: Davaras 1976a, 349, pl. 301:3.

138 (HCH 72; Fig. 13). Bridge/open-spouted jar or teapot, rim sherd. D. of rim 10.0 cm. A pale fabric (reddish yellow, 7.5YR 7/6). Straight, slightly upturned rim, probably had two horizontal handles on the shoulder and a bridged or open spout (now missing). Dark slip on interior of rim and on exterior. Added white missing. *Comments*: from Rooms 3–4 (1983 season). *Date*: MM IB–II.

Jars

139 (HNM 12,438; Fig. 13; Pl. 1). Jar, complete. H. 9.6; d. of rim 4.9; d. of base 4.7 cm. A pale fabric (reddish yellow, 5YR 7/6). Globular form, straight, slightly flaring rim, two horizontal handles with circular sections on lower shoulder. Dark slip on interior of rim and on exterior. Added white band at base of neck, sets of three diagonal lines extending from base of neck to center of body, above a frieze of dots bounded by white bands. *Comments*: from Room 4, upper levels (1983 season). *Parallels*: for the ornament of sets of white lines, see Zois 1969, pl. 21 (Gournes); Alexiou and Warren 2004, fig. 44:117 (Lebena). For the shape, see Alexiou and Warren 2004, fig. 6:1 (Lebena). *Date*: MM IB.

140 (HNM 12,427; Fig. 13; Pl. 2). Miniature jar, complete. H. 6.2; d. of rim 4.6; d. of base 3.6 cm. A pale fabric (pink, 5YR 7/4). Somewhat conical form, flaring rim, pronounced base, two horizontal handles with circular sections on shoulder. Dark slip on interior of rim and on exterior. Added white band at base of neck, lines extending from below base of neck to lower body. *Comments*: from Room 1 (1976 season). *Parallels*: Marinatos 1929, 112, fig. 9:8 (Krasi). *Date*: MM IB.

141 (HNM 13,827; Fig. 13; Pl. 2). Small jar, complete. H. 8.2 cm. A fine fabric (light red, 2.5YR 6/8). Conical form, flaring rim, two handles with circular sections on shoulder. Dark slip on interior of rim and on exterior. Added white bands on rim and at base of neck, groups of four diagonal lines on body. *Comments*: from Room 4, upper levels (1983 season). *Parallels*: for the ornament of sets of white lines, see Zois 1969, pl. 21 (Gournes). For the shape, see Marinatos 1929, 112, fig. 9:8 (Krasi). *Date*: MM I–II.

Pyxis

142 (HNM 14,066; Fig. 13; pl. 2). Pyxis, complete. H. 10.5; d. of rim 13.5; d. of base 16 cm. A pale fabric (surface slipped, pink, 5YR 6/4). Rim with inner ledge for lid, which is missing; almost straight walls; pronounced base; three small feet; four small holes on inner side of rim, probably to attach the lid. Dark slip: band on inside of rim, outer rim, and base; checkerboard on body. *Comments*: from Room 4, upper levels (1983 season). *Date*: MM II.

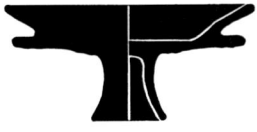

8

Diagonal Line Style Jugs and Other Small Jugs in Pale Fabrics, MM I

A group of handmade jugs stands apart from the majority of the Hagios Charalambos assemblage, which consists mostly of pottery designed for pouring and drinking. Most examples are miniature, although a few medium-sized jugs were also recovered. The small size and narrow openings of the majority of the pieces suggest other purposes. The clay fabric is usually reddish yellow, and many pieces may have come from Malia, but **162** is from Lasithi.

In terms of structure, the miniature jugs have small, perfectly circular openings in the necks, so they could easily have been sealed with stoppers made from sticks wrapped in a material such as leather. Once filled, the ability to seal their openings would allow for transport with the assurance that the contents would remain intact. In later periods, perfumed oil was a typical offering in tombs, and it was offered in small vessels with narrow necks. This shape allows liquid to be poured slowly and conservatively. In the Late Bronze Age, the vessels used for grave offerings of oil were stirrup jars,

and they were replaced by the alabastron and the lekythos after the end of the Bronze Age. Perhaps we may consider this class of miniature jug as a predecessor of the perfumed oil vessels of later periods because of the miniature size and the design of the neck opening (Betancourt 2008a, 28–29). Further, the consistent decorative scheme of diagonal bands and variations on this motif may have made these miniature jugs readily recognizable.

Diagonal Line Style Jugs

The Diagonal Line Style is the cave's largest group of small jugs in pale fabrics. Decoration is either in dark-on-light or light-on-dark bands applied in a diagonal manner across the bodies of the jugs. There is little variation in the composition, and the decorative motifs are limited. The diagonal line motif often consists of a pair of diagonal designs with the highest parts under the spout (Betancourt 2011a). With narrow mouths and slightly raised delicate spouts, the jugs may have held a precious oil or unguent and were most likely deposited in the primary burials as offerings.

These special vessels have parallels at Malia (Demargne 1945, pl. 28:8598, 8599, 8600, 8601), the Trapeza Cave (Pendlebury, Pendlebury, and Money-Coutts 1935–1936, fig. 17:643 [somewhat different]), the Kyparissi Cave (Serpetsidaki 2006, 257, fig. 3:3, 4), Archanes (Sakellarakis and Sapouna-Sakellaraki 1997, 387, fig. 229, lower right); and Krasi (Marinatos 1929, 113, fig. 10:gamma).

143 (HCH 113; Fig. 14). Small jug, body sherd. Rest. h. ca. 6.5 cm. A pale fine fabric (reddish yellow, 5YR 6/6 to 7/6); burnished. Rounded profile. Dark slip: diagonal lines flanking a zigzag motif. *Comments*: from Rooms 1–4 (1976–1983 seasons). *Date*: MM IA. *Bibl.*: Betancourt 2011a, 27, fig. 1:1.

144 (HNM 13,829; Fig. 14; Pl. 2). Small jug, almost complete. Rest. h. ca. 7; d. of base 3.7 cm. A pale fine fabric (white, 10YR 8/2); burnished. Rounded profile, raised spout, handle with circular section attached at rim and shoulder. Dark slip: band at base of neck, diagonal lines flanking a zigzag motif, one band on lower body at widest part. *Comments*: from Rooms 1–3 (1976–1983 seasons). *Date*: MM IA. *Bibl.*: Betancourt 2011a, 27, fig. 1:2.

145 (HNM 13,826; Fig. 14; Pl. 2). Small jug, almost complete. Rest. h. ca. 8; d. of base 3.4 cm. A pale fine fabric (reddish yellow, 5YR 7/6); burnished. Rounded profile, handle with circular section attached at rim and shoulder. Dark slip: band at base of neck, diagonal lines flanking a zigzag motif, one band on lower body at widest part. *Comments*: from Rooms 1–3 (1976–1983 seasons). *Date*: MM IA. *Bibl.*: Langford-Verstegen 2008, 555, fig. 9:9; Betancourt 2011a, 27, fig. 1:3.

146 (HCH 130; Fig. 14). Small jug, body sherd from the shoulder. Max. dim. 5.9 cm. A pale fine fabric (pink, 7.5YR 7/6); burnished. Rounded profile. Dark slip: diagonal lines flanking a zigzag motif. *Comments*: from Rooms 1–3 (1983 season). *Date*: MM IA. *Bibl.*: Betancourt 2011a, 27, fig. 1:4.

147 (HCH 115; Fig. 14). Small jug, body sherd from the shoulder. Max. dim. 3.4 cm. A pale fine fabric (reddish yellow, 7.5YR 7/6); burnished. Rounded profile. Dark slip: diagonal lines flanking a zigzag motif. *Comments*: from Rooms 1–3 (1983 season). *Date*: MM IA. *Bibl.*: Betancourt 2011a, 27, fig. 1:5.

148 (HCH 57; Fig. 14). Small jug, body sherd from the shoulder. Max. dim. 2.6 cm. A pale fine fabric (pink, 5YR 8/4); burnished. Rounded profile. Dark slip: diagonal lines flanking a band of filled triangles. *Comments*: from Rooms 1–3 (1983 season). *Date*: MM IA. *Bibl.*: Betancourt 2011a, 27, fig. 1:6.

149 (HCH 92; Fig. 14). Small jug, body sherd with handle. Max. dim. 5.2 cm. A pale fine fabric (reddish yellow, 5YR 7/6); burnished. Rounded profile, handle with circular section. Dark slip: diagonal lines flanking a band of crosshatched lines. *Comments*: from Rooms 1–3 (1983 season). *Date*: MM IA. *Bibl.*: Betancourt 2011a, 27, fig. 1:7.

150 (HCH 112; Fig. 14). Small jug, body sherd with handle. Max. dim. 5.2 cm. A pale fine fabric (pink, 5YR 7/4); burnished. Rounded profile, handle with circular section. Dark slip: diagonal lines flanking a band of crosshatched lines. *Comments*: from Rooms 1–3 (1983 season). *Date*: MM IA. *Bibl.*: Betancourt 2011a, 27, fig. 1:8.

151 (HCH 114; Fig. 14). Small jug, body sherd. Max. dim. 4.1 cm. A pale fine fabric (reddish yellow, 5YR 6/6); burnished. Rounded profile. Dark slip: diagonal lines flanking an incomplete (unidentifiable) motif. *Comments*: from Rooms 1–3 (1983 season). *Date*: MM IA. *Bibl.*: Betancourt 2011a, 27, fig. 1:9.

152 (HCH 59; Fig. 14). Small jug, body sherd. Max. dim. 3.2 cm. A pale fine fabric (reddish yellow, 5YR 7/6); burnished. Rounded profile. Dark slip: diagonal lines flanking a band of groups of short diagonal lines. *Comments*: from Rooms 1–3 (1983 season). *Date*: MM IA. *Bibl.*: Betancourt 2011a, 27, fig. 1:10.

153 (HCH 91; Fig. 14). Small jug, upper part. Max. dim. 6.8; rest. h. ca. 12 cm. A pale fine fabric (reddish yellow, 7.5YR 8/6); burnished. Rounded profile, small raised spout, handle with circular section. Dark slip: diagonal lines on tip of spout, two bands at base of neck, diagonal lines flanking bands of chevrons. *Comments*: from Rooms 1–3 (1983 season). *Date*: MM IA. *Bibl.*: Betancourt 2011a, 27, fig. 1:11.

154 (HCH 164; Fig. 14). Small jug, body sherd from shoulder. Max. dim. 3.4; rest. h. ca. 12 cm. A pale fine fabric (reddish yellow, 5YR 7/6); burnished. Rounded profile. Dark slip: diagonal lines flanking a band of chevrons. *Comments*: from Rooms 1–3 (1983 season). *Date*: MM IA. *Bibl.*: Betancourt 2011a, 27, fig. 1:12.

155 (HCH 181; Fig. 14). Small jug, lower part. Rest. h. ca. 12; d. of base 4.0 cm. A pale fine fabric (between pink, 7.5YR 7/4, and reddish yellow, 5YR 7/6); burnished. Rounded profile. Dark slip: diagonal lines flanking a band of diagonal lines. *Comments*: from Rooms 1–3 (1982 season). *Date*: MM IA. *Bibl.*: Betancourt 2011a, 27, fig. 1:13.

156 (HNM 13,809; Fig. 14). Small jug, almost complete (missing upper part of spout and part of handle). Rest. h. ca. 7.5; d. of base 3.7 cm. A pale fine fabric, reddish yellow (7.5YR 7/5); burnished. Rounded profile, small raised spout, handle with circular section attached at rim and on shoulder. Dark slip: horizontal lines on neck, diagonal lines on body, horizontal line on lower body. *Comments*: from Rooms 1–4 (1976–1983 seasons). *Date*: MM IA. *Bibl.*: Betancourt 2011a, 27, fig. 1:14.

157 (HNM 13,807; Fig. 14). Small jug, almost complete (missing top of spout and handle). H. 9; d. of base 3.4 cm. A pale fine fabric (pink, 7.5YR 7/4); burnished. Rounded profile, handle with circular section attached below rim and on shoulder. Dark slip: horizontal line at base of neck, sets of two diagonal lines on body, two horizontal bands on body. *Comments*: from Rooms 1–4 (1976–1983 seasons). *Date*: MM IA. *Bibl.*: Betancourt 2011a, 27, fig. 1:15.

158 (HNM 13,828; Fig. 14). Small jug, almost complete (missing handle and top of spout). Rest. h. ca. 10; d. of base 3.7 cm. A pale fine fabric (pink, 7.5YR 7/4). Rounded profile, small raised spout, handle with circular section attached at rim and on shoulder, rounded profile. Dark slip on exterior. Added white band at base of neck, diagonal lines flanking a chevron motif, and two bands on lower body. *Comments*: from Rooms 1–4 (1976–1983 seasons). *Parallels*: for the light-on-dark variation at Malia, see Demargne 1945, pl. 28:8600. *Date*: MM IA. *Bibl.*: Betancourt 2011a, 27, fig. 1:16.

159 (HCH 04-350; Fig. 14). Medium-sized jug, lower part. Rest. h. ca. 17–18; d. of base 8.2 cm. A pale fine fabric (reddish yellow, 5YR 7/6); burnished. Rounded profile. Dark slip: diagonal lines flanking zigzag band, double band around body. *Comments*: from Room 4, upper levels (1983 season). *Date*: MM IA. *Bibl.*: Betancourt 2011a, 27, fig. 1:17.

160 (HCH 05-452; Fig. 14). Medium-sized jug, body sherd. Rest. h. ca. 17–18; max. dim. 4.0 cm. A red fabric (light red, 2.5YR 6/8); burnished. Rounded profile. Dark slip: diagonal lines flanking band of groups of short diagonal lines. *Comments*: from Rooms 1–4 (1983 season). *Date*: MM IA. *Bibl.*: Betancourt 2011a, 27, fig. 1:18.

161 (HCH 04-358; Fig. 15). Medium-sized jug, rim sherd with handle. Rest. h. ca. 17–18; d. of rim 6 cm. A pale fine fabric (reddish yellow, 5YR 7/6); burnished. Rounded profile, small raised spout, handle with circular section attached below rim and on shoulder. Dark slip: horizontal band at base of neck, diagonal lines flanking band of chevrons on body. *Comments*: from Room 4, upper levels (1983 season). *Date*: MM IA. *Bibl.*: Betancourt 2011a, 27, fig. 1:19.

162 (HCH 02-56; Fig. 15). Medium-sized jug, almost complete. H. ca. 12.2; d. of base 6.2 cm. A fine fabric (red, 2.5YR 5/8). Rounded profile, small raised spout, handle with circular section attached below rim and on shoulder, pair of lugs on sides of spout. Dark slip (red, 10R 5/8) on exterior. Added white band on rim, band at base of neck, two bands at center of body, band on lower body, diagonal band on shoulder consisting of thin bands flanking another motif, which is not preserved (possibly chevrons). *Comments*: from Room 4, upper levels (1983 season). Analyzed by ceramic

petrography, sample no. HCH 04/11, Hagios Charalambos Fabric Group 2b (Lasithi Red Fabric Group). *Date*: MM IA. *Bibl.*: Betancourt 2011a, 27, fig. 1:20.

Jugs with Bosses on Body

163 (HNM 13,819; Fig. 15). Jug, almost complete. Rest. h. ca. 10; d. of base 3.2 cm. A fine pale fabric (pink, 5YR 8/4). Piriform shape, handle with circular section. Five bosses on upper shoulder. Dark bands on upper shoulder, triple diagonal bands on body. *Comments*: from Rooms 1–4 (1976–1983 seasons). *Parallels*: for the small jug with large bosses, see Sakellarakis and Sapouna-Sakellaraki 1997, 383, fig. 335, second from left (Archanes). *Date*: MM I–IIB.

164 (HCH 03-216; Fig. 15; Pl. 2). Jug, half complete. Rest. h. ca. 9–10; d. of base 6 cm. A fine pale fabric (pink, 7.5YR 8/4). Piriform shape, handle with thick oval section attached at rim and shoulder. Three or four bosses on shoulder. Three dark bands on upper shoulder, three bands on lower body. *Comments*: from Rooms 1–4 (1976–1983 seasons). *Parallels*: Pendlebury, Pendlebury, and Money-Coutts 1935–1936, 74, fig. 17:662 (Trapeza Cave); Momigliano 1991, pl. 31:27 (Knossos). *Date*: MM I–IIB. *Bibl.*: Betancourt et al. 2014, 29, fig. 17:f.

165 (HCH 19; Fig. 15). Jug, base and body sherd. Rest. h. ca. 7–9; d. of base 3.3 cm. A fine pale fabric (reddish yellow, 5YR 7/6). Rounded profile. Four bosses, two on shoulder and two on lower body. Paint missing or never present. *Comments*: from Rooms 1–4 (1976–1983 seasons). *Parallels*: see **164**. *Date*: MM I–IIB.

166 (HCH 03-218; Fig. 15). Jug, body sherd from the neck and shoulder. Max. dim. 6.4; rest. h. ca. 10–15 cm. A fine fabric (reddish yellow, 5YR 7/6); burnished. Rounded profile. Bosses at center of body. Dark slip on exterior. Added white bands on neck and shoulder. *Comments*: from Room 5, area 5, level 2 (unit HCH 02-3-5-2). *Date*: MM I–IIB.

Various Classes of Small Jugs

167 (HNM 13,813; Fig. 15). Small jug, complete. H. 7.4; d. of base 4.5 cm. A pale fine fabric (pink, 7.5YR 7/4). Rounded profile, small low spout, handle with elliptical section attached at rim and shoulder. Dark slip: horizontal lines on spout, sets of three diagonal lines on body, vertical line on handle. *Comments*: from Rooms 1–4 (1976–1983 seasons). *Date*: MM I.

168 (HNM 12,413; Fig. 15). Small jug, complete. H. 7; d. of base 3.1 cm. A pale fine fabric (pink, 5YR 7/4). Rounded profile, slim shape, small low spout, handle with circular section attached at rim and shoulder. Dark slip: line on spout, horizontal band at base of neck, sets of four diagonal lines on body placed in alternate directions. *Comments*: from Room 4, upper levels (1983 season). *Date*: MM I.

169 (HCH 04-304; Fig. 15). Small jug, half complete. Rest. h. ca. 7–9; d. of base 4.2 cm. A pale fine fabric (reddish yellow, 5YR 7/6); burnished. Rounded profile, handle attached at rim and shoulder. Surface in poor condition. *Comments*: mended from the Room 4/5 Entrance (unit HCH 02-2/3 Ent) and from Room 4, upper levels (1983 season). *Date*: MM I.

170 (HCH 02-114; Fig. 15). Small jug, upper part. Pres. h. 6.8; rest. h. ca. 8 cm. A pale fine fabric (light red, 2.5YR 6/6); burnished. Rounded profile. Dark slip: diagonal lines on tip of spout, horizontal band at base of neck, traces of diagonal lines at upper part of shoulder, with the remainder of the decoration missing. *Comments*: from the 4/5 Entrance, level 3 (unit HCH 02-2/3 Ent-3). *Date*: MM I.

171 (HCH 128; Fig. 15). Small jug, almost complete. Rest. h. ca. 9–10; d. of base ca. 6 cm. A fine pale fabric (reddish yellow, 5YR 7/6). Rounded profile. Dark slip: bands on neck. *Comments*: from Rooms 1–4 (1976–1983 seasons). *Date*: MM I–IIB.

172 (HCH 177; Fig. 15). Small jug, upper part. Rest. h. ca. 15 cm. A fine pale fabric (reddish yellow, 5YR 7/6); burnished. Conical shoulder and small handle. Dark slip: horizontal bands on upper part of shoulder and body. *Comments*: from Rooms 1–3 (1982 season). *Date*: MM I–IIB.

173 (HNM 12,408; Fig. 15). Small jug, almost complete. Rest. h. ca. 8 cm. A fine pale fabric (pink, 7.5YR 7/6). Rounded profile, small slightly raised spout, handle with circular section attached at rim and shoulder. Paint missing or never present. *Comments*: from Room 4, upper levels (1983 season). *Date*: MM I–IIB.

174 (HNM 12,461; Fig. 15; Pl. 2). Small jug, complete. H. 9.6 cm. A fabric tempered with phyllite and chaff (light red, 2.5YR 6/8). Rounded profile, slim shape, small slightly raised spout, tiny handle attached at rim and on shoulder. Paint missing or never present. *Comments*: from Rooms 1–4 (1976–1983 seasons). *Date*: EM II–MM I.

175 (HNM 13,815; Fig. 15). Small jug, lower part. Rest. h. ca. 10–12 cm. A fine fabric (red, 2.5YR 5/6 with pale surface). Rounded profile. Paint missing or never present. *Comments*: from Room 4, upper levels (1983 season). *Date*: MM I.

176 (HCH 02-54, in HNM; Fig. 16). Small jug, almost complete (handle and spout missing). Rest. h. ca. 10–12 cm. A fine fabric (reddish yellow, 7.5YR 6/6). Rounded profile. Dark slip: traces. Added white traces of hatched area. *Comments*: from Room 5, area 5, level 2 (unit HCH 02-3-5-2). *Date*: MM I.

177 (HCH 02-99, in HNM; Fig. 16). Small jug, almost complete. Rest. h. ca. 10–12; d. of base 4 cm. A fine fabric (light brown, 7.5YR 6/6). Rounded profile, small spout, handle with circular section attached at rim and shoulder. Dark slip: crosshatched band on shoulder, bands on rim, neck, and lower body. *Comments*: from the Room 4/5 Entrance, level 3 (unit HCH 02-2/3 Ent-3). *Date*: MM I.

178 (HCH 02-20, in HNM; Fig. 16). Small jug, complete except for handle. H. 11.3; d. of base 4.2 cm. A fine fabric (reddish yellow, 7.5YR 7/6). Rounded profile. Dark slip: traces. *Comments*: from Room 4, upper levels (unit HCH 02-2-2). *Date*: MM I–IIB.

179 (HCH 58; Fig. 16). Small jug, upper body sherd. Max. dim. 5.3 cm. A pale fine fabric (reddish yellow, 5YR 6/6); burnished. Small jug with rounded profile. Dark slip: vertical bands. *Comments*: from Rooms 1–3 (1982 season). *Date*: MM I.

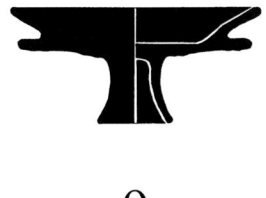

9

Jugs with Bosses in Barbotine Style, MM I–II

Added plastic decoration, such as bosses, is the defining characteristic of the Middle Minoan Barbotine Style (Kaiser 1976; Foster 1982, 1–52). At Hagios Charalambos, this class is made up of several fragments of jugs with horizontal rows of small bosses. They are cataloged separately because they do not join, but it is likely that a few may be from the same vessel. A fabric sample of this class was analyzed, and it is classified as Hagios Charalambos Fabric Group 8, which may be from the area of Malia.

It is interesting to note that the Barbotine Style was decidedly unpopular at Hagios Charalambos, as few examples are present, and some fragments may join. However, if indeed the fabric of this class is of Malian manufacture (more rare than the hallmark Barbotine vessels from South Crete), this would mean that the examples here, albeit few, contribute to the significant increase at Hagios Charalambos of MM II pottery from Malia that belies an economic or political shift, which influenced the Lasithi-Pediada region during the Middle Bronze Age.

Jugs in Barbotine Style

The decoration on these jugs consists of three horizontal bands of small bosses and dark-painted areas that are sometimes across the barbotine. Vessels with this type of decoration occur at sites near the northern coast of Central Crete (Demargne and Gallet de Santerre 1953, pl. 51:g [Malia]; Zois 1969, pl. 31; Foster 1982, pl. 22 [Gournes]; Sakellarakis and Sapouna-Sakellaraki 1997, 415, fig. 380, left and right [Archanes]; Momigliano 1991, pl. 31:40a [Knossos]).

180 (HCH 04-320; Fig. 16). Wide-mouthed jug, complete profile. H. 9.8; d. of rim ca. 7.5; d. of base 4.6 cm. A pale fabric (reddish yellow, 7.5YR 7/6). Rounded profile, handle with circular section attached below rim and on shoulder. Plastic bosses: three horizontal rows at mid body. Dark slip on exterior (grayish brown 10R 5/2). Added white traces? *Comments*: from Room 4 (1983 season). Barbotine Style. *Parallels*: for the form, which has been recorded from Malia, Lasithi, and elsewhere in Central Crete, see Walberg 1983, 173, type 130. *Date*: MM IB–IIA. *Bibl.*: Langford-Verstegen 2008, 555, fig. 9:7.

181 (HCH 137; Fig. 16). Jug, body sherd. Max. dim. 4.5 cm. A pale fabric (reddish yellow, 5YR 5/6). Rounded profile. Plastic bosses, probably in horizontal rows. Dark slip on exterior. Added white two horizontal bands. *Comments*: from Rooms 1–4 (1983 season). Barbotine Style. *Date*: MM IB–IIA.

182 (HCH 176; Fig. 16). Jug, body sherd. Max. dim. 3.1 cm. A pale fabric (light reddish brown, 5YR 6/4). Rounded profile. Plastic bosses: three horizontal or diagonal rows. Traces of dark slip on exterior. *Comments*: from Rooms 1–4 (1983 season). Barbotine Style. Analyzed by ceramic petrography, sample no. HCH 04/45 (Hagios Charalambos Fabric 8). *Date*: MM IB–IIA

183 (HCH 138; Fig. 16). Jug, body sherd. Max. dim. 5 cm. A pale fabric (light red, 2.5YR 6/8). Rounded profile. Plastic bosses: two horizontal rows. Dark slip on exterior (light reddish brown, 5YR 6/3–6/4). Added white traces? *Comments*: from Rooms 1–4 (1983 season). Barbotine Style. *Date*: MM IB–IIA.

184 (HCH 188; Fig. 16). Jug, body sherd. Max. dim. 2.9 cm. A pale fabric (pink, 5YR 7/4). Rounded profile. Plastic bosses. Dark slip on exterior. Added white two horizontal bands. *Comments*: from Rooms 1–4 (1983 season). Barbotine Style. *Date*: MM IB–IIA.

185 (HCH 187; Fig. 16). Jug, body sherd. Max. dim. 8 cm. A pale fabric (reddish yellow, 5YR 7/8). Rounded profile. Plastic bosses. Traces of dark slip on exterior. *Comments*: from Rooms 1–4 (1983 season). Barbotine Style. *Date*: MM IB–IIA.

186 (HCH 02-110; Fig. 16). Jug, body sherd. Max. dim. 4.7 cm. A pale fabric (reddish yellow, 5YR 7/6). Rounded profile. Plastic bosses. Dark slip on exterior. Added white horizontal bands. *Comments*: from the Room 4/5 Entrance, level 3 (unit HCH 02-3/4 Ent-3). Barbotine Style. *Date*: MM IB–IIA.

187 (HCH 184; Fig. 16). Jug, body sherd. Max. dim. 5.6 cm. A medium coarse fabric, (reddish brown, 2.5 YR 5/4), with phyllite. Rounded profile. Band of large plastic bosses. Dark slip on exterior. Added white bands. *Comments*: from Rooms 1–4 (1983 season). Barbotine Style. Analyzed by ceramic petrography, sample no. HCH 04/46 (Hagios Charalambos metamorphric fabric, loner). *Date*: MM IB–IIA.

10

Knossian Pottery, EM II–MM II

Early imports from Knossos are few, but significant; they make up a small, representative set of pouring and drinking vessels. Two examples of goblets are constructed of a tall cup attached to a foot. Two jars, one of which has a bridged spout, are slipped and decorated with bands. The spouted jar has parallels from the North Central Court at Knossos (Momigliano 1991, 177, fig. 5:1; 2000, 79, fig. 8:24, 25, 26).

Goblets

Knossian goblets are taller than the shape favored farther east (Momigliano 1990; 1991, 247–248; Macdonald and Knappet 2007, fig. 3.1). They consist of a small conical foot supporting a tall conically shaped cup with a slightly rounded profile. This form also occurs elsewhere in North-Central Crete (Zois 1969, pl. 30:2 [Gournes]; Sakellarakis and Sapouna-Sakellaraki 1997, 391, fig. 342, upper right [Archanes]).

188 (HCH 44; Fig. 17). Goblet, base sherd. D. of base 6 cm. A pale fine fabric (light red, 2.5YR 6/6). Footed goblet. Dark slip on exterior. *Comments*: from Rooms 1–4 (1983 season). Identified as Knossian, pers. comm. D. Wilson and N. Momigliano. *Date*: EM IIB or EM II–MM I.

189 (HCH 45; Fig. 17). Goblet, base sherd. D. of base 4.6 cm. A medium coarse pale fabric (pink, 5YR 7/6). Footed goblet with tall cup. Dark slip on exterior (light reddish brown, 5YR 6/4). *Comments*: from Rooms 1–4 (1983 season). Cup attached directly to thin base with rounded edges. Identified as Knossian, pers. comm. D. Wilson and N. Momigliano. *Date*: EM IIB or EM III–MM I. *Bibl.*: Betancourt et al. 2008, 166, fig. 16:g; Langford-Verstegen 2008, 555, fig. 9:6.

190 (HCH 03-204; Fig. 17). Goblet, base sherd. D. of base 4.6 cm. A fine pale fabric (reddish yellow, 5YR 7/6), burnished. Footed goblet with tall cup. *Comments*: from Room 5, area 5, level 1 (unit HCH 02-3-5-1). *Date*: EM IIB or EM III–MM I.

Spouted Jar

191 (HCH 179; Fig. 17). Spouted jar, rim sherd with neck. D. of rim 10 cm. A pale fabric (pink, 7.5YR 7/4); burnished. Rounded profile, slightly outturned rim. Dark slip on interior of rim, band at rim. Added white pairs of horizontal thin bands. *Comments*: from Rooms 1–3 (1982 season). *Parallels*: Knossos, North Central Court (Momigliano 1991, 177, fig. 5:1; 2000, 79, fig. 8:24, 25, 26). *Date*: MM I–II.

Wide-Mouthed Vessel with Bridged Spout

192 (HCH 03-166; Fig. 17). Bridge-spouted vessel, spout and rim sherds. D. of rim 14 cm. A pale fine fabric (between pink, 7.5YR 7/4, and reddish yellow, 7.5YR 7/6). Highly burnished. Convex profile, everted rim; handles do not survive. Two bands at interior of rim, two bands at exterior of rim, four bands on spout and top of rim. *Comments*: mended from sherds found in Room 5, area 3, level 4 and area 5, level 2 (units HCH 02-3-5-2 and HCH 03-5-3-4), Room 7, surface (unit HCH 03-7-1), and the 1976–1983 dump (unit HCH-Dump). *Date*: MM I–II.

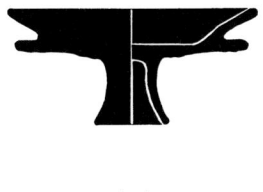

11

Fine Red Fabrics with Dark Red Slip, Mostly MM IB

Several different shapes were imported into the site from production centers that made pottery with thin walls in red-firing clay and finished them with a darker red or black-firing slip. Tumblers predominate in this class, but a few other shapes also exist, all small and delicate. Besides the tumbler, open shapes include a goblet, a kantharos, and a carinated cup. Closed shapes include a miniature pithos and jugs. The fabric is dark red (10R 4/8 to 2.5YR 5/8), and the slip of the red pieces is even darker red than the fabric. Among the members of this group are the vases from the Pediada region that have been discussed by Rethemiotakis and Christakis in their study of pottery from Galatas and Kastelli and called Kastelli Fabric (Rethemiotakis and Christakis 2004). Petrographic analyses of three samples from this group show that more than a single fabric is present in the class imported to Hagios Charalambos.

Tumbler

193 (HCH 67; Fig. 17). Tumbler, base sherd. D. of base 2 cm. A fine fabric (red, 2.5YR 4/8); burnished. Dark red slip on interior and exterior. Tall cup, pronounced base. *Comments*: from Room 4, upper levels (1983 season). *Date*: MM IB.

194 (HCH 68; Fig. 17). Tumbler, base sherd. D. of base 2.5 cm. A fine fabric (red, 2.5YR 6/6). Dark red slip (10R 5/6) on interior and exterior. Tall cup with straight profile. *Comments*: from Rooms 1–4 (1983 season). *Date*: MM IB.

195 (HCH 03-224; Fig. 17). Tumbler, base sherd. D. of base 2 cm. A fine fabric (light red, 2.5YR 5/6). Dark slip on exterior (red, 2.5YR 5/6). Tall cup with almost straight profile, pronounced base. *Comments*: from Room 5, area 5, level 2 (unit HCH 02-3-5-2). *Date*: MM IB.

196 (HCH 04-381; Fig. 17). Tumbler, base sherd. D. of base 3.8 cm. A fine fabric (red, 2.5YR 4/6–8); burnished. Dark red slip on exterior (2.5YR 4-5/6). Tall cup with straight profile. *Comments*: from Room 4, upper levels (1983 season). Analyzed by ceramic petrography, sample no. HCH 04/44 (Hagios Charalambos Fabric Group 5), considered to be from the coastal plain of Malia. *Date*: MM IB.

197 (HCH 05-418; Fig. 17). Tumbler, body sherd. Max. dim. 2.1 cm. A fine fabric (red, 2.5YR 5/8); burnished. Dark red slip on exterior. Tall cup with straight profile. *Comments*: from Room 4, upper levels (1983 season). Analyzed by ceramic petrography, sample no. HCH 04/10 (Hagios Charalambos Fabric Group 5), considered to be from the coastal plain of Malia. *Date*: MM IB.

Goblet

198 (HCH 55; Fig. 17). Goblet, lower part. D. of base 3.7 cm. A fine fabric (light red, 2.5YR 6/8). Dark red slip (5YR 4/8) on interior and exterior. Tall conical goblet. *Comments*: from Rooms 1–4 (1983 season). *Parallels*: for the shape, see MacGillivray 1998, pl. 1:55 (Knossos); 2007, 115, fig. 4.7:8 (Knossos). *Date*: MM IB.

Kantharos

199 (HCH 06-460; Fig. 17). Kantharos, rim sherd. D. of rim 8.8 cm. A fine fabric (red, 2.5YR 5/6). Dark red slip (5YR 4/8) on interior and exterior. Open vessel with two opposed handles. *Comments*: from Rooms 1–4 (1983 season). *Date*: MM IB.

Tall Carinated Cup

200 (HCH 22; Fig. 17). Tall carinated cup with raised handle, body, base, and handle sherds. Rest. h. ca. 8.5; d. of base 2.8 cm. Black slip. *Comments*: from Rooms 1–4 (1983 season). Analyzed by ceramic petrography, sample no. HCH 04/43 (Hagios Charalambos Micaceous Fabric, loner). A similar vase comes from Galatas (Rethemiotakis and Christakis 2004, 172, fig. 12.2:c, from MM IB). *Date*: MM IB.

Carinated Cup with Grooves

201 (HCH 03-212; Fig. 17). Carinated cup, body sherd. Max. dim. 3.4 cm. A fine fabric (red, 2.5YR 5/6). Horizontal incised lines on exterior. *Comments*: from Rooms 1–4 (1983 season). The carinated cup

begins in MM IB, and it continues into MM IIB. The class with horizontal grooves on the upper section is a "type fossil" for MM IIB in East-Cental and East Crete (Walberg 1976, 150–151, form 47; 1983, pls. 16, 17, nos. 257–260). *Parallels*: Hawes et al. 1908, pl. 2:11 (Gournia); Betancourt and Silverman 1991, 25–26, nos. 410–415 (Gournia); Pendlebury, Pendlebury, and Money-Coutts 1935–1936, 62 (Trapeza Cave); Demargne 1945, pl. 33:8657 (Malia); Coldstream and Huxley 1972, 94–95 (Kythera); Betancourt 1983, pls. 19, 20, nos. 257–260 (Vasiliki); Betancourt and Davaras, eds., 2003, 15, 46, 79, nos. 1.122, 4.61, 4.62, 9.64 (Pseira); Tsipopoulou and Wedde 2000, 377, fig. 14, lower left (Petras); Poursat and Knappett 2005, 224–225, nos. 720–765 (Malia). *Date*: MM IIB.

Miniature Pithos

202 (HCH 02-25; Fig. 17; Pl. 2). Miniature pithos, three quarters complete. H. 6.4; d. of rim 5; d. of base 4 cm. A fine fabric (red, 2.5YR 4/6–8); burnished. Dark red slip on exterior. Miniature jar with slightly everted rim, pronounced base, three sets of three vertical handles. *Comments*: from the 4/5 Entrance, surface (unit HCH 03-4/5 Ent-Surface). *Parallels*: Poursat 1996, pls. 22:a–g, 33:f (Malia). *Date*: MM IB.

Small Jug(?)

203 (HCH 04-380; Fig. 17). Jug(?), neck and shoulder sherd. Max. dim. 5.6 cm. A fine red (possibly Kastelli?) fabric (red, 2.5YR 5/8). Dark red slip on exterior (red, 2.5YR 5/6), burnished. *Comments*: from Rooms 1–4 (1983 season). *Date*: MM IB.

Jug with Incised Band on Body

204 (HCH 02-122; Fig. 17). Jug, spout, base, neck, and body sherds. D. of base 4.9; max. d. of spout and neck 8.1 cm. A fine, dark (burned?) fabric containing phyllite. Thin spout, rounded shoulder, straight base, strap handle attached at rim and shoulder. *Comments*: mended from sherds from Room 5, area 5, level 2 (unit HCH 02-5-5-2) and the 4/5 Entrance, level 3 (unit HCH 03-4/5 Ent-3). Analyzed by ceramic petrography, sample no. HCH 04/12 (Hagios Charalambos Fabric Group 2a, Lasithi Red Fabric Group). *Date*: MM IB.

12

Cups and Trefoil-Mouthed Jugs with Parallels from Galatas, MM IB

Open-mouthed jugs with trefoil spouts predominate in this class of pottery made from a fabric from the Pediada region. A sample was analyzed by ceramic petrography (sample no. HCH 04/6), and the fabric has been classified as metamorphic but different from Lasithi Fabric Group 2. The fabric is light red (2.5YR 6/8), and all examples have a light red, slightly lustrous surface that was probably covered with a slip made from clay that was similar to that used for the body. This fabric is one of the types used for pottery found at Galatas (Rethemiotakis and Christakis 2004, 171). One example has traces of added white decoration in the form of bands. Another example has plastic decoration consisting of small knobs.

Straight-Sided Cup

205 (HCH 5; Fig. 18). Straight-sided cup, base sherd with part of handle. D. of base 4.3 cm. A fine fabric (light red, 2.5YR 6/8); burnished. Red slip on exterior. Added white horizontal bands at center of cup and traces above bands. *Comments*: from Rooms 1–4 (1983 season). *Date*: MM IB.

Jugs with Trefoil Mouths

The attribution of this class to the Pediada is based on excavations at Galatas that uncovered many examples of the shape (Rethemiotakis and Christakis 2004, 172, fig. 12.2:c). The class has also been found at Kato Symi (Christakis 2013, 170, fig. 15.2), Knossos (Momigliano 1991, pl. 50:57), and Phaistos (Levi 1976, vol. I*, pl. 34:a, g, h). The vessel has relatively thin walls, a trefoil mouth, and a single vertical handle. The profile of the body varies from rounded to carinated.

206 (HNM 13,833; Fig. 18; Pl. 2). Trefoil-mouthed jug, complete except for rim and handle. Rest. h. ca. 14; d. of base 4.5 cm. A fine fabric (light red 2.5YR 6/5). Profile with rounded carination, slim shape, handle with circular section attached at rim and lower shoulder. Red slip on exterior. *Comments*: from Rooms 1–4 (1983 season). *Date*: MM IB.

207 (HNM 13,850; Fig. 18). Trefoil-mouthed jug, lower part. Rest. h. ca. 14; d. of base 5.4 cm. A fine fabric (red, 2.5YR 5/8, with a darker core). Profile with rounded carination, handle with circular section attached at lower shoulder. Red slip on exterior. *Comments*: from Room 4, upper levels (1983 season). *Date*: MM IB.

208 (HNM 13,852; Fig. 18). Trefoil-mouthed jug, almost complete. H. ca. 25 cm. A fine fabric (red, 2.5YR 5/6 to 10R 6/8). Body with rounded profile, handle with circular section attached at rim and lower shoulder. Red slip on exterior. *Comments*: from Rooms 1–4 (1976–1983 seasons). *Date*: MM IB.

209 (HNM 12,440; Fig. 18). Trefoil-mouthed jug, almost complete. H. 14.0 cm. A fine fabric (red, 2.5YR 5/6). Body with rounded profile, handle with circular section attached at rim and lower shoulder. Red slip on exterior. *Comments*: from Room 4 (1983 season). Burned. *Date*: MM IB.

210 (HNM 13,845; Fig. 18). Trefoil-mouthed jug, almost complete. H. 15.5; d. of base 7.5 cm. A fine red fabric (2.5YR 5/8). Body with rounded profile, handle with circular section attached at rim and lower shoulder. Red slip on exterior. *Comments*: from Room 3 (1982 season). *Date*: MM IB.

211 (HNM 13,853; Fig. 18). Trefoil-mouthed jug, almost complete, missing rim and handle. Rest. h. ca. 14; d. of base 5.1 cm. A fine fabric (red, 2.5YR 5/6) with a darker surface. Body with rounded profile. Red slip on exterior. *Comments*: from Room 4 (1983 season). *Date*: MM IB.

212 (HCH 33; Fig. 18). Trefoil-mouthed jug, rim sherd. Max. dim. 5.0 cm. A fine fabric (light red, 2.5YR 6/8). Red slip on exterior. *Comments*: from Room 4, upper levels (1983 season). *Date*: MM IB.

213 (HCH 02-64; Fig. 18). Trefoil-mouthed jug, upper part. Rest. h. ca. 14; d. of rim ca. 8.1 cm. A fine fabric (red, 2.5YR 5/6, with a darker core). Body with rounded profile, handle with circular section attached at rim and lower shoulder. Red slip on exterior. Possible traces of white. *Comments*: from the Room 4/5 Entrance, surface (unit HCH 02-2/3 Ent). *Date*: MM IB.

214 (HCH 02-46; Fig. 18). Trefoil-mouthed jug(?), lower part. D. of base 4.7 cm. A fine fabric (red, 2.5YR 4/8). Body with rounded profile, handle with circular section attached at lower shoulder. Red slip on exterior. *Comments*: from the Room 4/5 Entrance, surface (unit HCH 02-2/3 Ent). *Date*: MM IB.

215 (HCH 02-132; Fig. 19). Trefoil-mouthed jug, upper part. Pres. h. 16.3; d. of rim ca. 10 cm. A fine fabric (red, 2.5YR 4/4). Body with rounded profile, handle with circular section attached at rim and lower shoulder. Red slip on exterior. *Comments*: from the Room 4/5 Entrance, level 3 (unit HCH 02-2/3 Ent-3). *Date*: MM IB.

216 (HCH 02-55, Fig. 19). Trefoil-mouthed jug, complete profile. H. 11.1; d. of base 5.2 cm. A fine fabric

(between red, 2.5YR 6/8, and light red, 2.5YR 5/8). Body with rounded profile, handle with circular section attached at rim and lower shoulder. Red slip on exterior. *Comments*: from Room 5, area 5, level 2 (unit HCH 02-3-5-2). Burned. *Date*: MM IB.

217 (HCH 05-414; Fig. 19). Trefoil-mouthed jug, rim sherd. Max. dim. 5.3 cm. A fine fabric (light red, 2.5YR 6/8). Red slip on exterior. *Comments*: from Room 4, upper levels (1983 season). Analyzed by ceramic petrography, sample no. HCH 04/6 (Hagios Charalambos Metamorphic Fabric). *Date*: MM IB.

Side-Spouted Jug or Cup

218 (HNM 13,832; Fig. 19). Side-spouted jug or cup, almost complete. H. 9.7; d. of rim 8.3; d. of base 5.5 cm. A fine fabric (light red, 2.5YR 6/6). Wide-mouthed vessel with rounded profile and slightly flaring rim, with open shallow spout at the side, vertical handle with circular section attached at rim and widest part of body, two pairs of bosses on opposite sides of body, and one boss under the spout. Red slip on exterior. *Comments*: from Room 4, upper levels (1983 season). *Date*: MM IB.

219 (HCH 04-306; Fig. 19). Side-spouted jug or cup, body sherd. Max. dim. 8.5 cm. A fine fabric (light red, 2.5YR 6/8). Wide-mouthed vessel with rounded profile and slightly flaring rim, probably with open shallow spout at the side, with small knobs on body. Red slip on exterior. *Comments*: from Room 4, upper levels (1983 season). *Date*: MM IB.

220 (HNM 12,442; Fig. 19; Pl. 2). Side-spouted jug or cup, complete. H. 9.4; d. of rim 8.1; d. of base 5.7 cm. A fine fabric (light red, 2.5YR 6/6). Wide-mouthed vessel with rounded profile and slightly flaring rim, with open shallow spout at the side, vertical handle with circular section attached at rim and widest part of body. Red slip on exterior. *Comments*: from Room 4, upper levels (1983 season). *Date*: MM IB.

221 (HNM 12,443; Fig. 19). Side-spouted jug or cup, complete. H. 10.2; d. of rim 7.1; d. of base 5.1 cm. A fine fabric (light red, 2.5YR 6/6). Wide-mouthed vessel with rounded profile and slightly flaring rim, with open shallow spout at the side, vertical handle with circular section attached at rim and widest part of body. Red slip on exterior. *Comments*: from Room 4, upper levels (1983 season). Burned. *Date*: MM IB.

222 (HNM 12,435; Fig. 19). Side-spouted jug or cup, complete. H. 10.2; d. of rim 8.4; d. of base 6 cm. A fine fabric (light red, 2.5YR 6/6). Wide-mouthed vessel with rounded profile and slightly flaring rim, with open shallow spout at the side, vertical handle with circular section attached at rim and widest part of body. Red slip on exterior. *Comments*: from Room 4, upper levels (1983 season). Burned. *Date*: MM IB. *Bibl.*: Betancourt et al. 2014, 29, fig. 17:g.

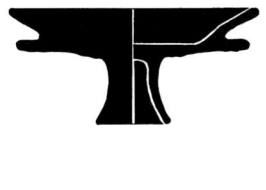

13

Kamares Ware, MM IB–IIB

Kamares Ware is a Middle Minoan style of polychrome pottery highly decorated with dynamic geometric motifs as well as designs drawn from the natural world (Walberg 1976). The Early Kamares Ware pottery, from the initial period of the tradition in MM IB to MM IIA, has light-on-dark decoration with added white and red paint in the form of bands and other details that are not as elaborate as the ornament from the mature phase in MM IIB. The decoration of the pieces found in the cave is by no means as sophisticated or developed as the finest examples from the palaces, but one cannot be certain if the date of these fragments is early or late in the sequence. Shapes here include cups, jugs, and jars.

Straight-Sided Cup

223 (HCH 05-447; Fig. 20). Straight-sided cup, base sherd with part of handle. D. of base 8.0 cm. A fine, pale fabric (reddish yellow, 5YR 7/6). Handle with elliptical section. Dark slip on exterior and interior. Added white: horizontal bands flanking a red band. Added red: crossing pairs of diagonal bands. *Comments*: from Rooms

1–4 (1983 season). Classical Kamares Ware. *Parallels*: MacGillivray 1998, pl. 38:12 (Knossos). *Date*: MM IIA–IIB. *Bibl.*: Betancourt et al. 2008b, 166, fig. 28:h.

Semiglobular Cup

224 (HCH 71; Fig. 20). Semiglobular cup, rim sherd. D. of rim 12 cm. A pale, fine fabric (pink, 5YR 8/4 to 7/4). Dark slip: traces on interior rim and on exterior. Added white: twin horizontal and diagonal bands. Added red: horizontal and diagonal bands. *Comments*: from Rooms 1–4 (1983 season). Kamares Ware. *Date*: MM IB–IIA.

Carinated Cup

225 (HCH 165; Fig. 20). Carinated cup, rim sherd. D. of rim 8 cm. A pale fabric (pink, 5YR 7/4). Dark slip on interior and exterior (black, 5YR 2.5/1). Added white: horizontal bands on exterior and diagonal cross-hatched lines. Added red: band below rim and two diagonal lines. *Comments*: from Rooms 1–4 (1983 season). Kamares Ware. *Parallels*: for cross-hatching on the upper part of a carinated cup, see Levi 1976, I**, color pl. LIIIc (Phaistos). *Date*: MM IIB.

Jug

226 (HNM 13,818; Fig. 20). Jug, almost complete. H. 10.3; d. of base 4.5 cm. A pale, fine fabric (reddish yellow, 7.5 YR 7/6). Rounded profile. Dark slip in traces. Added white and red traces. *Comments*: from Rooms 1–4 (1976–1983 seasons). *Parallels*: compare Macdonald and Knappett 2007, fig. 3.27:499 (Knossos). *Date*: MM I–IIB.

227 (HCH 04-370; Fig. 20). Jug, spout sherd with handle. Opening of spout 2.3; max. dim. 6.2 cm. A pale fabric (reddish yellow, 7.5YR 7/6). Convex profile, handle with circular section. Dark slip on interior of spout and exterior. Added white: thin horizontal bands on spout and neck. Added red: band at neck and shoulder. *Comments*: from the Room 4/5 Entrance (unit HCH 02-2/3 Ent). Kamares Ware. *Date*: MM IB–IIA.

228 (HCH 04-388; Fig. 20). Jug, neck sherd. Max. dim. 5.2; d. of neck 3.8 cm. A pale, fine fabric (pink, 5YR 7/4). Dark slip on exterior. Added white: pair of vertical petals. Added red: vertical band. *Comments*: from Room 1 (1982 season). Classical Kamares Ware. *Parallels*: MacGillivray 1998, pl. 50:170 (Knossos). *Date*: MM IIA–IIB. *Bibl.*: Langford-Verstegen 2008, 555, fig. 9:8.

229 (HNM 13,801; Fig. 20). Jug, spout sherd. Max. dim. 4 cm. A pale fabric (reddish yellow, 7.5YR 7/6). Dark slip on exterior. Added white: thin horizontal bands on rim and neck. Added red: horizontal band at neck. *Comments*: from Rooms 1–4 (1976–1983 seasons). *Date*: MM IB–IIA.

230 (HCH 163; Fig. 20). Jug(?), body sherd. Max. dim. 5.4 cm. A pale fine fabric (reddish yellow, 7.5YR 7/6). Convex profile. Dark slip on exterior. Added white: thin horizontal bands on the body (flanking the red band) and groups of diagonal lines on lower body. Added red: horizontal band between two sets of two white bands. *Comments*: from Rooms 1–4 (1983 season). Kamares Ware. *Date*: MM IB–IIA.

231 (HCH 40; Fig. 20). Jug, body sherd. Max. dim. 3.4 cm. A pale fine fabric (pink, 5YR 7/4). Convex profile. Dark slip on exterior. Added white thin diagonal lines hatched with thin added red lines. *Comments*: from Rooms 1–4 (1983 season). Kamares Ware. *Date*: MM IB–IIA.

Jar

232 (HCH 186; Fig. 20). Jar(?), handle sherd. Max. dim. 3.4 cm. A pale fabric (pink, 7.5YR 7/4). Vessel (probably a small jar) with twisted handle (two coils, twisted together to form a handle). Dark slip. *Comments*: from Rooms 1–4 (1983 season). Kamares Ware. *Date*: MM IB–IIA.

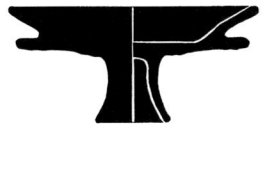

14

Chamaizi Pots, MM IIB

A small vessel with a piriform body and a cylindrical rim/neck with no separate spout is called a Chamaizi pot after the site where the class was first found. All the known examples are made from pale-firing clay. Incised decoration on the cylindrical upper part is usual, though examples with a plain rim are also known (Poursat and Knappett 2005, 36–37). Inscriptions in the Cretan hieroglyphic script occur on some of the pieces. The class can be placed in MM IIB by its presence in the destruction level of Quartier Mu at Malia (Poursat 1996, pl. 30:d, e; Poursat and Knappett 2005, 236, nos. 1123–1139, pl. 36, upper 2 rows). Other parallels include the following: Xanthoudides 1906, pl. 9:1–3 (Chamaizi); Hall 1914, 117, fig. 67 (Vrokastro); Chapouthier and Demargne 1942, pl. 7:g (Malia); Demargne 1945, pls. 31.2:8649, 8650, 32.2:8602 (Malia); Demargne and Gallet de Santerre 1953, pl. 8:8507 (Malia); Zois 1969, pl. 24:2 (Gournes); Pendlebury, Pendlebury, and Money-Coutts 1935–1936, 81, fig. 17:665 (Trapeza Cave).

Only a few examples of the Chamaizi pot occur in the cave. They vary somewhat in size, in the shape of the upper part, and in the type of incised decoration. Only one example (**233**) is complete enough to discern the shape of the body (Walberg 1983, 175, form 25, type 149). No inscriptions occur from this location. Analysis of the fine, pale fabric (petrographic sample no. HCH 04/8, called Hagios Charalambos Fabric Group 10) does not match any of the petrographic samples known from Malia.

233 (HNM 12,390; Fig. 20; Pl. 2). Chamaizi pot, almost complete. H. 6.2; d. of rim 2.5; d. of base 3.1 cm. A pale, fine fabric (reddish yellow, 7.5YR 7/6). Incised diamonds on neck. *Comments*: from Room 3 (1976–1983 seasons). *Date*: MM IIB. *Bibl.*: Betancourt 2007, 215, fig. 25.5:B; Betancourt et al. 2014, 29, fig. 17:i.

234 (HNM 13,811; Fig. 20). Chamaizi pot, upper part. Rest. h. ca. 7; d. of rim 2.7 cm. A pale, fine fabric (pink, 7.5YR 7/4). Incised crosshatching on neck, widely spaced. *Comments*: from Room 3 (1976–1983 seasons). *Date*: MM IIB. *Bibl.*: Langford-Verstegen 2008, 556, fig. 10:14.

235 (HCH 05-416; Fig. 20). Chamaizi pot, upper part. Rest. h. ca. 12; d. of rim ca. 4 cm. A pale, fine fabric (reddish yellow, 7.5YR 7/6). Incised crosshatching on neck, closely spaced. *Comments*: from Room 4, upper levels (1976–1983 seasons). Analyzed by ceramic petrography, sample no. HCH 04/8 (Hagios Charalambos Fabric Group 10). *Date*: MM IIB.

236 (HCH 42; Fig. 20). Chamaizi pot, upper part. Rest. h. ca. 9–10; d. of rim ca. 2.5 cm. A pale, fine fabric (pink, 5YR 7/4). Incised crosshatching on neck, closely spaced. *Comments*: from Rooms 1–4 (1983 season). *Date*: MM IIB.

237 (HCH 135; Fig. 20). Chamaizi pot, neck sherd. Rest. h. ca. 10–12; d. of rim ca. 2.6 cm. A pale, fine fabric (pink, 5YR 7/4). Incised crosshatching on neck, closely spaced. *Comments*: from Rooms 1–4 (1976–1983 seasons). *Date*: MM IIB.

238 (HCH 136; Fig. 20). Chamaizi pot, rim sherd. Rest. h. ca. 12; d. of rim ca. 4 cm. A pale, fine fabric (between pink, 7.5YR 7/4, and reddish yellow, 7.5YR 7/6). Incised crosshatching on neck, widely spaced. *Comments*: from Room 4, upper levels (1983 season). *Date*: MM IIB.

239 (HCH 04-357; Fig. 20). Chamaizi pot, upper part. Rest. h. ca. 12; d. of rim ca. 2.4 cm. A pale, fine fabric (pink, 7.5YR 7/4). Incised crosshatching on neck, widely spaced. *Comments*: from Room 4, upper levels (1983 season). *Date*: MM IIB.

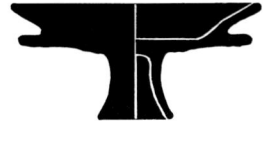

15

Imports into Lasithi with Parallels from Malia, MM I–IIB

In MM II the amount of pottery from production centers that are related to those at Malia increases while the percentage of imports from elsewhere decreases. This fact may contribute to a greater understanding of the Malian political and economic sphere of influence in the Middle Minoan period. Objects from the cave with published parallels include: jugs with incised bands on the shoulder; tall, open-spouted jugs; two-handled, spouted cups with painted and incised decoration; carinated vessels with undulating rims; carinated cups with grooves; and several other shapes. As in the other classes, many of the vessels are designed for pouring and drinking. Petrographic analysis on samples has shown matches with a Malian pale fabric.

Spouted Two-Handled Cup, Incised

240 (HNM 13,982; Fig. 21). Spouted two-handled cup, almost complete. D. of rim 7.8; d. of base 5.4 cm. A pale, fine fabric (reddish yellow, 5YR 7/6). Straight rim with open spout at the rim, two opposed vertical handles with circular section. Band of incised diagonal lines, in groups, alternating directions. Dark slip on upper part and lower part of exterior. Added white chevrons and bands. *Comments*: mended from sherds found in Room 5, area 5, level 2 (unit HCH 02-5-5-2) and Room 4/5 Entrance, lower levels (unit HCH 02-4/5 Ent). *Parallels*: Deshayes and Dessenne 1959, pl. 53:c (Malia); Marinatos 1929, 113, fig. 10:epsilon (Krasi). *Date*: MM IB–IIA. *Bibl.*: Betancourt et al. 2008b, 166, fig. 28:1.

241 (HCH 04-288; Fig. 21). Spouted two-handled cup, rim and body sherds. D. of rim ca. 8 cm. A pale, fine fabric (reddish yellow, 5YR 7/8). Straight rim with open spout at the rim, two opposed vertical handles with circular section. Band of incised diagonal lines. Dark slip on upper part and lower part of exterior. Added white: three bands, chevrons. *Comments*: from Room 4, upper levels (1983 season). *Parallels*: see **240**. *Date*: MM IB–IIA. *Bibl.*: Betancourt 2007, fig. 25.5:A; Langford-Verstegen 2008, 556, fig. 10:12.

242 (HCH 04-289; Fig. 21). Spouted two-handled cup, handle and body sherds. Estimated d. of rim ca. 8–10 cm. A pale, fine fabric (pink, 5YR 7/4). Shape probably like **240**, two opposed vertical handles with circular section. Band of incised diagonal lines. Dark slip (light red, 2.5 YR 6/6) below and probably above incised band. Added white not preserved. *Comments*: from Room 4, upper levels (1983 season). *Parallels*: see **240**. *Date*: MM IB–IIA.

243 (HCH 05-419; Fig. 21). Spouted two-handled cup, body sherd. Max. dim. 1.8 cm. A pale, fine fabric (between light reddish brown, 5YR 6/4, and reddish yellow, 5 YR 6/6). Band of incised diagonal lines. *Comments*: from Room 4, upper levels (1983 season). Analyzed by ceramic petrography, sample no. HCH 04/21 (Hagios Charalambos Fabric Group 8). *Parallels*: see **240**. Date: MM IB–IIA.

244 (HCH 02-51; Pl. 2). Spouted two-handled cup, fragments. D. of rim 10 cm. A fine fabric (pink, 7.5YR 7/4), burnished. Thin straight rim, two vertical handles attached just below rim and on incised band. Dark slip on interior of rim and on rim above incised band and below incised band. Added white: two thin bands above and below incised band of chevrons. *Comments*: from Room 3, area 5, level 2 (unit 02-3-5-2). *Date*: MM I–II.

Jugs with Incised Lines

Small jugs are common at the cave. These examples have incised horizontal lines on the shoulders. They belong to a class of trefoil-mouthed jug with parallels from Malia and elsewhere in North-Central Crete: Poursat 1996, pls. 22:a–g, 33:f (Malia); Poursat and Knappett 2005, 61, fig. 14, lower row right, pl. 23:559, 581, 582, 625 (Malia); Zois 1969, pl. 8:6942 (Gournes).

245 (HCH 122; Fig. 21). Jug with incised lines, rim and body sherds at shoulder. D. of rim 4.8 cm. Convex profile. A fine, light red (2.5YR 6/6), very hard fabric. Three incised horizontal lines on upper shoulder. *Comments*: from Rooms 1–4 (1983 season). *Date*: MM IIB. *Bibl.*: Langford-Verstegen 2008, 555, fig. 9:10.

246 (HCH 84; Fig. 21). Jug with incised lines, sherds from the shoulder. D. of neck 4.5 cm. A fine, light red (2.5YR 6/6), very hard fabric. Four incised horizontal lines preserved on upper shoulder. *Comments*: from Rooms 1–4 (1983 season). *Date*: MM IIB.

247 (HCH 04-290; Fig. 21). Jug with incised lines, sherds from the shoulder. D. of neck 4.5 cm. A fine, light red (2.5YR 6/8), very hard fabric. Five incised horizontal lines preserved on upper shoulder. *Comments*: from Room 3 (1983 season). *Date*: MM IIB

Carinated Jug with Wide Mouth

248 (HCH 04-349; Fig. 21). Open-spouted jug with carinated shoulder, almost complete. D. of rim ca. 12; d. of base 6.2; h. with handle ca. 11 cm. A pale, fine fabric (reddish yellow, 5YR 7/6). Dark slip on exterior and interior of rim. *Comments*: from Room 4 (1983

season). *Parallels*: Poursat 1996, pl. 33, 1; Poursat and Knappett 2005, 221:639–645, pl. 24 (Malia); Hawes et al. 1908, pl. 6:17 (Gournia); Tsipopoulou and Wedde 2000, 377, fig. 13, right (Petras). *Date*: MM I–IIB. *Bibl.*: Langford-Verstegen 2008, 555–556, fig. 10:11.

Semiglobular Cup with Grooves

249 (HCH 03-211; Fig. 21). Semiglobular cup with grooves, body sherd. Max. dim. 5.3 cm. A pale, fine fabric (pink, 5YR 7/4). Dark slip on interior and exterior. *Comments*: from Room 4/5 Entrance, level 4 (unit HCH 03-4/5Ent-4). *Date*: MM IIB.

Carinated Cup without Grooves

250 (HCH 03-175; Fig. 21). Carinated cup, rim sherd. D. of rim ca. 10 cm. A pale, fine fabric (pink, 5YR 8/4). No grooves. Dark slip on interior and exterior. *Comments*: from Rooms 1–4 (from the dump outside the cave, unit HCH 03-11-1). *Parallels*: Demargne and Gallet de Santerre 1953, pl. 8:8407 (Malia). *Date*: MM IIB.

Carinatd Cups with Grooves, Straight Rim

Numerous examples of this class of cup are known. The standard shape consists of a cylindrical upper part and a lower section with a rounded profile, with the two parts divided by a carination. Grooves are made on the upper part on the potter's wheel. Most examples are monochrome with an overall coat of dark slip.

Parallels include the following, among many others: Pendlebury, Pendlebury, and Money-Coutts 1935–1936, 61 (Trapeza Cave); Chapouthier and Charbonneaux 1928, pl. 27:4 (Malia); Demargne 1945, pl. 33:8657 (Malia); Poursat and Knappett 2005, 70, fig. 19 (Malia).

251 (HCH 66; Fig. 21). Carinated cup with grooves, rim sherd with handle. D. of rim 10; max. dim. 4.4 cm. A pale, fine fabric (pink, 5YR 7/4). Dark slip on interior (light red, 2.5YR 6/6) and exterior (varies from black to light red, 2.5YR 6/6). Straight rim, flat handle with elliptical section. *Comments*: from Room 4, upper levels (1983 season). *Date*: MM IIB. *Bibl.*: Betancourt 2007, 215, fig. 25.5:A; Langford-Verstegen 2008, 556, fig. 10:13.

252 (HCH 03-174; Fig. 21). Carinated cup with grooves, body sherd. Max. dim. 3 cm. A pale, fine fabric (reddish yellow, 7.5YR 7/6). Dark slip traces. *Comments*: from cave exterior, trench 12, level 3 (unit HCH 03-12-3). *Date*: MM IIB.

253 (HCH 03-214; Fig. 21). Carinated cup with grooves, rim sherd. D. of rim 12 cm. A pale, fine fabric (pink, 7.5YR 8/4). Dark slip on interior and exterior (very dark gray, 7.5YR N3). *Comments*: from Room 5, area 5, level 2 (unit HCH 02-5-2). *Date*: MM IIB.

Carinated Cup or Kantharos with Undulating Rim

254 (HCH 02-15). Carinated cup or kantharos, body sherd. Max. dim. 6.9 cm. A pale, fine fabric (pink, 5YR 7/4). Lower part of the undulating rim. Dark slip on interior and exterior (very dark gray, 7.5 YR N3). *Comments*: from Room 2, level 2 (unit HCH 02-2-2). *Date*: MM IIB.

255 (HCH 03-209; Fig. 21). Carinated cup or kantharos, rim sherd. Max. dim. 3.4 cm. A pale, fine fabric (pink, 7.5YR 9/4). Undulating rim. Dark slip on interior and exterior (very dark gray, 7.5YR N3). *Comments*: from Room 5, area 5, level 2 (unit HCH 02-5-5-2). *Parallels*: for the carinated shape with an undulating rim, see Poursat and Knappett 2005, 73 (Malia). These parallels all have two opposed handles, so the shape would be called a kantharos. *Date*: MM IIB.

256 (HCH 02-129; Fig. 21). Carinated cup or kantharos, rim sherd. D. of rim 6–8; max. dim. 3.4 cm. A pale, fine fabric (reddish yellow, 5YR 7/6). Undulating rim. Dark slip on exterior and interior (very dark gray, 7.5 YR N3). *Comments*: from Room 4/5 Entrance, level 3 (unit HCH 02-4/5Ent-3). *Date*: MM IIB.

Carinated Cup with Grooves, Unknown Type of Rim

257 (HCH 03-217; Fig. 21). Carinated cup with grooves, body sherd. Max. dim. 4.9 cm. A pale, fine fabric, pink (5YR 7/4). Dark slip on interior and exterior, very dark gray (7.5YR N3). *Comments*: from Room 5, area 5, level 2 (unit HCH 02-5-5-2). *Date*: MM IIB.

258 (HCH 16; Fig. 21). Carinated cup with grooves, rim sherd. D. of rim 6–8 cm. A pale, fine fabric (reddish yellow, 5YR 7/6). Traces of dark slip on interior and exterior. *Comments*: from Rooms 1–4 (1983 season). *Date*: MM IIB.

259 (HCH 02-123; Fig. 21). Carinated cup with grooves, body sherd. Max. dim. 2.5 cm. A pale, fine fabric (pink, 7.5YR 8/4). Traces of dark slip on interior and exterior. *Comments*: from Room 4/5 Entrance, level 3 (unit HCH 02-4/5-3). *Date*: MM IIB.

260 (HCH 05-417; Fig. 22). Carinated cup with grooves, body sherd. D. of rim 10.8 cm. A pale, fine fabric (pink, 5YR 7/4). Dark slip on interior and exterior. *Comments*: from Room 4 (1976–1983 seasons). Analyzed by ceramic petrography, sample no. HCH 04/9 (Hagios Charalambos Fabric Group 8). *Date*: MM IIB. *Bibl.*: Betancourt et al. 2014, 29, fig. 17:h.

Straight-Sided Cup with Grooves

261 (HCH 41; Fig. 22). Straight-sided cup with grooves, lower part. D. of base 3.4 cm. A pale, fine fabric (pink, 5YR 7/4). Dark slip on interior and exterior. *Comments*: from Rooms 1–4 (1983 season). *Parallels*: Pendlebury, Pendlebury, and Money-Coutts 1935–1936, 60, fig. 14:536, 537, 548 (Trapeza Cave); Demargne and Gallet de Santerre 1953, pl. 8:8520 (Malia). *Date*: MM IIB.

Scoop

262 (HCH 195; Fig. 22). Scoop, body sherd with handle. D. of rim ca. 8–10 cm. A fine fabric (light red, 2.5YR 6/8). Rounded cup, handle inside vessel. *Comments*: from Rooms 1–4 (1983 season). *Parallels*: Poursat and Knappett 2005, 224:704–716 (Malia); Hawes et al. 1908, pl. 6:32, 33 (Gournia). For discussion of the shape and a list of other examples, see Floyd 1999. *Date*: MM IIB.

Basket-Shaped Vessel

263 (HCH 193; Fig. 22). Basket-shaped vessel, rim sherd with part of handle. D. of rim ca. 10 cm. A fine fabric (reddish yellow, 5YR 7/6). Elliptical rim, horizontal handle raised above the rim. Dark slip traces. *Comments*: from Rooms 1–4 (1983 season). *Parallels*: Poursat and Knappett 2005, 238:1175 (Malia). *Date*: MM IIB.

16

Various Nonlocal Fabrics, EM III–MM IIB

Among the pottery vessels and fragments found at the Hagios Charalambos Cave are a number of pieces that were imported into the Lasithi Plain from other parts of Crete. Most of these vessels were made in pale fabrics. Only a few of these vases can be assigned to specific places of origin.

Shallow Bowls

264 (HCH 02-128; Fig. 22). Shallow bowl, rim sherd. D. of rim ca. 28–34 cm. Mirabello Fabric (red, 2.5YR 5/8). Conical shape. *Comments*: from the Room 4/5 Entrance, level 3 (unit HCH 02-2/3-3). From the Gulf of Mirabello region. *Parallels*: Hall 1912, 46, fig. 20 (Sphoungaras); Betancourt 2006, 91, fig. 5.7:90 (Chrysokamino). *Date*: EM III–MM I.

265 (HCH 110; Fig. 22). Shallow bowl, rim sherd. D. of rim ca. 18 cm. A pale fabric (pink, 7.5YR 7/4). Conical shape. Dark slip on interior and exterior. Added white traces on rim. *Comments*: from Room 4, upper levels (1983 season). *Parallels*: Hall 1912, 46, fig. 20 (Sphoungaras); Betancourt 2006, 91, fig. 5.7:87 (Chrysokamino). *Date*: EM III–MM I.

Rounded Cup

266 (HCH 05-444; Fig. 22). Rounded cup, rim sherd with handle. Rest. h. 6; d. of rim 7.4 cm. A pale fabric (reddish yellow, 5YR 5/6). Dark band on rim on both interior and exterior. *Comments*: from Rooms 1–4 (1976–1983 seasons). This is a common cup from EM III in the region of the Gulf of Mirabello. *Parallels*: Hall 1912, 55, fig. 28:a, b, e (Sphoungaras); Betancourt 2006, 85, fig. 5.6:78–81 (Chrysokamino). *Date*: EM III–MM IA. *Bibl.*: Betancourt et al. 2008b, 166, fig. 28:d.

Conical Cup

267 (HCH 02-37; Fig. 22; Pl. 2). Conical cup, almost complete. H. 6.2; d. of rim 10; d. of base 3.9 cm. A pale fabric (light red, 2.5YR 6/6). Conical shape, straight rim, straight walls. No slip preserved. *Comments*: from the Room 4/5 Entrance, surface level (unit HCH 02-2/3 Ent). This is the only unpainted conical cup without a handle from the deposit. *Date*: MM I–II.

Cups

268 (HCH 98; Fig. 22). Cup, rim sherd with part of handle. D. of rim 8 cm. A fine, pale fabric (reddish yellow, 5YR 7/6). Straight rim; handle with circular section. Dark slip (red) on interior and exterior. *Comments*: from Room 4, upper levels (1983 season). *Date*: EM III–MM IA.

269 (HCH 99; Fig. 22). Cup with pronounced base, base sherd. D. of base 4.4 cm. A fine, pale fabric (reddish yellow, 7.5YR 7/6). Dark slip on interior and exterior. *Comments*: from Room 4, upper levels (1983 season). *Date*: EM III–MM IB.

270 (HCH 97; Fig. 22). Cup(?), body sherd. Max. dim. 3.4 cm. A fine fabric (reddish yellow, 5YR 7/8). Almost straight profile. Pinkish-white (5YR 8/2) slip on interior and exterior. *Comments*: from Room 4, upper levels (1983 season). Analyzed by ceramic petrography, sample HCH 04/47 (metamorphic fabric, loner). The cup may be from the Pediada (see App. A). *Date*: MM IA–IB.

271 (HCH 47; Fig. 22). Cup, rim sherd with handle. D. of rim ca. 10 cm. A fine, pale fabric (pinkish white, 7.5YR 8/2). Straight rim, handle with thin oval section. *Comments*: from Room 4, upper levels (1983 season). *Date*: MM IIB.

Bowl with Tab Handles

272 (HCH 04-347; Fig. 22). Bowl with tab handles on the rim, complete profile. D. of rim 16; d. of base 8 cm. A medium-coarse fabric (light red, 2.5YR 6/8), slipped (reddish brown, 2.5YR 5/4). Ledge rim, rounded shoulder. *Comments*: from Room 4, upper levels (1983 season). *Date*: MM IB.

Miniature Tripod Vessel

273 (HCH 81; Fig. 22). Miniature tripod vessel, leg sherd. Length of leg 3.3 cm. A pale fabric (reddish yellow, 5YR 7/8). Probably a conical shape, leg with elliptical section. *Comments*: from Room 4, upper levels (1983 season). *Date*: EM III–MM II.

Jugs

274 (HCH 02-106; Fig. 22). Jug, spout sherd with handle scar. D. of opening 2.6 cm. A pale fabric (pink to reddish brown, 5YR 6–7/4). Rounded profile, tall raised spout, handle with circular section. Three dark bands on rim, two bands on base of neck, slip on shoulder. *Comments*: from Room 4/5 Entrance, level 3 (unit HCH 02-2/3 Ent-3). *Parallels*: Hazzidakis 1934, pl. 18:j (Tylissos). *Date*: MM I–II.

275 (HCH 03-207; Fig. 22). Jug, neck and shoulder sherd. D. of opening 5.2 cm. A pale fabric (reddish yellow, 5YR 7/6). Raised band with incised lines at junction between neck and shoulder. No slip preserved. *Comments*: from Room 5, area 5, level 3 (unit HCH 02-3-5-3). *Date*: MM I–II.

276 (HCH 64; Fig. 22). Jug, neck and upper shoulder sherd. D. of opening 4.6 cm. A pale fabric (reddish yellow, 5YR 7/6). Rounded profile, raised band with incised lines at junction between neck and shoulder. Two dark bands on base of neck, traces of dark slip on shoulder. *Comments*: from Room 4, upper levels (1983 season). *Date*: MM I–II.

277 (HCH 90; Fig. 22). Jug, neck sherd with handle and upper shoulder. Max. dim. 10.7 cm. A pale fabric (reddish yellow, 5YR 7/8). Raised spout, handle with circular section, two raised bands with incising on upper shoulder. Dark slip on exterior. Added white horizontal bands on neck and shoulder, vertical band on handle. *Comments*: from Room 4, upper levels (1983 season). *Date*: MM I–II.

278 (HCH 03-205; Fig. 23). Jug, upper part with most of spout missing. Rest. h. ca. 15–18 cm. A pale fabric (reddish yellow, 5YR 7/6, to pink, 5YR 7/4). Rounded profile, raised spout, handle with circular section attached at rim and shoulder. Dark bands on spout, wide band on neck, lines on handle. *Comments*: from Room 5, area 5, level 1 (Unit HCH 02-3-5-1). *Date*: MM I–II.

279 (HCH 03-206; Fig. 23). Jug, upper part with most of spout missing. Rest. h. ca. 12–14 cm. A pale fabric (pink, 7.5YR 8/4). Rounded profile, raised spout, handle with circular section attached below rim and on shoulder. Dark bands on spout, wide band on upper shoulder, diagonal lines on body. *Comments*: from Room 5, area 5, level 1 (unit HCH 02-3-5-1). *Date*: MM I–II.

280 (HCH 05-413; Fig. 23). Jug, spout sherd. Rest. h. ca. 20 cm. A fine pale fabric (pink, 7.5YR 7–8/4). Small raised spout. Three dark bands on spout, band on lower part of neck. *Comments*: analyzed by ceramic petrography, sample no. HCH 04/5 (Hagios Charalambos Fabric Group 8). *Parallels*: Hazzidakis 1934, pl. 18:j (Tylissos). *Date*: MM I–II.

281 (HCH 18; Fig. 23). Jug, spout sherd. Rest. h. ca. 16 cm. A pale fabric (pink, 7.5YR 7/4). Small raised spout, handle attached at rim and shoulder. Dark band on spout, band on lower part of neck, wide band on shoulder. *Comments*: from Room 4, upper levels (1983 season). *Parallels*: Hazzidakis 1934, pl. 18:j (Tylissos). *Date*: MM I–II.

282 (HCH 02-69; Fig. 23). Jug, spout sherd with handle. D. of opening 2.8 cm. A pale fabric (light brown, 7.5YR 6/4). Small raised spout, handle with circular section attached below rim. Two dark bands on rim, with third band on spout. *Comments*: from Room 5, area 5, level 2 (unit HCH 02-3-5-2). *Parallels*: Hazzidakis 1934, pl. 18:j (Tylissos). *Date*: MM I–II.

283 (HCH 05-434; Fig. 23). Jug, spout and shoulder sherd. Rest. h. ca. 20 cm. A pale fabric (reddish yellow,

5YR 7/6). Rounded profile, small raised spout. Molding at junction of neck and shoulder. Dark bands on rim and at junction of neck and shoulder. *Comments*: from Room 7, washed in from Room 5 (unit HCH 03-7-1). *Date*: MM I–II.

284 (HNM 12,415; Fig. 23). Jug with no spout (flagon), complete. H. 15.7; d. of rim 3.4; d. of base 8.1 cm. A pale fabric (reddish yellow, 7.5YR 7/6). Globular body, straight rim with no spout, handle with circular section attached below rim and on shoulder. Pale slip on exterior. *Comments*: from Room 4, upper levels (1983 season). *Date*: MM I–II.

285 (HCH 39; Fig. 23). Jug(?), base sherd. D. of base 6.5 cm. A pale fabric (reddish yellow, 5YR 7/6). Dark traces on lower body. *Comments*: from Room 4, upper levels (1983 season). *Date*: MM I–II.

Wide-Mouthed Vessel with Tube Spout

286 (HCH 04-354; Fig. 23). Side-spouted vessel, rim sherd with part of tube spout. D. of rim 16 cm. A medium-coarse fabric (red, 2.5YR 5/8). Outturned rim; tube spout (d. 1.9 cm) below the rim, attached at base of shoulder; handles do not survive, so the shape is unknown. Traces of dark slip. *Comments*: from Room 4, upper levels (1983 season). *Date*: EM III–MM IA.

Closed Vessel

287 (HCH 03-221; Fig. 23). Closed vessel, body sherd. Max. dim. 4.3 cm. A pale very hard fabric (pink, 7.5YR 8/4, with very pale brown surface, 10YR 8/3). *Comments*: from Room 4, upper levels (1983 season). Cataloged because of the unusual fabric. Analyzed by ceramic petrography, sample no. HCH 04/48 (Hagios Charalambos Fabric Group 13). *Date*: MM I–II.

Bridge-Spouted Jars

288 (HCH 101; Fig. 23). Bridge-spouted jar, body sherd with handle. Max. dim. 6.2 cm. A medium textured fabric (reddish yellow, 7.5YR 7/6). Dark slip on the exterior. *Comments*: from Room 4, upper levels (1983 season). Dated by the characteristic slip and thin handle. *Date*: MM I–II.

289 (HCH 60; Fig. 23). Bridge- or side-spouted jar, rim sherd. D. of rim 12 cm. A medium textured fabric (light red, 2.5YR 6/6). Straight, inturned rim, no handles or spout preserved. Dark band on interior of rim, possible traces on exterior. *Comments*: from Room 4, upper levels (1983 season). *Date*: EM III–MM II.

Side-Spouted Jug

290 (HCH 04-345; Fig. 24). Side-spouted jug, rim sherd with edge of tube spout. D. of rim ca. 9.5 cm. A fine fabric (very pale brown, 10YR 7/4). Wide mouth, slightly everted rim, handle opposite a tube or teapot-type spout. *Comments*: from Room 4, upper levels (1983 season). *Parallels*: Xanthoudides 1906, pls. 9:6, 11:19 (Chamaizi); Poursat and Knappett 2005, pl. 49:680 (Malia); Betancourt, Dierckx, and Reese 2003, fig. 21:1.1 (Pseira). *Date*: MM I–IIA.

Spouted Two-Handled Cups

291 (HCH 04-376; Fig. 24). Spouted two-handled cup, many sherds. D. of rim ca. 10 cm. A fine fabric (pink, 5YR 7/3). Wide mouth, slightly inturned rim, two opposed vertical handles with circular sections, open rim-spout. Dark slip on exterior. Added white: band of isolated chevrons below rim, frieze of C-spiral motifs joined by diagonal lines on upper shoulder, bands on lower body. *Comments*: from Room 4, upper levels (1983 season). *Parallels*: for the band of C-spirals at Malia, see Poursat and Knappett 2005, 122, fig. 35:3. *Date*: MM IIB.

292 (HCH 168; Fig. 24). Spouted two-handled cup, rim sherd. D. of rim ca. 10–12 cm. A fine fabric (pink, 5YR 7/3). Wide mouth, slightly inturned rim. Dark slip on exterior. Added white: band of double quirks below rim, two bands below rim, frieze of spirals joined by diagonal lines on upper shoulder. *Comments*: from Room 3 or 4, upper levels (1983 season). *Date*: MM I–II.

293 (HCH 169; Fig. 24). Spouted two-handled cup, rim sherd with spout. D. of rim ca. 10–12 cm. A fine fabric (pink, 5YR 7/4). Wide mouth, slightly inturned rim, open rim-spout. Dark slip on exterior. Added white: band of spirals below rim, two bands below rim, frieze of C-spirals joined by diagonal lines on upper shoulder, band below frieze. *Comments*: from Room 3 or 4, upper levels (1983 season). Analyzed by ceramic petrography, sample no. HCH 04/41 (Hagios Charalambos Fabric Group 12). *Parallels*: for the band of C-spirals at Malia, see Poursat and Knappett 2005, 122, fig. 35:3. *Date*: MM I–II.

294 (HCH 04-280; Fig. 24). Spouted two-handled cup, rim sherd with handle. D. of rim ca. 10 cm. A pale fabric (pink, 5YR 7/4). Straight rim. Dark slip on interior of rim and on exterior. Added white: two bands on rim and bands on body, frieze of quirks, possible traces of bands on handle. *Comments*: from Room 4, upper levels (1983 season). *Parallels*: for the band of quirks at Malia, see Poursat and Knappett 2005, 121, fig. 34:2. *Date*: MM I–II.

295 (HCH 04-281; Fig. 24). Spouted two-handled cup, rim sherd. D. of rim ca. 10–12 cm. A pale fabric (pink, 5YR 7–8/4). Straight rim. Dark slip on exterior and interior of rim. Added white: two bands on rim. *Comments*: from Room 4, upper levels (1983 season). Analyzed by ceramic petrography, sample no. HCH 04/21 (Hagios Charalambos Fabric Group 7a,

Granodiorite Fabric). The fabric shows that this vessel is an import from the Gulf of Mirabello. *Date*: MM I–II.

296 (HCH 121; Fig. 24). Spouted two-handled cup, rim sherd. D. of rim ca. 10–12 cm. A fine fabric (light red, 2.5YR 6/8). Wide mouth, slightly inturned rim. Dark slip on exterior. Added white: two bands below rim, frieze of spirals joined by diagonal lines on upper shoulder. *Comments*: from Room 3 or 4, upper levels (1983 season). *Date*: MM I–II.

297 (HCH 70; Fig. 24). Spouted two-handled cup, rim sherd. Rest. h. ca. 10; d. of rim ca. 8 cm. A pale fabric (pink, 7.5YR 7/4). Straight rim. Dark slip on interior of rim and on exterior. Added white traces. *Comments*: from Room 4, upper levels (1983 season). *Date*: MM I–II.

298 (HCH 02-142; Fig. 24). Cup or two-handled cup, rim sherd. Rest. h. ca. 8; d. of rim 8 cm. A fine, pale fabric (pink, 7.5YR 8/4). Straight-sided cup? Dark slip on exterior and interior of rim. Added white: zigzag band at rim, two bands above spiral motif. *Comments*: from Room 4, upper levels (1983 season). *Date*: MM I–II.

299 (HCH 174; Fig. 24). Spouted two-handled cup, rim sherd with edge of spout. D. of neck ca. 5 cm. A pale fabric (light red, 2.5YR 6/8). Dark slip on exterior. Added white: bands and spiral on rim. *Comments*: from Rooms 1–4 (1983 season). *Date*: MM I–II.

Jar

300 (HCH 118; Fig. 24). Small jar(?), body sherd. D. of rim 6 cm. A fine fabric (pink, 5YR 7/4). Vertical, straight profile. Dark slip on the interior of the rim and on the exterior. *Comments*: from Room 4, upper levels (1983 season). *Date*: EM III–MM II.

Jar with Thickened Rim

The presence of several sherds from storage jars in the cave (mostly made in the Lasithi Red Fabric Group, see Ch. 17) demonstrates that storage containers were present in the cave in small numbers. Considering the presence of larnax sherds in the assemblage, it is possible that the jars were used as burial containers. Like the larnakes, they survive only in sherds.

301 (HCH 34; Fig. 24). Jar, rim sherd. D. of rim ca. 26 cm. A pale fabric (reddish yellow, 5YR 7/6). Thickened rim. Dark slip on top of rim and on exterior. *Comments*: from Room 4, upper levels (1983 season). *Parallels*: Pendlebury, Pendlebury, and Money-Coutts 1935–1936, 86–88, fig. 20:902–910 (Trapeza Cave); Betancourt 2006, 80, 95, figs. 5.4:43, 5.9:106 (Chrysokamino). *Date*: MM I–II.

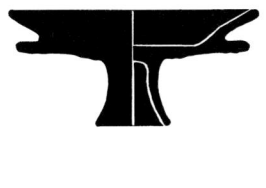

17

The Lasithi Red Fabric Group, EM II–MM II

The largest group of vessels from the cave is of local or regional production using red-firing clays containing varying amounts of phyllite (App. A, Fabrics 2 and 11). Many examples are coarsely textured and have thicker walls than comparable vessels from many other parts of Crete. The overall coat of slip used for many of these vases is red, light red, or reddish yellow rather than the black produced elsewhere in Crete.

Open Shapes in Lasithi White-on-Dark Ware

A pottery tradition using both shapes and decorations in a manner imitating East Cretan White-on-Dark Ware, but also with ties to the Mesara, has been identified in the assemblage found in the cave (for East Cretan examples and for White-on-Dark Ware in general, see Betancourt 1984). East Cretan White-on-Dark Ware flourished in EM III–MM IA, with its main region of production extending along the northern coast of eastern Crete from Malia to Palaikastro. Similarities with pottery in the Mesara can be seen at Moni Odigitria (Vasilakis and Branigan 2010, figs. 50:P282, P283, 51:P307) and Phaistos (Levi 1976, vol. I*, pl. 33:c). The Lasithi imitation consists mostly of cups, shallow bowls, and jugs. Most numerous are cups that are rounded or conical in form with one vertical handle (with round or elliptical section), attached at or across the rim and on the lower body of the vessel. The cups have a dark slip that is applied to the exterior and to the inside of the rim or to the exterior and entire interior. Decoration in white is added over the dark undercoat.

The local Lasithi production uses the phyllite-tempered Fabric Group 2 (Lasithi Red Fabric Group). Many of the vases are coarsely textured (including the cups). The East Cretan production uses Mirabello Fabric, a pale-firing clay tempered with granodiorite (for the description, see Myer, McIntosh, and Betancourt 1995; here, it is Fabric 7).

The Lasithi production employs a more limited repertoire of decorative motifs than is used in the main region of White-on-Dark Ware. Spirals, circles, chevrons, and quirks were common motifs used in East Crete, but they were not used here. Instead, ornamentation is simpler; motifs are both rectilinear and curvilinear. The most common decoration is diagonal lines from which a series of lines are joined at an angle. The triangle motif of the East Cretan variety of this ware is a common design that is sometimes filled with crosshatching, dots, and other motifs (Betancourt 1984, 22). Here, the triangle is only filled with parallel lines that converge at an angle (hatched triangles). Two phases of imitation are illustrated by cups with, first, paint on the interior of the rim and, secondly, cups with paint and decoration on the whole interior. In the Gulf of Mirabello area, the earlier phase of decoration is dated to EM III–MM IA, and the second phase is MM IB.

Shallow Conical Bowls

Shallow bowls have diameters between 20 and 35 cm and smaller bases, creating a shallow conical form. They were surely placed in the original tombs as containers for something else (probably food?) rather than for their own sake. Several parallels are known: Pendlebury, Pendlebury, and Money-Coutts 1935–1936, 50, fig. 11:303 (Trapeza Cave); Hall 1912, 46, fig. 20 (Sphoungaras); Betancourt 2006, 90, 92, figs. 5.7:87–90, 5.8:91–95 (Chrysokamino).

302 (HCH 76; Fig. 25). Shallow bowl, rim sherd. D. of rim 26 cm. Lasithi Red Fabric Group (between red, 2.5YR 5/6, and reddish brown, 2.5YR 4/4). Outturned rim. Dark slip: traces on interior and band on exterior of rim. Added white not preserved. *Comments*: from Room 4, upper levels (1983 season). *Date*: EM III–MM IA.

303 (HCH 120; Fig. 25). Shallow bowl, rim sherd. D. of rim ca. 20 cm. Lasithi Red Fabric Group (red, 10R 4/8). Traces of dark slip. Added white diagonal bands at rim on interior and traces on exterior. *Comments*: from Room 4 (1976–1983 seasons). Analyzed by ceramic petrography, sample no. HCH 04/4 (Hagios Charalambos Fabric Group 2a, Lasithi Red Fabric Group). *Date*: EM III–MM IA. *Bibl.*: Langford-Verstegen 2008, 554, fig. 8:3.

304 (HCH 2; Fig. 25). Shallow bowl, rim sherd. D. of rim 26 cm. Lasithi Red Fabric Group (red, 2.5YR 6/8). Outturned rim. Dark slip: traces on interior and exterior. Added white not preserved. *Comments*: from Room 4, upper levels (1983 season). *Date*: EM III–MM IA.

305 (HCH 3; Fig. 25). Shallow bowl, rim sherd. D. of rim ca. 20–25 cm. Lasithi Red Fabric Group (dark red, 2.5YR 3/6). Traces of dark slip on exterior. Added white not preserved. *Comments*: from Room 4, upper levels (1983 season). *Date*: EM III–MM IA.

306 (HCH 25; Fig. 25). Shallow bowl, rim sherd. D. of rim 26 cm. Lasithi Red Fabric Group (red, 2.5YR 5/8). Dark slip on interior and exterior. Added white not preserved. *Comments*: from Room 4, upper levels (1983 season). *Date*: EM III–MM IA.

307 (HCH 93; Fig. 25). Shallow bowl, rim sherd. D. of rim 34 cm. Lasithi Red Fabric Group (red, 2.5YR 5/8). Straight profile. Traces of dark slip on interior and exterior. Added white not preserved. *Comments*: from Room 4, upper levels (1983 season). *Date*: EM III–MM IA.

308 (HCH 04-403; Fig. 25). Shallow bowl, rim sherd. D. of rim 25 cm. Lasithi Red Fabric Group (light red, 2.5YR 6/6). Almost straight profile. Surface not preserved. *Comments*: from Rooms 4, upper levels (1983 season). *Date*: EM III–MM IA.

Shallow Conical Bowls with Tab Handle

309 (HCH 106; Fig. 25). Shallow bowl, rim sherd. D. of rim 34 cm. Lasithi Red Fabric Group (reddish yellow, 5YR 6/6). Outturned rim, tab handle. Dark slip on interior and exterior. Added white diagonal lines on interior and exterior. *Comments*: from Room 4, upper levels (1983 season). *Date*: EM III–MM IA.

310 (HCH 172; Fig. 25). Shallow bowl, rim sherd. D. of rim ca. 30 cm. Lasithi Red Fabric Group (red, 2.5R 4/6). Straight rim, tab handle. Dark slip on rim and handle. No added white preserved. *Comments*: from Room 4, upper levels (1983 season). *Date*: EM III–MM IA.

311 (HCH 05-412). Shallow bowl, rim sherd. D. of rim 30 cm. Lasithi Red Fabric Group (red, 10YR 4/8). Dark slip on exterior. Added white missing. *Comments*: from Room 4, upper levels (1983 season). *Date*: EM III–MM IA.

Rounded Cup with Lugs

This is one of the shapes used for White-on-Dark Ware, but this version of the rounded cup is much less common than the type with a vertical handle. Similar cups from EM III to MM IA in a pale colored fabric are known from Vasiliki (Seager

1906–1907, fig. 4) and the Alatzomouri rock shelter (Apostolakou, Betancourt, and Brogan 2011, 40, fig. 4:2). Date: EM III–MM IA.

312 (HNM 12,433; Fig. 25). Rounded cup with lugs, complete. H. 5.3; d. of rim 8.9; d. of base 5.7 cm. Lasithi Red Fabric Group (light reddish brown, 2.5YR 6/4). Rounded shape, straight rim, three small lugs low on body. Dark slip on interior and exterior. Added white not preserved. Comments: from Rooms 1–4 (1976–1983 seasons). Date: EM III–MM IA.

Straight-Sided Cups, Dark Paint on Interior of Rim and Exterior

This is one of the most popular classes of vessel in the Lasithi White-on-Dark Ware. The decoration is often pendent hatched triangles, a motif borrowed from the production of this class in northeastern Crete where it occurs in hundreds of examples from many sites (Betancourt 1984, 22, 24).

313 (HCH 02-145; Fig. 26). Straight-sided cup with conical shape, three-quarters complete. H. 6; d. of rim 11; d. of base 6 cm. Lasithi Red Fabric Group (light red, 2.5YR 6/8). Handle missing. Dark slip on interior of rim and on exterior. Added white pendent triangles with hatched upper corners. Comments: from Room 5, area 5, level 2 (unit HCH 02-3-5-5). Date: EM III–MM IA.

314 (HCH 02-148; Fig. 26). Straight-sided cup with conical shape, complete profile. H. 6.2; d. of rim 12; d. of base 6 cm. Lasithi Red Fabric Group (red, 2.5YR 4/8, with a darker core, dusky red, 2.5YR 3/2). Handle with circular section attached at rim and center of body. Dark slip on interior of rim and on exterior. Added white pendent triangles with hatched upper corners. Comments: from the Room 4/5 Entrance, level 2 (unit HCH 02-2/3-2). Date: EM III–MM IA.

315 (HCH 02-149; Fig. 26). Straight-sided cup with conical shape, half complete. H. 5.5; d. of rim 11; d. of base 6.0. Lasithi Red Fabric Group (red, 2.5YR 5/8). Handle with elliptical section attached at rim and center of body. Dark slip on interior of rim and on exterior. Added white pendent triangles with hatched upper corners. Comments: from Room 5, area 5, level 2 (unit HCH 02-3-5-5). Date: EM III–MM IA. Bibl.: Betancourt et al. 2014, 29, fig. 17:e.

316 (HCH 02-68; Fig. 26; Pl. 2). Straight-sided cup with conical shape, complete profile. H. 6.0; d. of rim 11; d. of base 6 cm. Lasithi Red Fabric Group (red, 2.5YR 6/6). Handle with circular section attached at rim and below center of body. Dark slip on interior of rim and on exterior. Added white pendent triangles with hatched upper corners. Comments: mended from sherds from the Rooms 4/5 Entrance, level 3 and Room 5, area 5, level 2 (units HCH 02-2/3Ent-3 and HCH 02-3-5-2). Date: EM III–MM IA. Bibl.: Betancourt et al. 2008b, 166, fig. 28:e.

317 (HCH 94; Fig. 26). Straight-sided cup with conical shape, rim sherd, handle missing. D. of rim ca. 16 cm. Lasithi Red Fabric Group (reddish yellow, 5YR 6/6, to yellowish red, 5YR 5/6). Dark slip on interior of rim and on exterior. Added white pendent triangles with hatched upper corners. Comments: from Rooms 1–4 (1976–1983 seasons). Date: EM III–MM IA.

318 (HCH 03-219; Fig. 26). Straight-sided cup with conical shape, fragments. H. 6.3; d. of rim 12; d. of base 5.5 cm. Lasithi Red Fabric Group (red, 2.5YR 5/8). Handle with circular section attached at rim and center of body. Dark slip on interior of rim and on exterior. Added white crosshatching. Comments: from Room 5, area 5, level 2 (unit HCH 02-3-5-2). Date: EM III–MM IA.

319 (HNM 12,434; Fig. 26; Pl. 2). Straight-sided cup with conical shape, complete. H. 6.3; d. of rim 13.7; d. of base 6.6 cm. Lasithi Red Fabric Group (red, 2.5YR 6/6). Handle with circular section attached at rim and below center of body. Dark slip on interior of rim and on exterior. Added white pendent triangles with hatched upper corners. Comments: from Room 4, upper levels (1983 season). Date: EM III–MM IA. Bibl.: Langford-Verstegen 2008, 554, fig. 9:1.

320 (HCH 02-150; Fig. 26). Straight-sided cup with conical shape, rim sherd. Rest. h. 6; d. of rim 10 cm. Lasithi Red Fabric Group (red, 2.5YR 5/8). Handle with circular section attached at rim and center of body. Dark slip on interior of rim and on exterior. Added white missing. Comments: from Room 5, area 3, below surface (unit HCH 02-3-3-Below Surf). Date: EM III–MM IA.

321 (HCH 53; Fig. 26). Straight-sided cup with conical shape, rim sherd. Rest. h. 6.0; d. of rim 12.0 cm. Lasithi Red Fabric Group (red, 2.5YR 6/8). Handle with circular section attached at rim and center of body. Dark slip on interior of rim and on exterior. Added white missing. Comments: from Rooms 1–4 (1983 season). Date: EM III–MM IA.

322 (HCH 173; Fig. 26). Straight-sided cup with conical shape, rim sherd. D. of rim 12.0 cm. Lasithi Red Fabric Group (light red, 2.5YR 6/8). Straight walls, small lug or tab handle at rim. Dark slip on interior of rim and on exterior. Added white missing. Comments: from Rooms 1–4 (1983 season). Date: EM III–MM IA.

323 (HNM 12,457; Fig. 26). Straight-sided cup with conical shape, almost complete. H. 6.5; d. of rim 10.9; d. of base 6 cm. Lasithi Red Fabric Group (light red, 2.5YR 6/6). Straight walls, handle with circular section attached on outside of rim and below center of body. Dark slip (light reddish brown, 2.5YR 6/4) on interior of rim and on exterior. Added white traces. Comments:

from Room 4, upper levels (1983 season). *Date*: EM III–MM IA.

324 (HCH 02-30; Fig. 26). Straight-sided cup with conical shape, almost complete. H. 5.5; d. of rim 10.5; d. of base 6.0 cm. Lasithi Red Fabric Group (varies from dark red, 2.5YR 6/6, to red, 2.5YR 3/6). Handle with circular section attached at outside of rim and center of body. Dark slip on interior of rim and on exterior. Added white traces of pendent semicircles. *Comments*: from the Room 4/5 Entrance (unit HCH 02-2/3 Ent-Surface). Burned. *Date*: EM III–MM IA.

325 (HCH 05-411; Fig. 26). Straight-sided cup with conical shape, rim sherd. D. of rim ca. 16 cm. Lasithi Red Fabric Group (varies from reddish yellow, 5YR 6/6, to yellowish red, 5YR 5/6). Handle not preserved. Dark slip on interior of rim and on exterior. Added white hatched triangles. *Comments*: from Room 4, upper levels (1983 season). Analyzed by ceramic petrography, sample no. HCH 04/03, Hagios Charalambos Fabric Group 2b (Lasithi Red Fabric Group). *Date*: EM III–MM IA.

326 (HCH 02-70; Fig. 26). Straight-sided cup with conical shape, complete profile. H. 5.5; d. of rim 12; d. of base 8 cm. Lasithi Red Fabric Group (red, 2.5YR 5/6). Almost straight walls, handle not preserved. Dark slip on interior of rim and on exterior. Added white traces, probably pendent hatched triangles. *Comments*: from Room 5, area 5, level 2 (unit HCH 02-3-5-2). *Date*: EM III–MM IA.

327 (HCH 03-226; Fig. 26). Straight-sided cup with conical shape, complete profile. H. 5.4; d. of rim 12; d. of base 6 cm. Lasithi Red Fabric Group (yellowish red, 5YR 5/6). Almost straight walls, handle missing. Dark slip on interior of rim and on exterior. Added white pendent hatched triangles. *Comments*: from Room 5, area 1, level 2 (unit HCH 03-5-1-2). *Date*: EM III–MM IA.

328 (HCH 63; Fig. 26). Straight-sided cup with conical shape, complete profile. H. 6.2; d. of rim ca. 12; d. of base 10 cm. Lasithi Red Fabric Group (reddish brown, 2.5YR 5/4). Conical shape, straight walls, handle with circular section attached at top of rim and center of body. Dark slip on interior of rim and traces on exterior. Added white bands on interior of rim, exterior not preserved. *Comments*: from Room 4, upper levels (1983 season). *Date*: EM III–MM IA.

Rounded Cups, Dark Paint on Interior of Rim and on Exterior

This is an extremely common shape at the cave. A continuous series exists between rounded and almost straight walls. Both the shape and the style of the decoration are a local or regional version of East Cretan White-on-Dark Ware (Betancourt

1984, 44, fig. 4-3:C; Apostolakou, Betancourt, and Brogan 2011, 40, fig. 4:2). Similar examples were found at the Trapeza Cave (Pendlebury, Pendlebury, and Money-Coutts 1935–1936, 56). The dark slip is light red to red to reddish yellow unless stated otherwise.

329 (HCH 02-113; Fig. 27). Rounded cup, rim sherd with handle. D. of rim ca. 12 cm. Lasithi Red Fabric Group (reddish brown, 5YR 4/4). Handle with circular section attached just below rim and at center of body. Dark slip on interior of rim and on exterior. Added white traces of pendent hatched triangles(?). *Comments*: from the Rooms 4/5 Entrance, level 3 (unit HCH 02-2/3 Ent-3). *Date*: EM III–MM IA.

330 (HCH 03-186; Fig. 27). Rounded cup, complete profile, with handle missing. H. 5.3; d. of rim ca. 9 cm. Lasithi Red Fabric Group (strong brown, 7.5YR 4/6). Dark slip on interior of rim and on exterior. Added white traces of pendent hatched triangles(?). *Comments*: from Room 7, surface (unit HCH 02-7-1). *Date*: EM III–MM IA.

331 (HCH 02-136; Fig. 27). Rounded cup, almost complete profile. D. of rim ca. 12 cm. Lasithi Red Fabric Group (light red, 2.5YR 6/8). Thickened rim, rounded shape, handle with elliptical section attached at rim and center of body. Dark slip on interior of rim and on exterior. Added white lines on handle and traces on exterior. *Comments*: from Room 5, area 5, level 2 (unit HCH 02-3-5-2). *Date*: EM III–MM IA.

332 (HNM 12,453; Fig. 27; Pl. 2). Rounded cup, complete profile. H. 6.9; d. of rim 9.2; d. of base 5.6 cm. Lasithi Red Fabric Group (reddish yellow, 5YR 7/6). Handle with circular section attached on outside of rim and center of body. Dark slip on interior of rim and on exterior. Added white traces. *Comments*: from Room 4, upper levels (1983 season). Burned. *Date*: EM III–MM IA.

333 (HNM 13,817; Fig. 27). Rounded cup, three-quarters complete, with handle missing. H. 5.7; d. of rim 9–10; d. of base 4.6 cm. Lasithi Red Fabric Group (red, 2.5YR 6/6). Dark slip on interior of rim and on exterior. Added white traces. *Comments*: from Room 4, upper levels (1983 season). Misshapen. *Date*: EM III–MM IA.

334 (HCH 03-165; Fig. 27). Rounded cup, most of a profile. D. of rim 8 cm. Lasithi Red Fabric Group (red, 5YR 6/8). Handle with circular section attached at rim and below center of body. Dark slip on interior of rim and on exterior. Added white not preserved. *Comments*: from Room 4, upper levels (1983 season). Burned. *Date*: EM III–MM IA.

335 (HCH 02-57; Fig. 27). Rounded cup, complete profile. H. 5.6; d. of rim 12; d. of base 6 cm. Lasithi Red Fabric Group (red, 2.5YR 4/6). Handle with thick oval section attached at rim and below center of body. Traces of dark slip. *Comments*: from Room 5, area 5, level 2 (unit HCH 02-3-5-2). *Date*: EM III–MM IA.

336 (HNM 12,403; Fig. 27). Rounded cup, complete. H. 5.5; d. of rim 8.4; d. of base 4.2 cm. Lasithi Red Fabric Group (red, 2.5YR 5/8). Handle with circular section attached at rim and center of body. Dark slip on interior of rim and on exterior. Added white not preserved. *Comments*: from Rooms 1–3 (1976–1983 seasons). *Date*: EM III–MM IA. *Bibl.*: Betancourt et al. 2008b, 166, fig. 28:i.

337 (HCH 03-227; Fig. 27). Rounded cup, rim sherd with part of handle. Max. dim. 5.1 cm. Lasithi Red Fabric Group (red, 2.5YR 5/8). Handle with circular section attached at rim. Dark slip missing. *Comments*: from the top of the original entrance to the cave, found in a small deposit of the original soil at the cave's mouth along with animal bones with cut marks on them. *Date*: EM III–MM IA.

338 (HNM 12,404; Fig. 27). Rounded cup, complete. H. 5.6; d. of rim 9.1; d. of base 6.1 cm. Lasithi Red Fabric Group (red, 2.5YR 6/6). Handle with circular section attached at rim and below center of body. Dark slip on interior of rim and on exterior. Added white traces. *Comments*: from Room 4, upper levels (1983 season). *Date*: EM III–MM IA.

339 (HNM 12,407; Fig. 27). Rounded cup, complete. H. 4.9; d. of rim 7.8; d. of base 4.7 cm. Lasithi Red Fabric Group (red, 2.5YR 6/8). Handle with circular section attached at rim and below center of body. Dark slip on interior of rim and on exterior. Added white not preserved. *Comments*: from Room 4, upper levels (1983 season). *Date*: EM III–MM IA.

340 (HNM 12,430; Fig. 27). Rounded cup, complete. H. 5.6; d. of rim 8.6; d. of base 5.5 cm. Lasithi Red Fabric Group (red, 2.5YR 5/6). Handle with circular section attached at rim and below center of body. Dark slip on interior of rim and traces on exterior. Added white not preserved. *Comments*: from Room 4, upper levels (1983 season). *Date*: EM III–MM IA.

341 (HNM 13,822; Fig. 27). Rounded cup with spout, almost complete. H. 6.5; d. of rim 9.5; d. of base 5.1 cm. Lasithi Red Fabric Group (red, 2.5YR 5/8). Handle with circular section attached at rim and center of body. Traces of dark slip. Added white not preserved. *Comments*: from Room 4, upper levels (1983 season). *Date*: EM III–MM IA.

342 (HCH 05-432; Fig. 27). Rounded cup, rim sherd with handle. D. of rim 8.0 cm. Lasithi Red Fabric Group (reddish yellow, 2.5YR 7/6). Handle with circular section attached at rim and center of body and extending well above the rim. Dark slip on interior of rim and on exterior. Added white traces. *Comments*: from Room 4, upper levels (1983 season). *Date*: EM III–MM IA.

343 (HNM 13,824; Fig. 27). Rounded cup, almost complete. H. 7.5; d. of rim 9.5; d. of base 5.8 cm. Lasithi

Red Fabric Group (red, 2.5YR 4/8). Handle with circular section attached on top of rim and center of body. Dark slip on interior of rim and on exterior. Added white traces. *Comments*: from Room 4, upper levels (1983 season). *Date*: EM III–MM IA.

344 (HNM 13,834; Fig. 28). Rounded cup, complete. H. 6.5; d. of rim 10.3; d. of base 5.5 cm. Lasithi Red Fabric Group (red, 2.5YR 5/6). Handle with circular section attached at rim and center of body. Dark slip on interior of rim and on exterior. Added white not preserved. *Comments*: from Room 4, upper levels (1983 season). *Date*: EM III–MM IA.

345 (HNM 12,455; Fig. 28; Pl. 2). Rounded cup, complete. H. 5.6; d. of rim 9.5; d. of base 5.9 cm. Lasithi Red Fabric Group (red, 2.5YR 5/6). Handle attached at outside of rim and center of body. *Comments*: from Room 4, upper levels (1983 season). *Date*: EM III–MM IA.

346 (HCH 151; Fig. 28). Rounded cup, complete profile. H. 5.8; d. of rim 12; d. of base 6.6 cm. Lasithi Red Fabric Group (red, 2.5YR 5/8). Handle attached at outside of rim and center of body. Dark slip on interior of rim and traces on exterior. Added white not preserved. *Comments*: from Room 4, upper levels (1983 season). *Date*: EM III–MM IA.

347 (HNM 13,821; Fig. 28). Rounded cup, complete profile. H. 6.4; d. of rim 10.6; d. of base 7.7 cm. Lasithi Red Fabric Group (red, 2.5YR 6/6). Handle with circular section attached across the rim and below center of body. Dark slip on interior of rim and on exterior. Added white missing. *Comments*: from Room 4, upper levels (1983 season). Burned. *Date*: EM III–MM IA.

Rounded Cup, Plain or Paint Missing

The paint was probably once present on many or all of these cups, but it often did not adhere well to the vessels. The problem is caused by differences in the amount of shrinkage between the clay recipe used for the body of the vessel in comparison with the slip, plus the large amount of water that apparently flowed through this cave over the centuries. If the cup was unpainted, it may have been made in EM II.

348 (HNM 12,450; Fig. 28). Rounded cup, almost complete. H. 6.7; d. of rim 12.3; d. of base 6.0 cm. Lasithi Red Fabric Group (red, 2.5YR 5/6). Handle attachment missing at rim, handle with circular section attached below center of body. *Comments*: from Room 4, upper levels (1983 season). Burned on one side. *Date*: EM II–MM I.

349 (HNM 12,428; Fig. 28). Rounded cup, complete. H. 7.4; d. of rim 10.9; d. of base 6.2 cm. Lasithi Red Fabric Group (varies from red, 10R 5/8, to black). Handle with circular section attached across the rim and below center of body. Surface not preserved. *Comments*: from Room 4, upper levels (1983 season). Burned. *Date*: EM II–MM I.

350 (HNM 12,448; Fig. 28; Pl. 2). Rounded cup, three-quarters complete. H. 4.8; d. of rim 7.2; d. of base 4.2 cm. Lasithi Red Fabric Group (red, 2.5YR 5/8). Handle with circular section attached at outside of rim and below center of body. Traces of dark slip. *Comments*: from Room 4, upper levels (1983 season). *Date*: EM II–MM I.

351 (HNM 12,446; Fig. 28). Rounded cup, almost complete. H. 5.6; d. of rim 8.5; d. of base 4.8 cm. Lasithi Red Fabric Group (brown, 2.5YR 6/4). Handle with circular section attached at outside of rim and below center of body. *Comments*: from Room 4, upper levels (1983 season). *Date*: EM II–MM I.

352 (HCH 03-170; Fig. 28). Rounded cup, rim sherd with handle. D. of rim ca. 8.5 cm. Lasithi Red Fabric Group (varies from red to light red, 2.5YR 6/8). Rounded shape, handle with thick oval section attached at outside of rim. *Comments*: from Room 4, upper levels (1983 season). *Date*: EM II–MM I.

353 (HCH 27; Fig. 28). Rounded cup, rim sherd with handle. D. of rim ca. 8.5 cm. Lasithi Red Fabric Group (varies from red to light red, 2.5YR 6/8). Handle with circular section attached at outside of rim and below center of body. Surface not preserved. *Comments*: from Room 4, upper levels (1983 season). *Date*: EM II–MM I.

354 (HNM 12,405; Fig. 28). Rounded cup, complete. H. 7.3; d. of rim 10.9; d. of base 5.2 cm. Lasithi Red Fabric Group (light red, 2.5YR 6/6). Handle with circular section attached across rim and below center of body. *Comments*: from Room 4, upper levels (1983 season). *Date*: EM II–MM I.

355 (HNM 13,805; Fig. 28). Rounded cup, half complete. H. 7.8; d. of rim 9.5; d. of base 7 cm. Lasithi Red Fabric Group (reddish brown, 5YR 6/4, with a darker core). Handle with circular section attached across the rim and below center of body. Surface not preserved. *Comments*: from Room 4, upper levels (1983 season). Burned. *Date*: EM II–MM I.

356 (HNM 13,849; Fig. 28). Rounded cup, almost complete. H. 8.2; d. of rim 12.5; d. of base 6.6 cm. Lasithi Red Fabric Group (varies from red, 2.5YR 5/6, to black). Handle with circular section attached across the rim and at bottom of body. Surface not preserved. *Comments*: from Room 4, upper levels (1983 season). Burned. *Date*: EM II–MM I.

357 (HCH 03-157; Fig. 28). Rounded cup with almost straight profile, almost complete. H. 8.2 including the handle; d. of rim 10; d. of base 8 cm. Lasithi Red Fabric Group (varies from red, 2.5YR 5/8, to light red, 2.5YR 6/8). Handle with circular section attached across the rim and at base of body. Surface not preserved. *Comments*: from Room 5, area 3, level 4 (unit HCH 03-5-3-4). Burned. *Date*: EM II–MM I.

358 (HCH 26; Fig. 29). Rounded cup, rim sherd with handle. D. of rim ca. 16 cm. Lasithi Red Fabric Group (red, 2.5YR 6/6). Handle with circular section attached below rim and at center of body, open spout. Surface not preserved. *Comments*: from Rooms 1–4 (1983 season). *Date*: EM II–MM I.

359 (HCH 03-208; Fig. 29). Rounded cup, complete profile. D. of rim ca. 12 cm. Lasithi Red Fabric Group (reddish yellow, 5YR 6/6). Handle with circular section attached at rim and center of body. Surface not preserved. *Comments*: from Room 5, area 5, level 3 (unit HCH 02-3-5-3). *Date*: EM II–MM I.

360 (HNM 12,421; Fig. 29). Rounded cup, almost complete. H. 5.9; d. of rim 8.6; d. of base 7.3 cm. Lasithi Red Fabric Group (red, 2.5YR 6/6). Handle attached at outside of rim and center of body. *Comments*: from Room 4, upper levels (1983 season). *Date*: EM II–MM I.

361 (HCH 147; Fig. 29). Rounded cup with spout, fragments, handle missing. D. of rim 14.0 cm. Lasithi Red Fabric Group (reddish brown, 5YR 4/4). Rounded shape. Traces of dark slip on interior and exterior. Added white traces of bands in interior. *Comments*: from Room 4, upper levels (1983 season). *Date*: EM II–MM I.

362 (HCH 02-49; Fig. 29). Rounded cup, complete profile. H. 11.9; d. of rim ca. 14; d. of base 8.8 cm. A fabric with phyllite inclusions (red, 5YR 5/6). Handle with circular section attached below rim and at center of body. *Comments*: from Room 5, area 5, level below surface (unit HCH 02-3-5-Below Surface). Analyzed by ceramic petrography, sample no. HCH 04/18 (Hagios Charalambos Metamorphic Fabric). *Date*: EM II–MM I.

363 (HNM 12,420; Fig. 29). Rounded cup, almost complete. H. 6.9; d. of rim 11.1; d. of base 6.7 cm. Lasithi Red Fabric Group (red, 2.5YR 6/6). Handle with circular section attached at outside of rim and below center of body. *Comments*: from Room 4, upper levels (1983 season). *Date*: EM II–MM I.

364 (HNM 12,429; Fig. 29). Rounded cup, almost complete. H. 7.7; d. of rim 11.75; d. of base 7.3 cm. Lasithi Red Fabric Group (red, 2.5YR 5/6). Handle with circular section attached at outside of rim and below center of body. *Comments*: from Room 4, upper levels (1983 season). Burned. *Date*: EM II–MM I.

365 (HNM 13,830; Fig. 29). Rounded cup, almost complete, most of handle missing. H. 7.2; d. of rim 11; d. of base 7.2 cm. Lasithi Red Fabric Group (red, 2.5YR 5/6). Handle with circular section attached at outside of rim and below center of body. *Comments*: from Room 4, upper levels (1983 season). Burned. *Date*: EM II–MM I.

366 (HNM 13,831; Fig. 29). Rounded cup, almost complete, with handle missing. H. 7.8; d. of rim 10.7; d. of base 5.2 cm. Lasithi Red Fabric Group (red, 2.5YR

5/6). *Comments*: from Rooms 1–4 (1976–1983 seasons). Burned. *Date*: EM II–MM I.

367 (HNM 13,835; Fig. 29). Rounded cup, rim sherd with handle. H. ca. 7; d. of rim 9 cm. Lasithi Red Fabric Group (red, 2.5YR 5/6). Handle with circular section attached at outside of rim and center of body. *Comments*: from Room 4 (1983 season). *Date*: EM II–MM I.

368 (HNM 13,840; Fig. 29). Rounded cup, almost complete, most of handle missing. H. 6.9; d. of rim 8.8; d. of base 4.8 cm. Lasithi Red Fabric Group (red, 2.5YR 6/6). Handle with circular section attached at outside of rim and below center of body. *Comments*: from Rooms 1–4 (1976–1983 seasons). *Date*: EM II–MM I.

369 (HNM 13,844; Fig. 29). Rounded cup, almost complete, with most of handle missing. H. 6; d. of rim 9.2; d. of base 7.0 cm. Lasithi Red Fabric Group (red, 2.5YR 4/6). Handle with circular section attached at outside of rim and center of body. *Comments*: from Room 3 (1976–1983 seasons). *Date*: EM II–MM I.

370 (HNM 12,394; Fig. 30). Rounded cup, complete profile, with most of handle missing. H. 6.1; d. of rim 9.7; d. of base 6.7 cm. Lasithi Red Fabric Group (red, 2.5YR 5/6). *Comments*: from Room 3 (1976–1983 seasons). *Date*: EM II–MM I.

371 (HNM 12,419; Fig. 30; Pl. 2). Rounded cup, three-quarters complete, with handle missing. H. 6.2; d. of rim 9.6; d. of base 6 cm. Lasithi Red Fabric Group (red, 2.5YR 5/6). *Comments*: from Room 4, upper levels (1983 season). Burned. *Date*: EM II–MM I.

372 (HCH 03-213; Fig. 30). Rounded cup, complete profile, with handle missing. H. 6.1; d. of rim 12; d. of base 6.6 cm. Lasithi Red Fabric Group (red, 2.5YR 5/6). *Comments*: from Room 5, area 5, level 3 (unit HCH 02-3-5-3). *Date*: EM II–MM I.

373 (HCH 03-153; Fig. 30). Rounded cup, complete profile, with handle missing. H. 5.7; d. of rim ca. 8–10; d. of base 6 cm. Lasithi Red Fabric Group (red, 2.5YR 6/6). *Comments*: from Room 5, area 3, level 3 (unit HCH 03-5-3-3). *Date*: EM II–MM I.

374 (HNM 12,396; Fig. 30). Rounded cup, almost complete, with handle missing. H. 8.9; d. of rim 12.1; d. of base 8.6 cm. Lasithi Red Fabric Group (varies from black to red, 2.5YR 5/6, with a darker surface). *Comments*: from Room 3 (1983 season). *Date*: EM II–MM I.

375 (HNM 13,847; Fig. 30). Rounded cup, two thirds complete. H. 8.8; d. of rim 13.5; d. of base 8.1 cm. Lasithi Red Fabric Group (red, 2.5YR 5/6, to black). Handle with thick oval section attached at outside of rim and bottom of body. *Comments*: from Rooms 1–4 (1976–1983 seasons). *Date*: EM II–MM I.

376 (HNM 13,854; Fig. 30). Rounded cup, two-thirds complete. H. 11.0; d. of base 8 cm. Lasithi Red Fabric Group (red, 2.5YR 5/8). Handle with circular section attached at outside of rim and center of body. *Comments*: from Rooms 1–4 (1976–1983 seasons). *Date*: EM II–MM I.

377 (HCH 02-60; Fig. 30; Pl. 3). Rounded cup, three-quarters complete. H. 6.6; d. of rim ca. 9; d. of base 5.7 cm. Lasithi Red Fabric Group (red, 2.5YR 4/8). Out-turned rim with spout, handle with circular section attached at outside of rim and below center of body. *Comments*: from Rooms 1–4 (1976–1983 seasons). *Date*: EM II–MM I.

378 (HCH 02-151; Fig. 30). Rounded cup, complete profile. D. of rim 10; d. of base 6 cm. Lasithi Red Fabric Group (red, 2.5YR 5/8). Handle with circular section attached at outside of rim and below center of body. *Comments*: from Rooms 1–4 (1976–1983 seasons). *Date*: EM II–MM I.

379 (HCH 4; Fig. 30). Rounded cup, rim sherd with handle. D. of rim ca. 10–12 cm. Lasithi Red Fabric Group (red, 2.5YR 4/8). Handle with circular section attached at rim and center of body. Surface not preserved. *Comments*: from Rooms 1–4 (1983 seasons). Burned. *Date*: EM II–MM I.

380 (HCH 132; Fig. 31). Rounded cup, complete profile. H. 5.6; d. of rim 12.0; d. of base 6 cm. Lasithi Red Fabric Group (red, 2.5YR 2.5/4). Handle with circular section attached at top of rim and center of body. *Comments*: from Rooms 1–4 (1976–1983 seasons). *Date*: EM II–MM I.

381 (HCH 03-154; Fig. 31). Rounded cup, complete. H. 6.1; d. of rim 8.7; d. of base 5.2 cm. Lasithi Red Fabric Group (red, 2.5YR 4/8). Handle with circular section attached across top of rim and below center of body. *Comments*: from Room 5, area 3, level 3 (unit HCH 03-5-3-3). *Date*: EM II–MM I.

382 (HCH 02-62; Fig. 31). Cup with almost straight profile, complete profile. H. 6.1; d. of rim 7.6; d. of base 5.4 cm. Lasithi Red Fabric Group (red, 2.5YR 4/6). Rounded shape, handle with circular section attached across top of rim and below center of body. *Comments*: from Room 5, area 4, level 2 (unit HCH 02-3-5-2). *Parallels*: for the shape, see Pendlebury, Pendlebury, and Money-Coutts 1935–1936, 57, fig. 13:522 (Trapeza Cave). *Date*: EM II–MM I.

Cylindrical Cup, Dark Paint on Interior and Exterior

The cup with a vertical handle and a cylindrical shape is uncommon. The only preserved example is in poor condition.

383 (HNM 12,456; Fig. 31; Pl. 3). Straight-sided cup, complete. H. 5.7; d. of rim 7.7; d. of base 6.4 cm. Lasithi Red Fabric Group (between light reddish brown, 2.5YR 6/4, and reddish brown, 2.5YR 5/4). Cylindrical shape, straight walls, handle with circular section attached at rim and below center of body. Dark slip on interior and exterior. Added white traces of band on the rim. *Comments*: from Room 4, upper levels (1983 season). *Date*: MM I–II.

Rounded Cup, Dark Paint on Interior and Exterior

In the East Cretan White-on-Dark Ware of the Gulf of Mirabello and other places on Crete's northeastern coast, which was the region that acted as the inspiration for this version of the style, the practice of painting the entire interior of a cup began in MM IB (all the sherds from the North Trench at Gournia, the largest deposit for the earlier EM III/MM IA phase, added only a band to the interior of the rim; see Betancourt 1984, 11, fig. 2-5). The detail is of special interest because it demonstrates that those who made the pottery in the Lasithi version of White-on-Dark Ware were informed of continuing developments in the tradition. The cups continued to be handmade.

384 (HCH 02-77; Fig. 31). Rounded cup, complete profile. H. 6.3; d. of rim 10; d. of base 6 cm. Lasithi Red Fabric Group (red, 10R 4/6). Handle with circular section attached at rim and center of body. Dark slip on interior and exterior. *Comments*: from Room 5, area 5, level 3 (unit HCH 02-3-5-3). Analyzed by ceramic petrography, sample no. HCH 04/16, Hagios Charalambos Fabric Group 2b (Lasithi Red Fabric Group). *Date*: MM I–II.

385 (HCH 02-83; Fig. 31). Rounded cup, rim sherd with handle. D. of rim ca. 8 cm. Lasithi Red Fabric Group (red, 2.5YR 6/6). Handle with circular section attached at rim and below center of body. Dark slip on interior and exterior. No added white survives. *Comments*: from Rooms 1–4 (1983 season). *Date*: MM I–II.

386 (HCH 02-90; Fig. 31). Rounded cup, complete profile. H. 5.7; d. of rim 10; d. of base 6.0 cm. Lasithi Red Fabric Group (red, 2.5YR 4/6). Handle with thick oval section attached at or across rim and at center of body. Dark slip on interior and exterior. *Comments*: from Room 5, area 4, level 3 (unit HCH 02-3-5-3). *Date*: MM I–II.

387 (HNM 13,816; Fig. 31). Rounded cup, almost complete. H. 6.9; d. of rim 8.6; d. of base 7.4 cm. Lasithi Red Fabric Group (red, 2.5YR 5/6). Handle with circular section attached across rim and at center of body. Dark slip on interior and exterior. Added white traces. *Comments*: from Room 2 (1982 season). *Date*: MM I–II.

388 (HNM 12,452; Fig. 31). Rounded cup, complete. H. 5.8; d. of rim 10.3; d. of base 4.9 cm. Lasithi Red Fabric Group (light red, 2.5YR 6/6). Handle with circular section attached at rim and below center of body. Dark slip on interior and exterior. Added white missing. *Comments*: from Rooms 1–4 (1976–1983 seasons). *Date*: MM I–II.

389 (HNM 12,447; Fig. 31). Rounded cup, three-quarters complete. H. 5.4; d. of rim 9.13; d. of base 5.8 cm. Lasithi Red Fabric Group (red, 10R 5/6). Handle missing. Dark slip (black) on interior and exterior. Added white missing. *Comments*: from Room 4, upper levels (1983 seasons). *Date*: MM I–II.

390 (HCH 03-163; Fig. 31). Rounded cup, complete. H. 5.4; d. of rim 8.9; d. of base 6.0 cm. Lasithi Red Fabric Group (red, 2.5YR 5/8). Handle with circular section attached at rim and below center of body. Dark slip on interior and exterior. Added white missing. *Comments*: from Rooms 1–4 (1976–1983 seasons). *Date*: MM I–II.

391 (HCH 03-156; Fig. 31). Rounded cup, almost complete. H. 5.3; d. of rim 9.3; d. of base 6.0 cm. Lasithi Red Fabric Group (reddish yellow, 5YR 6/6). Handle attachment missing. Dark slip on interior and exterior. Added white missing. *Comments*: from Room 5, area 3, level 4 (unit HCH 03-5-3-4). *Date*: MM I–II.

392 (HCH 02-139; Fig. 31). Rounded cup, rim sherd with part of handle. D. of rim ca. 10–12 cm. Lasithi Red Fabric Group (reddish yellow, 5YR 7/6). Handle with circular section attached at rim. Dark slip on interior and exterior. Added white missing. *Comments*: from the Room 4/5 Entrance, level 3 (unit HCH 02-2/3-3). *Date*: MM I–II.

393 (HCH 88; Fig. 31). Rounded cup, complete profile. H. 5.2 including handle; d. of rim 12; d. of base 5.7 cm. Lasithi Red Fabric Group (light red, 2.5YR 6/8). Handle with circular section attached at rim and below center of body. Dark slip on interior and exterior. Added white band on lower body with rest missing. *Comments*: from Rooms 1–3 (1976–1983 seasons). *Date*: MM I–II.

394 (HNM 12,431; Fig. 31). Rounded cup, complete profile. H. 7.8; d. of rim 11.9; d. of base 7.0 cm. Lasithi Red Fabric Group (red, 2.5YR 5/6). Handle with circular section attached across rim and below center of body. Dark slip on interior and exterior. Added white traces. *Comments*: from Room 4, upper levels (1983 season). *Date*: MM I–II.

Straight-Sided Cups with Conical Shape

Conical cups with one vertical handle made in the Lasithi White-on-Dark Ware were common offerings in MM IB. They were often decorated with pendent hatched triangles or pendent arcs, which are both motifs that were borrowed from the tradition in eastern Crete (Betancourt 1984, 24–32). In comparison with the style of East Crete made in Mirabello Fabric, the shapes here are larger and less well made, and their dark slip is usually red, not black.

395 (HCH 124; Fig. 32). Straight-sided cup with conical shape, complete profile. H. 5.8; d. of rim 14; d. of base 6.0 cm. Lasithi Red Fabric Group (red, 2.5YR 5/6). Straight walls and rim, handle with circular section attached at rim and below center of body. Dark slip on interior and exterior. Added white pendent arcs on exterior, frieze of diagonal lines above bands on interior. *Comments*: from Rooms 1–4 (1983 season). Analyzed by ceramic petrography, sample no. HCH 04/13, Hagios Charalambos Fabric Group 2b (Lasithi Red Fabric Group). *Date*: MM IA–IB.

396 (HNM 13,810; Fig. 32). Straight-sided cup with conical shape, complete profile. H. 6.2; d. of rim 14; d. of base 6 cm. Lasithi Red Fabric Group (light red, 2.5YR 6/6). Straight walls and rim, handle with circular section attached across rim and at center of body. Dark slip on interior and exterior. Added white pendent arcs on interior, hatched triangles on exterior *Comments*: from Room 4, upper levels (1983 season). *Date*: MM IB. *Bibl.*: Langford-Verstegen 2008, 554, fig. 9:2.

397 (HNM 12,399; Fig. 32). Straight-sided cup with conical shape, half complete. H. 6.0; d. of rim 12.6; d. of base 6.0 cm. Lasithi Red Fabric Group (red, 2.5YR 5/6). Straight walls and rim, handle with circular section attached at rim and below center of body. Dark slip on interior and exterior. Added white pendent triangles with hatched upper corners on exterior. *Comments*: from Room 2 (1982 season). *Date*: MM IB.

398 (HCH 125; Fig. 32). Straight-sided cup with conical shape, almost complete. H. 6; d. of rim 13.5; d. of base 5.8 cm. Lasithi Red Fabric Group (light red, 2.5YR 6/8). Straight walls and rim, handle with circular section attached at rim and below center of body. Dark slip on interior and exterior. Added white pendent arcs on interior, pendent hatched triangles on exterior. *Comments*: from Rooms 1–4 (1983 season). *Date*: MM IB.

399 (HCH 78; Fig. 32). Straight-sided cup with conical shape, complete profile. H. 6.0; d. of rim 12.0; d of base 6.9 cm. Lasithi Red Fabric Group (red, 5YR 6/3). Straight walls and rim, handle with circular section attached at rim and below center of body. Dark slip on interior and exterior. Added white pendent arcs on interior, exterior surface missing. *Comments*: from Rooms 1–4 (1983 season). *Date*: MM IB.

400 (HNM 12,389; Fig. 32; Pl. 3). Straight-sided cup with conical shape, almost complete. H. 5.6; d. of rim 11.1; d. of base 6.2 cm. Lasithi Red Fabric Group (light red, 2.5YR 6/8). Straight walls and rim, handle with circular section attached at rim and below center of body. Dark slip on interior and exterior. Added white pendent arcs on interior, exterior surface missing. *Comments*: from Room 3 (1983 season). *Date*: MM IB. *Bibl.*: Betancourt et al. 2008b, 166, fig. 28:f.

401 (HCH 02-67; Fig. 32). Straight-sided cup with conical shape, complete profile with handle. H. 6.2; d. of rim 12; d. of base 7 cm. Lasithi Red Fabric Group (red, 2.5YR 4/8). Straight walls and rim, handle with circular section attached at rim and center of body. Surface not preserved. *Comments*: from Room 5, area 5, level 2 (unit HCH 02-3-5-2). Burned inside, and probably used as a lamp. *Date*: EM III–MM IA.

402 (HCH 150; Fig. 32). Straight-sided cup with conical shape, rim sherd, handle missing. D. of rim 10 cm. Lasithi Red Fabric Group (red, 2.5YR 5/8). Almost straight walls and rim. Dark slip (red) on exterior. Added white pendent hatched triangles. *Comments*: from Rooms 1–4 (1983 season). *Date*: EM III–MM IA.

403 (HCH 89; Fig. 32). Straight-sided cup with conical shape, rim sherd, handle missing. D. of rim 10 cm. Lasithi Red Fabric Group (red, 10R 5/8). Almost straight walls and rim. Dark slip on interior and exterior. Added white pendent hatched triangles. *Comments*: from Rooms 1–4 (1983 season). *Date*: EM III–MM IA.

404 (HCH 77; Fig. 32). Straight-sided cup with conical shape, rim sherd, handle missing. D. of rim 10 cm. Lasithi Red Fabric Group (light red, 2.5YR 6/6). Straight walls and rim. Dark slip on interior and exterior. Added white: crosshatched lines on rim, two horizontal bands, other traces. *Comments*: from Rooms 1–4 (1983 season). *Date*: EM III–MM IA.

405 (HNM 12,411; Fig. 32). Straight-sided cup with conical shape, almost complete, handle missing. H. 7.8; d. of rim 11.2; d. of base 5.3 cm. Lasithi Red Fabric Group (red, 2.5YR 5/8). Straight walls and rim with slightly concave walls, handle missing. Dark slip on interior and exterior. Added white not preserved. *Comments*: from Room 4, upper levels (1983 season). *Date*: MM IB.

406 (HCH 02-144; Fig. 32). Miniature straight-sided cup with conical shape, complete profile, handle missing. H. 3.4; d. of rim ca. 6; d. of base 4 cm. Lasithi Red Fabric Group (light red, 2.5YR 6/6). Straight walls. Dark slip on interior and exterior. Added white diagonal lines from rim to base on exterior. *Comments*: from Room 5, feature 1 (unit HCH 02-3-Feature 1). *Date*: MM IB.

407 (HNM 13,803; Fig. 32). Straight-sided cup with conical shape, complete profile, handle missing. H. 5.8; d. of rim 11.3; d. of base 5.8 cm. Lasithi Red Fabric Group (red, 2.5YR 6/6). Straight walls and rim. Surface missing. *Comments*: from Rooms 1–4 (1976–1983 seasons). *Date*: EM III–MM IA.

408 (HNM 13,839; Fig. 33). Straight-sided cup with conical shape, almost complete. H. 6.2; d. of rim 12.2; d. of base 6.5 cm. Lasithi Red Fabric Group (red, 2.5YR 4/6). Straight walls and rim, handle with elliptical section attached at rim and center of body. Surface missing. *Comments*: from Rooms 1–4 (1976–1983 seasons). Burned. *Date*: MM IB.

409 (HNM 13,848; Fig. 33). Straight-sided cup with conical shape, almost complete. H. 5.5; d. of rim 13.0; d. of base 5.7 cm. Lasithi Red Fabric Group (red, 2.5YR 5/8). Straight walls and rim, handle with elliptical section attached at rim and below center of body. Dark slip on interior and exterior. Added white traces of sets of three pendent arcs in interior, added white traces on exterior. *Comments*: from Room 3 (1976–1983 seasons). *Date*: MM IB.

410 (HNM 13,837; Fig. 33). Straight-sided cup with conical shape, almost complete. H. 5.6; d. of rim 10.8; d. of base 6.1 cm. Lasithi Red Fabric Group (red, 2.5YR 5/6). Straight walls and rim, handle with elliptical section attached at rim. Surface missing. *Comments*: from Rooms 1–4 (1976–1983 seasons). *Date*: MM IB.

411 (HCH 02-98, in HNM; Fig. 33). Straight-sided cup with conical shape, almost complete, handle missing. H. 5.4; d. of rim 10.9; d. of base 6.1 cm. Lasithi Red Fabric Group (red, 2.5YR 5/8). Straight walls and rim, handle with elliptical section attached at rim. Traces of dark slip on interior and exterior. *Comments*: from the Room 4/5 Entrance, level 3 (unit HCH 02-2/3-3). Burned in the interior (perhaps used as a lamp). *Date*: MM IB.

412 (HCH 02-14; Fig. 33). Straight-sided cup with conical shape, rim sherd with part of handle. D. of rim 12–14 cm. Lasithi Red Fabric Group (light red, 2.5YR 6/8). Straight walls and rim, handle with elliptical section attached at rim and center of body. *Comments*: from Room 4, lower levels (unit HCH 02-2-2). *Date*: MM IB.

Miscellaneous Open Shapes

413 (HCH 30; Fig. 33). Cup with angular carination low on body, rim sherd with handle. H. over 10 cm; d. of rim not measurable. A coarse fabric (red, 2.5YR 5/6). Conical shape, handle attached across rim and below center of body. *Comments*: from Room 4, upper levels (1983 season). Burned. *Date*: EM III–MM IA.

414 (HNM 12,432; Fig. 33). Bell-shaped spouted cup, complete. H. 7.7; d. of rim 10.7; d. of base 5.1 cm. Lasithi Red Fabric Group (red, 2.5YR 6/6). Rounded to conical shape, slightly outturned rim, handle with circular section attached at rim and below center of body. *Comments*: from Room 4, upper levels (1983 season). Burned. *Date*: EM III–MM IA.

415 (HCH 52; Fig. 33). Bell-shaped cup, rim sherd with handle. D. of rim 10–12 cm. Lasithi Red Fabric Group (light red, 2.5YR 6/8). Rounded to conical shape, slightly outturned rim, handle with circular section attached at top of rim and below center of body. Dark slip on interior and exterior. Added white missing. *Comments*: from Room 4, upper levels (1983 season). *Date*: EM III–MM IA.

416 (HCH 03-220; Fig. 33). Miniature tumbler, base sherd. D. of base 2.85 cm. Lasithi Red Fabric Group (light red, 2.5YR 6/6). Tall conical shape, no handles preserved. Dark slip on exterior. *Comments*: from Rooms 1–4 (unit HCH 02-Dump). *Parallels*: Zois 1969, pl. 29 (Gournes). *Date*: EM III–MM II.

417 (HCH 06-459; Fig. 33). Brazier, base and handle sherds. Rest. h. ca. 10; d. of base 8.0 cm. Lasithi Red Fabric Group (red, 2.5YR 4–5/8). Slightly rounded shape of profile on handle side, with pulled rim opposite handle, horizontal frying pan handle. Dark slip (red, 2.5YR 4/6) on exterior. *Comments*: from Room 4 (1983 season). *Date*: EM III–MM II.

418 (HCH 170; Fig. 33). Wide-mouthed cup, rim sherd. D. of rim 14 cm. Lasithi Red Fabric Group (light red, 2.5YR 6/8). Heavily burnished on interior, exterior surface missing. *Comments*: from Rooms 1–4 (dump from 1976–1983 seasons, unit HCH 03-12-1). *Date*: EM III–MM IA.

419 (HNM 12,451; Fig. 33; Pl. 3). Deep rounded cup, almost complete. H. 6.2; d. of rim 8.2; d. of base 4.6 cm. Lasithi Red Fabric Group (light red, 10R 6/6). Tall rounded cup, no handles. Dark slip (red, 2.5YR 6/6) on interior and exterior. Added white missing. *Comments*: from Room 4, upper levels (1983 season). *Date*: EM III–MM IA.

420 (HCH 43; Fig. 34). Vessel with loop handles, rim sherd with handle. D. of rim 8 cm. A red fabric (light red, 2.5YR 6/8). Horizontal loop handle rising above rim. Dark slip on interior and exterior. Added white not preserved. *Comments*: from Room 4, upper levels (1983 season). *Date*: EM III–MM IA.

421 (HCH 02-44; Fig. 34). Miniature cup, complete. H. 3.5; d. of rim 5.3; d. of base 2.6 cm. Lasithi Red Fabric Group (red, 2.5YR 5/6, with a white surface). Rounded shape, handle with circular section attached at rim and on body. *Comments*: from Room 4, lower levels (unit HCH 02-2-1). *Date*: MM I–II.

Cooking Vessels

Tripod Cooking Vessels

Tripod cooking pots are common Minoan cooking vessels (Betancourt 1980; Martlew 1988; Filippa-Touchais 2000). The earliest examples from EM I have very wide legs, a rounded profile, and an everted rim (Shank 2005, 105, pl. 11:f). The pieces from the cave are much later, with straighter profiles and legs that have elliptical to circular sections. The dates are uncertain, and for **424–426** the shape of the vessel is unknown.

422 (HCH 155; Fig. 34). Tripod brazier, base sherd with part of legs. D. of base 8 cm. Lasithi Red Fabric Group (dark red, 2.5YR 3/6). Straight-sided bowl supported on three legs, leg with elliptical section, horizontal handle with circular section. *Comments*: from Rooms 1–4 (1983 season). *Date*: MM I–II.

423 (HCH 37; Fig. 34). Tripod cooking pot, base sherd with part of leg. D. of base 18 cm. Lasithi Red Fabric Group (dark red, 2.5YR 3/6), partly burned black. Straight-sided bowl supported on three legs, leg with elliptical section. *Comments*: from Rooms 1–4 (1983 season). *Parallels*: Tripod cooking pots are discussed by Betancourt (1980) and Martlew (1988). *Date*: EM II–MM I.

424 (HCH 02-130; Fig. 34; Pl. 3). Tripod vessel, leg sherd. Pres. length 6.2 cm. Lasithi Red Fabric Group (light red, 2.5YR 6/6). Leg with thick oval section. *Comments*: from the Room 4/5 Entrance, level 3 (unit HCH 02-2/3-3). *Date*: MM I–II.

425 (HCH 02-131; Fig. 34). Tripod vessel, leg sherd. Pres. length 6.2 cm. Lasithi Red Fabric Group (light reddish brown, 2.5YR 5/6). Leg with thick oval section. Slipped. *Comments*: from the Room 4/5 Entrance, level 3 (unit HCH 02-2/3-3). *Date*: MM I–II.

426 (HCH 03-210; Fig. 34). Tripod vessel, leg sherd. Pres. length 7.0 cm. Lasithi Red Fabric Group (light red, 2.5YR 6/6). Leg with almost circular section. *Comments*: from the Room 4/5 Entrance, level 3 (unit HCH 02-2/3-3). *Date*: MM I–II.

427 (HCH 161; Fig. 34). Tripod cooking pot(?), leg sherd. Pres. length 8.8 cm. Lasithi Red Fabric Group (light red, 2.5YR 6/6). Leg with elliptical section. *Comments*: from Rooms 1–4 (1983 season). *Date*: EM II–MM I.

Cooking Vessel without Legs

428 (HCH 06-458; Fig. 34). Cooking pot, non-joining rim and base sherds with handles. Rest. h. ca. 23–25; d. of rim ca. 18; d. of base 17 cm. Lasithi Red Fabric Group (red, 2.5YR 4/6). Slightly outturned rim, rounded profile, no legs, two vertical handles with circular sections attached below rim and at shoulder. *Comments*: mended from Rooms 2 and 4 (1982 and 1983 seasons). *Date*: MM I–II.

Other Shapes

Small Tripod Bowl

429 (HCH 02-116; Fig. 34). Small tripod vessel, base sherd with part of leg. Max. dim. 8.2 cm. Lasithi Red Fabric Group (red, 2.5YR 4/6). Straight-sided or rounded bowl supported on three short legs, handle not preserved. *Comments*: from the Room 4/5 Entrance, level 3 (unit HCH 02-2/3 Ent-3). *Date*: MM I–II.

Jug or Kantharos(?)

430 (HCH 06-456; Fig. 35). Jug or kantharos, body sherd with handle. Max. dim. 6.6 cm. Lasithi Red Fabric Group (core dark reddish brown, 5YR 3 /4). Dark slip (light red, 2.5YR 6/6) on exterior. Handle with elliptical section. *Comments*: from Room 5, area 5, level 2 (unit HCH 02-3-5-2). Burned. *Date*: MM I–II.

Jugs

431 (HNM 12,416; Fig. 35; Pl. 3). Piriform jug, complete. H. 20.5; d. of base 6.7 cm. Lasithi Red Fabric Group (red, 2.5YR 5/8). Piriform shape, tall raised spout, handle with thick oval section attached below rim and on shoulder. Dark slip on exterior. Added white

not preserved. *Comments*: from Room 4 (1983 season). *Date*: EM IIB.

432 (HNM 12,414; Fig. 35). Jug, lower part. Rest. h. ca. 19–22; d. of base 7.5 cm. Lasithi Red Fabric Group (light red, 2.5YR 6/8). Globular shape, handle with circular section attached at rim and shoulder. Dark slip (red) on exterior. *Comments*: from Room 1 (1982 season). *Date*: EM IIB.

433 (HCH 28; Fig. 35). Jug, upper part. Pres. h. 8 cm. Lasithi Red Fabric Group (light red, 10R 6/8). Small raised spout. *Comments*: from Room 4 (1983 season). *Date*: EM II–III.

434 (HNM 12,454; Fig. 35). Jug, complete. H. 26.5; d. of base 8.5 cm. Lasithi Red Fabric Group (red, 2.5YR 5/6). Piriform shape, small raised spout, handle with circular section attached at rim and shoulder. Dark slip (red) on exterior. *Comments*: from Room 4, upper levels (1983 season). Burned. *Date*: EM IIB.

435 (HCH 02-71; Fig. 35). Jug, upper part. Pres. h. 8.5 cm. Lasithi Red Fabric Group (light red, 2,5YR 6/6). Rounded profile, small raised spout, handle with circular section attached at rim and shoulder. Dark band on rim, other traces. *Comments*: from Room 5, area 5, level 2 (unit HCH 02-3-5-2). *Date*: EM II–III.

436 (HNM 12,393; Fig. 35). Jug, almost complete. Pres. h. 13; d. of base 7.2 cm. Lasithi Red Fabric Group (red, 2.5YR 5/8). Globular shape, handle attached below rim and on shoulder. Dark slip on exterior. Added white missing. *Comments*: from Room 3 (1983 season). *Date*: EM II–III.

437 (HNM 12,391; Fig. 36). Jug, complete. H. 8.5; d. of base 5.2 cm. A light red fabric, light red (5YR 7/4). Trefoil mouth, handle with circular section attached below rim and on body. Unpainted. *Comments*: from Room 3 (1983 season). *Date*: MM IIB?

438 (HNM 12,459; Fig. 36; Pl. 3). Jug, almost complete. Rest. h. 9; d. of base 5.1 cm. Lasithi Red Fabric Group (red, 2.5YR 6/6). Piriform shape and smooth transition between neck and shoulder, handle missing. Unpainted. *Comments*: from Room 4, upper levels (1983 season). *Date*: MM I–IIB.

439 (HNM 12,406; Fig. 36; Pl. 3). Jug, almost complete. Rest. h. ca. 10; d. of base 5.3 cm. Lasithi Red Fabric Group (reddish yellow, 5YR 7/6). Piriform shape and smooth transition between neck and shoulder, handle missing. Unpainted. *Comments*: from Room 4, upper levels (1983 season). *Date*: MM I–IIB.

440 (HCH 02-61; Fig. 36). Jug, upper part and handle sherd. Rest. h. ca. 10; d. of rim 3.5 cm. Lasithi Red Fabric Group (between red and light red, 2.5YR 5–6/8). Piriform shape, horizontal spout, handle with circular section attached below rim and on body. Unpainted. *Comments*: from Room 5, area 5, level 2 (unit HCH 02-3-5-2). *Date*: MM I–IIB.

441 (HNM 12,392; Fig. 36). Jug, complete. H. 9.2; d. of rim 4.0; d. of base 5.0 cm. Lasithi Red Fabric Group (red, 2.5YR 5/8). Piriform shape and smooth transition between neck and shoulder, handle missing. Dark slip on exterior. Added white missing. *Comments*: from Room 3 (1983 season). *Date*: MM I–IIB?

442 (HCH 54; Fig. 36). Jug, complete profile. H. 6.4; d. of base 6.0 cm. Lasithi Red Fabric Group (light red, 2.5YR 6/8). Piriform shape and low position of maximum diameter, small low spout, two knobs on sides of spout. Unpainted. *Comments*: from Rooms 1–4 (1983 season). *Date*: MM I–II.

443 (HNM 13,846; Fig. 36). Jug, lower part. Rest. h. ca. 11–12; d. of base 6.5 cm. Lasithi Red Fabric Group (red, 2.5YR 5/6). Piriform shape and smooth transition between neck and shoulder, handle missing. Unpainted. *Comments*: from Room 4 (1983 season). *Date*: MM I–II.

444 (HCH 79; Fig. 36). Jug, upper part. Rest. h. ca. 14–16; d. of opening 3 x 2.3 cm. Lasithi Red Fabric Group (red, 2.5YR 5/8). Piriform shape and smooth transition between neck and shoulder, horizontal spout, handle with circular section attached at rim and center of body. *Comments*: from Room 4 (1983 season). *Date*: MM I–II.

445 (HCH 02-79; Fig. 36). Jug, upper part. Pres. h. 7.5 cm. Lasithi Red Fabric Group (red, 2.5YR 4/6). Rounded profile, low spout, handle with circular section attached below rim. Traces of dark slip. Added white band on rim and handle. *Comments*: from Room 4 (1983 season). *Date*: MM I–IIB.

Tube-Spouted and Bridge-Spouted Jugs

446 (HCH 06-455; Fig. 36). Bridge-spouted jug, many sherds. Rest. h. 11.5; d. of rim 11.8 cm. Lasithi Red Fabric Group (varies from reddish brown, 5YR 5/4, to reddish yellow, 5YR 7/8, with the core dark). Thickened, outturned rim, bridged spout at the shoulder, handle with thick oval section attached at rim and below shoulder. Dark slip (reddish brown, 5YR 5/3) on exterior. Black band under rim. *Comments*: from Room 4, upper levels (1983 season). Burned. *Date*: MM IB.

447 (HNM 13,820; Fig. 36; Pl. 3). Side-spouted or tube-spouted jug, almost complete. H. 8.3; d. of rim 6.5; d. of base 4.2 cm. Lasithi Red Fabric Group (reddish brown, 2.5YR 5/4). Straight rim, rounded body, bridged or tube spout at the shoulder, one vertical handle with circular section attached below rim and on shoulder opposite spout and two horizontal side-handles added at shoulder. Dark slip (red) on exterior. Added white missing. *Comments*: from Room 4, upper levels (1983 season). *Date*: MM IB.

448 (HCH 127; Fig. 37). Side-spouted jug or teapot, lower part. Rest. h. ca. 14; d. of base 10.8 cm. Lasithi Red Fabric Group (light red, 2.5YR 6/6). Either a tube spout or a teapot (spout missing), vertical handle with circular section. Dark slip on exterior. *Comments*: from Room 4, upper levels (1983 season). Burned. *Parallels*: the shape (except for the spout, whose shape is unknown) is similar to Betancourt and Davaras, eds., 2003, fig. 21:1.1 (Pseira). *Date*: EM IIB–MM I.

449 (HCH 02-127; Fig. 37). Bridge-spouted jar, rim sherd with spout. D. of rim 11–12 cm. Lasithi Red Fabric Group (weak red, 2.5YR 4/2). Thickened, inturned rim, rounded body, bridged spout at the shoulder. Traces of dark slip on exterior. Added white missing. *Comments*: mended from sherds from the Room 4/5 Entrance and Room 5, area 5, level 2 (units HCH 02-2/3-3 and HCH 02-3-5-2). Burned. *Date*: MM IB.

Closed Vessel

450 (HCH 02-126; Fig. 37). Closed vessel, rim sherd. D. of rim ca. 8 cm. Lasithi Red Fabric Group (light red, 2.5YR 6/6, with a darker core). Straight, slightly inturned rim. Dark slip on exterior. Added white cross-hatched area and diagonal band (motif mostly missing). *Comments*: from Room 5, area 5, level 2 (unit HCH 02-3-5-2). Burned. *Date*: MM I.

Jars

Large jars and pithoi in the assemblage are all preserved only in very small fragments, and not a single example is restorable. Fragments come from scattered locations. It is possible that these large shapes were originally used for jar burials.

Jar with Thickened Rim

451 (HCH 36; Fig. 37). Jar, rim sherd. D. of rim ca. 34 cm. Lasithi Red Fabric Group (light red, 2.5YR 6/8). Thickened rim. Dark slip on top of rim and on exterior. *Comments*: from Room 4, upper levels (1983 season). *Parallels*: Pendlebury, Pendlebury, and Money-Coutts 1935–1936, 86–88, fig. 20:902, 910 (Trapeza Cave); Betancourt 2006, 80, 95, figs. 5.4:43, 5.9:106 (Chrysokamino). *Date*: EM III–MM I.

Large Jar with Rope Decoration

452 (HCH 12; Fig. 37). Jar, rim sherd. D. of rim ca. 34 cm. Lasithi Red Fabric Group (light red, 2.5YR 6/8). Thickened rim. Ropework band below rim. Dark slip on top of rim and on exterior. *Comments*: from Room 4, upper levels (1983 season). *Parallels*: for the rim, see **451**; for both the rim and the molding below the rim, see Betancourt and Davaras, eds., 2003, fig. 28:2.50 (Pseira). *Date*: EM III–MM I.

Other Jars

453 (HCH 02-143; Fig. 37). Jar, rim sherd. D. of rim 32 cm. Lasithi Red Fabric Group (light red, 2.5YR 6/8). Thickened rim. Dark slip (red, 2.5YR 5/8) on top of rim and on exterior. *Comments*: from the Room 4/5 Entrance, surface level (unit HCH 02-2/3-Surface). *Parallels*: see **451**. *Date*: EM III–MM IIB.

454 (HCH 80; Fig. 37). Jar, lower part. D. of base 16 cm. Lasithi Red Fabric Group (light red, 2.5YR 6/8). Dark slip on exterior. Added white diagonal lines on body. *Comments*: from Room 4, upper levels (1983 season). *Date*: EM III–MM I.

455 (HCH 192; Fig. 37). Jar, body sherd with handle. Max. dim. 8.3 cm. Lasithi Red Fabric Group (red, 2.5YR 5/8). Wide vertical handle with elliptical cross section. Traces of dark slip. *Comments*: from Room 4, upper levels (1983 season). *Date*: EM III–MM IIB.

456 (HCH 35; Fig. 37). Jar, body sherd with handle. Max. dim. 10.0 cm. Lasithi Red Fabric Group (red, 2.5YR 5/8). Horizontal handle with circular cross section. *Comments*: from Room 4, upper levels (1983 season). *Date*: EM III–MM IIB.

457 (HCH 04-395; Fig. 37). Jar, body sherds. Max. dim. of largest sherd 11.0 cm. Lasithi Red Fabric Group (red, 2.5YR 5/8). Dark (reddish yellow, 5YR 7/6) horizontal bands on exterior. Added white traces of horizontal bands on the dark bands. *Comments*: from Room 4, upper levels (1983 season). *Date*: EM III–MM IIB.

458 (HCH 02-53). Jar with tripod legs, base sherd. Pres. h. 2.2; d. of base 27; length of leg 2.6 cm. Lasithi Red Fabric Group (light red, 2.5YR 6/8). Short legs. *Comments*: from Room 5, area 5, level 2 (unit 02-3-5-2). Analyzed by ceramic petrography, sample no. HCH 04/14 (Hagios Charalambos Fabric Group 2b). *Date*: MM I–II.

Lid

459 (HCH 83; Fig. 37). Lid, sherd. D. of rim 28 cm. Lasithi Red Fabric Group (red, 5YR 6/8). Flat lid. Dark slip (red). *Comments*: from Room 4, upper levels (1983 season). *Date*: EM II–MM IIB.

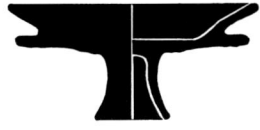

18

Conclusions

The pottery assemblage from the Hagios Charalambos Cave illustrates local production styles in the Lasithi region, and it also shows a taste for imported wares from elsewhere on the island of Crete. A total of 18,064 vases and sherds are present (App. B). This assemblage seems to represent a chronological continuum with no significant gaps. The earliest piece is a partial vessel from the Late Neolithic period (1). More pottery is present from FN, EM I, and EM II. Perhaps the settlement increased in size in EM III, because the quantity of pottery from this time includes many more pieces. Styles from MM I and MM II are well represented. The latest sherds from inside the cave are either from the very end of MM IIB or from slightly later. It is clear that the pottery represents a sequence of burials stretching from the Final Neolithic until MM IIB.

As is typical of pottery associated with burials, the vessels found in the cave were mostly used for the pouring and drinking of liquids. Cups and jugs are the dominant shapes of both imported and locally made objects. Minoan ritual drinking has received considerable discussion in recent years (for bibliography, see Hamilakis 1998; Tzedakis and Martlew, eds., 1999; Wright, ed., 2004; Halstead and Barrett, eds., 2004; Hitchcock, Laffineur, and Crowley, eds., 2008; Tzedakis, Martlew, and Jones, eds., 2008; Vasilakis and Branigan 2010, 255–261). Minoan drinking has been extensively documented both from domestic and funerary contexts. The evidence from this cave is important for the general discussion because it is possible to make some observations not only on the fact that eating and drinking were some of the activities represented by this assemblage, but also on the nature of those activities and how they changed through time.

In addition to vessels for the consumption of food and beverages, the cave also contained small vases that presumably held oil or a special unguent, sherds of jars that may have been burial containers, and vessels that can be classified as offering tables. Among these, the vases that may have contained oil or perfume are the most numerous. They suggest that the offering of oil to the deaceased was already a part of the ceremonial rituals associated with Early and Middle Bronze Age Crete.

The majority of the pottery from the site is made locally (App. B; Table B.19). The local pottery has a very distinctive fabric (called the Lasithi Red Fabric Group), and it also can be recognized by its style. The Lasithi Red Fabric Group is made of medium-coarse, reddish clay with phyllite inclusions. Petrographic analysis provides a thorough description (App. A). The petrography demonstrates that the metamorphic phyllite fragments consist of micas, deformed quartz grains, feldspars, and a few minor constituents.

Offering tables compliment the assemblage of locally made vessels for pouring and drinking. A typology of offering tables consists of twelve variations in added decoration to a relatively standardized shape that consists of a shallow bowl with a flat base and vertical or slightly flaring sides supported on a cylindrical lower part. Four types of offering tables are represented in the Hagios Charalambos assemblage. The tables are evidence for early exports out of Lasithi. A pattern of distribution can be discerned, as parallels are known from elsewhere in Lasithi, the northern coast, the area of the Gulf of Mirabello, the Mesara, and the southeastern coast.

Imported pottery found in the cave ranges in date from Neolithic to MM IIB. Early Minoan imports are from the Mesara, the area of the Gulf of Mirabello, and Knossos. In MM IB, there are imports from elsewhere in the Pediada, including Kastelli and Galatas. In MM II the amount of pottery from production centers at, or in the region of, Malia increased. This pattern can be recognized in other classes of artifacts from the cave as well (Ferrence 2007, 2008). This increased contact contributes to a greater understanding of the Malian political and economic sphere of influence in the Middle Minoan period. Several authors have discussed the possibility of a Malia-Lasithi state in MM II (Cadogan 1990; 1995; Knappett 1999; Betancourt 2011b). Petrographic analysis for this project has shown matches between pottery from the cave and a Malian pale fabric (App. A).

The closest parallel for the Hagios Charalambos Cave is the Trapeza Cave. The two caves seem to have served the same purpose during the same time periods. They are in close proximity, one situated almost directly across the Lasithi Plain from the other. Their proximity strongly suggests that those utilizing one cave would have known of the other. They were used concurrently, so this is not a case of one being filled and then the other coming into use at a later time.

Production Centers

Several specific production centers can be recognized within the assemblage from the cave. They represent different workshops or closely related workshop groups that were furnishing ceramics to the towns and villages in Lasithi and perhaps the nearby parts of the Pediada. The style of the pottery from these production centers was sometimes derived originally from other geographic locations, and it was sometimes developed locally from the large general style of earlier Minoan ceramics.

Lasithi White-on-Dark Ware

A local or regional workshop group decorating pottery in a manner imitating East Cretan White-on-Dark Ware provided a large amount of the pottery in the cave (Ch. 17). The products that were

made locally but related to the Gulf of Mirabello workshops have a highly distinctive micaceous red fabric. They can be traced from EM IIB until MM IB. The EM IIB pottery (Ch. 6) imitates the shapes of Vasiliki Ware, including goblets, jugs with upright spouts, and a conical bowl. Beginning in EM III, the workshop began imitating the East Cretan White-on-Dark Ware that replaced Vasiliki Ware in the region of the Gulf of Mirabello at this time (Ch. 17). The local change in style that closely follows the development in the Gulf of Mirabello region underscores the debt of the local potters to the styles of this part of Crete. During this period, the potters in the Lasithi region mostly made cups, shallow bowls, and jugs with lower spouts than in EM IIB and decorated them in white applied over a red to brown background. Although the decoration was heavily indebted to the region of the Gulf of Mirabello, one of the most popular shapes was a large handmade rounded cup with a vertical handle, a shape that may have been borrowed from South Crete (Vasilakis and Branigan 2010, fig. 51:P307).

Cups with rounded or conical shapes formed the largest group in the Lasithi White-on-Dark Ware. They had a background slip that was either applied to the exterior and to the inside of the rim or to the exterior and entire interior. Decoration in white was probably always added over the dark undercoat (it frequently does not survive). The local Lasithi production used a more limited repertoire of decorative motifs than was used in the main region of this style near the Gulf of Mirabello on the northeastern coast. Two phases of imitation exist: one with paint on the interior of the rim and another with paint and decoration on the whole interior. In the region of the Gulf of Mirabello, the earlier phase of decoration was restricted to the white-painted pottery dated EM III–MM IA, and the second phase can be placed in MM IB (Betancourt 1984). The same chronological development probably existed in Lasithi.

Recent excavation of a large deposit of White-on-Dark Ware from Alatzomouri (near the Gulf of Mirabello) allows the EM III and MM IA styles to be distinguished from one another (Apostolakou, Betancourt, and Brogan 2011). This deposit shows that the style of EM III consisted entirely of simple white motifs, especially hatched triangles, chevrons, and simple bands. Only later, in MM IA, did more complex designs appear (including circles, spirals, and other curvilinear motifs). The painting of the entire interior of the vessels appeared in MM IB. This chronological scheme is very important for an understanding of the pottery from Lasithi. It suggests that the direct influence for the local production was in EM III when only the simple motifs existed, and the Lasithi production did not copy the motifs of MM IA. The local potters, however, were aware enough of other styles to adopt the practice of painting all of the interiors of their cups in MM IB (the practice made the vessels more watertight, so it resulted in a definite improvement in the end product).

Vasiliki Ware

Vasiliki Ware is a class of pottery manufactured in the Gulf of Mirabello region in EM IIB (Betancourt 1979). It is identified easily by its mottled decoration in red, black, and brown colors and its use of Mirabello Fabric. Only a few examples of the actual ware come from the cave (Ch. 6), but several local imitations indicate that it was popular at the site where the residents lived. Jugs and goblets were apparently the main shapes that were chosen for funerary use.

Diagonal Line Style Juglets

Several examples of a workshop production that consisted of small juglets with decoration of diagonal lines were present at Hagios Charalambos (Ch. 8). The juglets were made in two sizes. They always had a narrow cylindrical neck that would allow the liquid contents to be tightly stopped as well as a small spout for pouring. The date is restricted to MM IA, and all examples from the cave are handmade (Betancourt 2011a). The location of the workshop (perhaps Malia) is uncertain.

Barbotine Style Jugs

Barbotine is the term applied to Minoan pottery with three-dimensional clay patterns applied onto the surface. The main styles of Barbotine Ware from the Mesara production, manufactured from Cretan South Coast Fabric, are not found at the cave. A small number of jugs decorated with small knobs, however, can be recognized in the assemblage (Ch. 9). Their date is MM IA–IB. The knobs are applied in horizontal rows on the jugs' bodies, which is a style with several parallels from North Crete (see catalog in Ch. 9), and the production for this class was probably somewhere in the north-central part of the island.

Knossian Goblets and Jugs

Only a few Knossian jugs and goblets are in the assemblage (Ch. 10). They are recognizable by their fine clay fabric as well as their style of decoration. Their date is in EM III to MM I. The small amount of this class of pottery suggests that

Knossos was not a major supplier of ceramics for this region between EM III and MM II.

Dark Red Vessels from Kastelli and Elsewhere

A class of pottery using very dark red slip over a fine-textured red clay has been identified from Kastelli in the Pediada (Rethemiotakis and Christakis 2004) as well as elsewhere (Ch. 11). Vases that belong to these production centers have been tentatively recognized in the pottery assemblage from the cave. They are all small, finely made luxury pieces. Their date is MM IB.

Trefoil-Mouthed Jugs

A group of trefoil-mouthed wheelmade vessels with a wide mouth and a single handle can be assigned to MM IB (Ch. 12). Close parallels for both the shape and the fine-grained orange fabric come from Galatas. Examples also come from Kastelli. The vases are always covered with a solid red slip on the exterior, and they are expertly made and well fired.

Chamaizi Pots

A distinctive shape called the Chamaizi pot consists of a small closed vessel with a rounded body and a cylindrical neck decorated with crosshatched incisions (Ch. 14). The shape begins in MM IA with handmade examples, and it continues into MM IIB.

By the end of the tradition, the vessels are made on a potter's wheel. The place where these vases were manufactured is uncertain. They are made from well-levigated, pale colored clay (a petrographic description is in App. A). Only a few examples of this class of vessel come from Hagios Charalambos.

Vessels with Parallels from Malia

Vases in several shapes have parallels from Malia (Ch. 15), and it is likely that most of them were produced either at Malia or in workshops that were closely related to that site. The group, mostly from MM II, includes both vessels that are unslipped and those that are covered with dark slip. They are all made on the potter's wheel. This group includes the carinated cups, which are among the latest vases in the main body of the assemblage.

Vessels with Fine White Spiraliform Ornament

A small group of vases, especially two-handled cups, with closely related elegant white spiraliform motifs form a tight stylistic group (Ch. 16, **291–295**). The date is MM II. The vases are all fine elite drinking vessels with spouts and two opposed handles. Their decoration is applied in white on a black background. The location of this production is not known for certain, but more than one fabric was used for the manufacture.

Comments

In conclusion, the Hagios Charalambos assemblage brings to light several points. First, the residents of Lasithi were not isolated from other parts of Crete. They had their own local or regional pottery workshops, but they were also able to acquire vessels from many other parts of the island. Imports from different periods show trade preferences, but at no time did the local consumers have exclusive trade relations with any one location. Imports from the Early Minoan period include Hagios Onouphrios Ware from EM I, Vasiliki Ware from EM IIB, and Knossian goblets from EM III to MM I. These objects indicate early links with

both the southern and the northern coast. In EM IIB, the local Lasithi workshops began to use shapes copied from Vasiliki Ware, which suggests a close tie with the Gulf of Mirabello region. This influence on the local ceramic production increased greatly in EM III. Local imitations of pottery styles from elsewhere on the island (such as Vasiliki Ware and East Cretan White-on-Dark Ware) indicate tastes in local pottery production.

Early Minoan exports out of Lasithi speak of reciprocity in early trade relations. This reciprocity is well illustrated by the distribution of offering tables to many different sites in the Mesara, the northern coast, the Gulf of Mirabello region, the southern coast, and sites in the Lasithi Plain.

Foreign connections increased in the Middle Minoan period. Many of these vessels were containers that may have held luxury goods such as perfume. Items from the northern coast included the Diagonal Line Style juglets, the jugs with barbotine decoration, the Chamaizi pots, and a series of small jugs in various styles. Elite drinking vessels like two-handled cups, some of them attractively decorated with white spiraliform motifs, suggest that some wealthy and influential citizens were among those buried in the cave. The significant increase of Malian imports in MM II may contribute to our understanding of the Malian sphere of influence.

The strong parallels with the nearby Trapeza Cave help us better understand the nature of the society that occupied the Lasithi Plain. The artifacts from the two caves reflect similar trade patterns and related styles of manufacture. The pottery assemblages from the two locations both include a large number of vessels for pouring and consuming beverages, a smaller number of vessels for some liquid like oil that required a small closed vessel with a narrow neck, and an even smaller number of miscellaneous shapes. The presence of a few sherds from large closed vessels in both contexts suggests that the practice of burial in jars had already begun in this region when the caves were filled with objects. For Hagios Charalambos, the number of buried individuals can be estimated at probably over 1,000 people. The figure of 1,000 burials is based on the secure number of almost 400 individuals from the surviving bones plus the estimate that 60% were unstudied from the early excavations and 10% were left unexcavated. Like the pottery, the disarticulated bones in this ossuary were not stratified. If the time span for the vessels is over 1,000 years (from before 2800 B.C. to ca. 1750 B.C.), this duration is a little less than one burial per year. These statistics support the idea of a single community for the use of the cave rather than many communities.

Unlike the bones, all the pottery from the old excavations was saved. The statistics (App. B) show that a total of 18,064 pieces were excavated from the various campaigns and recorded after mending, with 18,055 sherds and vases found inside the cave, representing an estimated minimum of 1,991 vases. The minimum number of vessels was determined separately for each category, based on counting both restorable vessels and unique portions (like handles for jugs and cups or differences in fabrics). This would be an average of about two vessels offered per burial. The largest group is the local production made of Lasithi Red Fabric Group (Table B.19 plus a few additional vases), which accounts for about half of the ceramic offerings.

Eating and Drinking

The largest category is drinking vessels, with about 50 examples from before EM III (when many of the vases are large communal chalices) and 733 from EM III and later when the vessels are individual drinking vessels. For ca. 1,000 burials, 783 drinking vessels suggest offerings to the dead in the form of food and drink rather than communal toasts by groups of mourners. This conclusion is supported by the offerings of meat as shown by animal bones mixed throughout the secondary deposit (Betancourt et al. 2008b).

This conclusion means that one can divide the funerary eating and drinking into two parts. The evidence from inside the cave where the drinking and pouring vessels were mixed throughout the deposit can be assigned to meals offered to the dead at the primary location where the deceased were placed as primary burials. The evidence for consuming food and drink outside the cave, discussed in the first volume in this series (Betancourt 2014a, 31–35), represents the ceremonial activity that occurred either in connection with the secondary deposit of the disarticulated bones in the cavern or with subsequent visits to the location in commemoration of the deceased. The presence of a small amount of Late Minoan (LM) III pottery found outside the cave (App. B, Table B.21) indicates that this activity may have continued for some time after the burial by at least a small number of persons. Although the evidence from the bones was not retained for Trapeza, the similarity in other areas suggests that the social practices may have been generally similar as well.

What was not found among the pottery assemblage is as interesting as what was present, and it also helps us understand the nature of the funerary customs here. No large numbers of cooking pots are in the assemblage, a situation that contrasts with the tholos tomb cemetery at Moni Odigitria (for the final publication, see Vasilakis and Branigan 2010). Cooking pots break into very small fragments, and because until recently excavations in Crete did not retain and study all classes of pottery, we cannot be sure of the situation in cemeteries excavated many years ago. Keith Branigan has studied the presence of cooking vessels at the cemetery of Moni Odigitria (Branigan 2008), and he has noted that they are found not in the tombs themselves but in ritual areas near the places of burial, suggesting their use in non-funerary communal ceremonies that occurred after the time of burial, perhaps many years later. The removal of soil from the vicinity of the cave of Hagios Charalambos in the 1970s when the roadway at this location was being constructed prevents any serious knowledge of the area near the cave. What we can conclude from the evidence inside the cave itself is that (as was the case at Moni Odigitria), ceremonies involving the cooking of food were not practiced at the time of primary burial when the bodies and their associated artifacts were first laid to rest. If they were present at this site, they represented, instead, a return to the cemetery in later times for ceremonial activity.

Other classes of pottery are also missing. In comparison with the tholoi in the Mesara (see Branigan 1970, 1988, 1991, 1993) where conical cups were found in great numbers and were sometimes deliberately placed just outside of the tombs (see Blackman and Branigan 1982), only one conical cup was found at Hagios Charalambos. The conical cups found in great numbers outside of tholos tombs strongly suggest drinking ceremonies for large groups of mourners. Here, a statistical analysis of the assemblage shows that in relation to the number of individuals who were buried, the drinking sets may have been for individual consumption, rather than for the use of mourners in a drinking ceremony (see App. B). Also missing from the assemblage and worthy of note are rhyta (see Koehl 2006), human or animal figurines, and animal-shaped vases (see Branigan 1970b). These objects helped define the personality of the deceased in the tholos tombs. Although they did not have rhyta or animal-shaped vessels, the deceased whose remains were deposited at Hagios Charalambos were by no means without social markers. They were buried with a selection of pottery that included both locally made containers and those imported from elsewhere. The precious objects in stone, gold, silver, and other metals, the amulets in bone and ivory, and the sealstones speak strongly of status and taste.

The distinct practices and preferences for burials in the Lasithi region represent a unique situation. The pottery from the cave shows evidence for trade conducted in several directions outside of the plain starting in EM (and export out of the plain in EM IIA), so the differences in burial practices as evidenced by the material that is lacking are clearly not as a result of isolation. Rather, this new information contributes to an overall picture of the Early and Middle Bronze Age in the Lasithi Plain with regard to trade, political and economic spheres of influence, and most importantly, unique tastes in pottery production and the use of ceramics in the local Middle Minoan burial practices in Lasithi.

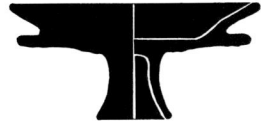

Appendix A

Petrographic Analysis of Selected Pottery Samples

Eleni Nodarou

The excavation of the burial cave of Hagios Charalambos and the study of the pottery assemblage established an array of ceramic vessels intended mainly for pouring and drinking, ranging in date from FN/EM I to MM IIB. The petrographic analysis involved 48 samples representative of the main vessel shapes and wares of the assemblage and of the chronological phases (for a complete catalogue of the pottery sampled, see Table A.1.). The aim of the analysis was the microscopic characterization of the fabrics according to their mineralogical composition as well as the investigation of continuity and change in the fabrics encountered in the cave. Issues of provenance and technology of production are also among this project's aims. As expected in a cave used for a long period and for purposes involving the ritual deposition of objects, some of them prestigious and exotic, it was expected that Hagios Charalambos would produce imported pottery. However, the main target of the analysis was to identify the local component of the Lasithi area and investigate its possible connection with Malia. In this respect the material from Karphi (Nodarou and Iliopoulos 2011) and the Pediada survey (the petrographic analysis is in the course of study, but preliminary reports on the survey can be found in Panagiotakis and Panagiotaki 2011; Panagiotakis et al. 2011) provided valuable comparative data, along with the data from the published assemblages from Quartier Mu at Malia (Poursat and Knappett 2005) and Sissi (Liard 2012). Fewer relations exist with West Crete (Nodarou 2003).

Petrography Number	Catalog Number	Excavation Number	Shape	Date
HCH 04/1	69	HCH 05-409	Offering table	EM IIA
HCH 04/2	79	HCH 04-276	Jug	EM IIB
HCH 04/3	325	HCH 05-411	Straight-sided cup	EM III–MM IA
HCH 04/4	303	HCH 120	Bowl	EM III–MM IA
HCH 04/5	280	HCH 05-413	Jug	MM I–II
HCH 04/6	217	HCH 05-414	Trefoil-mouthed jug	MM IB
HCH 04/7	not cataloged; not analyzed	HCH 05-415	Jug	MM I–II
HCH 04/8	235	HCH 05-416	Chamaizi pot	MM IIB
HCH 04/9	260	HCH 05-417	Carinated cup	MM IIB
HCH 04/10	197	HCH 05-418	Tumbler	MM IB
HCH 04/11	162	HCH 02-56	Jug	MM IA
HCH 04/12	204	HCH 02-122	Jug	MM IB
HCH 04/13	395	HCH 124	Straight-sided cup	MM IA–IB
HCH 04/14	458	HCH 02-53	Jar	MM I–II
HCH 04/15	377, not analyzed	HCH 02-60	Rounded cup	EM II–MM I
HCH 04/16	384	HCH 02-77	Rounded cup	MM I–II
HCH 04/17	vol. I (22)	HNM 14,280	Larnax	EM–MM I
HCH 04/18	362	HCH 02-49	Rounded cup	EM II–MM I
HCH 04/19	3	HCH 02-120	Closed vessel	FN–EM I
HCH 04/20	244, not analyzed	HCH 02-51	Two-handled cup	MM I–II
HCH 04/21	243	HCH 05-419	Two-handled cup	MM IB–IIA
HCH 04/22	37	HCH 04-404	Jug	EM I
HCH 04/23	271, not analyzed	HCH 47	Cup	MM IIB
HCH 04/24	132	HCH 05-410	Closed vessel	EM II–MM II
HCH 04/25	1	HCH 160	Carinated bowl	LN II
HCH 04/26	11	HCH 05-441	Cup	FN–EM I
HCH 04/27	24	HCH 03-202	Chalice	EM I
HCH 04/28	2	HCH 03-215	Open vessel	FN–EM I
HCH 04/29	12	HCH 05-442	Cup	FN–EM I
HCH 04/30	20	HCH 05-438	Open vessel	FN–EM I
HCH 04/31	23	HNM 13,856	Chalice	EM I
HCH 04/32	31	HCH 04-355	Chalice	EM I
HCH 04/33	33	HCH 05-443	Chalice	EM I
HCH 04/34	15	HCH 05-440	Open vessel (cup or bowl)	FN–EM I
HCH 04/35	41	HCH 75	Jug, Hagios Onouphrios Style	EM I
HCH 04/36	39	HCH 04-278	Jug, Hagios Onouphrios Style	EM IIA
HCH 04/37	63	HCH 04-390	Offering table	EM II
HCH 04/38	60	HCH 04-391	Offering table	EM II
HCH 04/39	70	HCH 05-439	Offering table	EM II

Table A.1. Concordance of sample, catalog, and excavation numbers of vessels that were petrographically analyzed.

Petrography Number	Catalog Number	Excavation Number	Shape	Date
HCH 04/40	102	HCH 73	Cup	MM IB–II
HCH 04/41	293	HCH 169	Two-handled cup	MM I–II
HCH 04/42	295	HCH 04-281	Two-handled cup	MM I–II
HCH 04/43	200	HCH 22	Tall carinated cup	MM IB
HCH 04/44	196	HCH 04-381	Tumbler	MM IB
HCH 04/45	182	HCH 176	Jug	MM IB–IIA
HCH 04/46	187	HCH 184	Jug	MM IB–IIA
HCH 04/47	270	HCH 97	Cup(?)	MM IA–IB
HCH 04/48	287	HCH 03-221	Closed vessel	MM I–II

Table A.1., cont. Concordance of sample, catalog, and excavation numbers of vessels that were petrographically analyzed.

Geological Setting

The cave is situated on the limestones and dolomites of Triassic-Jurassic Age. This series overlies the Phyllite Quartzite series of Permian-Triassic Age, which consists mainly of mica-carbonate schists and sandstones (IGME 1989). The Lasithi Plain is composed of alluvial deposits with terra rossa clayey material. Outcrops of flysch with shales and sandstones are located to the south of the plain (Fig. 38).

West of the Dicte Mountain the same geological environment is repeated in the area of the Pediada, extending from Kastelli to Thrapsano, the modern potting village. The alluvial plain of Kampos Kastelliou is surrounded by outcrops of limestones, phyllite-quartzites, and flysch, whereas the red deposits of Pleistocene age and fluviolacustrine origin of the Hagia Galini formation are to the south (area of Thrapsano). Alluvial deposits and Neogene marls extend also along the north coast of Crete near Chersonissos.

The repetitive geology that includes ophiolites, flysch, and metamorphic rocks and extends westward from Myrtos Ierapetra along the southern coast to the eastern Mesara impedes any secure assignment of provenance (Poursat and Knappett 2005; Nodarou and Rathossi 2008). The comparative material from Karphi, however, was used as an aid to establish the local component of the assemblage.

Petrographic Fabric Groups

The petrographic analysis resulted in the establishment of 13 fabrics, some comprising several samples, others a few, and some containing single samples (Table A.2). Detailed petrographic descriptions are provided at the end of the appendix.

Coarse and Semicoarse Fabrics

Fabric Group 1: Calcite Tempered

Samples HCH 04/22 (37), 04/27 (24), 04/28 (2)

This is a coarse fabric characterized by a reddish-brown firing clay matrix, which is optically active (Pl. 4:a). The dominant nonplastic component is the angular fragments of calcite. There are also a few fragments of quartz and rare fragments of quartzite. Textural concentration features are present in the form of rounded clay pellets. The shape and size as well as the bimodal size distribution of the calcite fragments indicate that they were intentionally added in the clay mix as temper. Evidence for tempering with organics is present in one sample (HCH 04/27) and is reflected in the elongate voids left

Fabric Group	Date	Shape-Ware	Possible Provenance
1. Calcite tempered	FN–EM II	Burnished	North Cretan coast
2A. Coarse phyllite	FN–EM IIA	Open vessels	Local
2B. Coarse phyllite	EM III–MM II	Drinking-serving	Local
3. Semicoarse with carbonate rocks	FN–EM I		Pediada
4. Calcareous grog-tempered	EM I	Chalices	North Cretan coast or Pediada
5. Red with quartz and clay pellets	MM IB	Tumbler	Malia
6. Red with quartz	EM I–MM II		Malia
7A. Granodiorite	MM I–II		Gournia/Kalo Chorio
7B. Granodiorite	EM IIB	Vasiliki Ware	Gournia/Kalo Chorio
8. Very fine with rare nonplastics	MM IA–IIB		Malia or Pediada
9. Fine calcareous with pellets	EM II–MM II		Import
10. Fine calcareous with biotite mica	MM IIB	Chamaizi vessel	Import
11. Fine with rounded sandstones	EM I	Dark-on-Light	Mesara
12. Very fine with pellets	MM I–II	White-on-Dark	Mesara or Pediada
13. Very fine with serpentine	MM I–II		South coast (Myrtos to Mesara)

Table A.2. Summary of fabric groups at Hagios Charalambos resulting from petrographic analysis.

after the organic material burned away. The vessels represented are a jug, a chalice, and an open vessel with burnished surfaces, ranging in date from the FN to the EM I period.

Tempering with calcite and vegetal matter is a common practice in the Neolithic and the Prepalatial period (mainly EM I–IIA). It occurs in ceramic assemblages across East Crete, for example, at Hagia Photia (Day, Wilson, and Kiriatzi 1998; Day et al. 2012), Petras Kephala and Petras Rock shelter (Papadatos 2008; Nodarou 2012), Mochlos (Day, Joyner, and Relaki 2003), Vrokastro and Priniatikos Pyrgos (Hayden 2003, 405; Molloy et al. 2014), and Kavousi (Day et al. 2005, 180; Haggis et al. 2007, 679–701), and in North Central Crete, for example, at Knossos (Tomkins, Day, and Kilikoglou 2004), Pyrgos, and Gournes (Galanaki 2006). On this basis, the calcite tempered fabric seems to represent a widely known recipe across the island, impeding any secure provenance assignment. Additionally, the matrix of the fabric is not diagnostic of origin; any red noncalcareous base clay could have been used. The constant presence of the clay concentrations in all the samples examined indicates incomplete clay mixing, but it is also indicative of a fairly consistent and systematic technique of manufacture, and possibly of common origin.

Fabric Group 2: Coarse Phyllite (Lasithi Red Fabric Group)

Group 2A, Samples HCH 04/1 (**69**), 04/4 (**303**), 04/12 (**204**), 04/26 (**11**), 04/29 (**12**), 04/37 (**63**), 04/38 (**60**), 04/39 (**70**)

Group 2B, Samples HCH 04/3 (**325**), 04/11 (**162**), 04/13 (**395**), 04/14 (**458**), 04/16 (**384**), 04/17 (Betancourt 2014b, 47, no. 22)

This is the most common fabric in the Hagios Charalambos assemblage. It is coarse and characterized by the presence of low grade metamorphic rock fragments, mainly phyllite and quartzite-schist. The color of the phyllite ranges from silvery gray to orangish brown, and it is composed of biotite mica and quartz and rarely contains muscovite (white) mica. The secondary nonplastic components include frequent fragments of quartz, a few carbonate rock fragments, chert, and siltstone.

Two subgroups were identified on the grounds of differentiation in color and optical activity of the matrix. For Group 2A the matrix is orangish brown and optically active (Pl. 4:b), whereas for Group 2B it ranges from reddish brown to dark brown and is optically inactive (Pl. 4:c). This differentiation is related to the firing temperature and is date specific. Group 2A is lower fired, it comprises black burnished vessels (offering tables, cups, and an open bowl), and the date range is from FN to EM IIA (with the exception of one MM IB sample). Group 2B is higher fired, and the date range is from EM IIA to MM II. The shapes represented are slipped drinking and serving vessels (cups, jugs).

As to provenance, considering the proximity of the Hagios Charalambos Cave to phyllitic outcrops and the abundance of this fabric in the assemblage, it could be suggested that this group represents broadly local production. This provenance assignment is reinforced by the material of the LM IIIC settlement at Karphi; the main coarse fabric of that assemblage is characterized by similar low grade metamorphic rock fragments in an optically active matrix (Nodarou and Iliopoulos 2011, 338–339). Despite the chronological distance between the two sites and the different function, it seems that the main coarse fabric in both of them reflects the Phylllite-Quartzite series of the Lasithi Plain. Possible connection with the production of the Malia plain cannot be established as the Maliot phyllitic fabric seems to be characterized by the regular presence of dark red mudstones, which are not seen at Hagios Charalambos (Poursat and Knappett 2005, 19; Liard 2012). It remains open for future research in settlement sites of the Lasithi Plain to investigate the existence and variability of local pottery production.

Small Groups and Loners with Metamorphic Rocks

Samples HCH 04/6 (**217**), 04/18 (**362**)

As was the case with Fabric Group 2, these samples are also characterized by the presence of metamorphic rock fragments (mainly phyllite) in a brown to dark brown matrix that is optically moderately active (Pl. 4:d). The secondary nonplastic components are quartz, quartzite, and chert. The presence of frequent clay pellets and the larger amount of quartz differentiates this group from the main phyllitic group (Fabric Group 2B). The clay pellets point toward an alluvial deposit where the phyllite was transported and deposited from the surrounding area.

The vessels represented are a trefoil mouthed jug of MM IB date and a rounded cup of EM II–MM IA date. In terms of provenance there is nothing to suggest a different environment than the phyllite outcrops of the area of Hagios Charalambos. However, the constant presence of the clay pellets might reflect a different workshop or a different technology of manufacture from Fabric Group 2B.

Sample HCH 04/30 (**20**)

This is a coarse fabric characterized by an orangish brown and optically active matrix (Pl. 4:e). The nonplastic components consist of low grade metamorphic rocks and frequent fragments of fractured epidote. The former are primarily fine-grained phyllites, quartzites, and biotite-mica schists. The secondary nonplastic components are quartz and plagioclase feldspar. This fabric is very different from the other metamorphic fabrics encountered at Hagios Charalambos. This rock and mineral suite indicates a metamorphic environment, but the presence of the epidote and the abundance of biotite mica in the clay matrix demonstrate a different source of raw material from that of the main metamorphic group.

The sample represented is an open vessel in scored ware of FN–EM I date, and it seems to be imported. The Scored Style is encountered in significant numbers in EM I contexts across West Crete, but this fabric does not correspond to any of the West Cretan recipes (Nodarou 2011). It is also found in the EM I assemblage of Knossos (Wilson and Day 2000, 39), and whether the Hagios Charalambos sample constitutes the product of a Central Cretan workshop is open to future research.

Sample HCH 04/46 (**187**)

This is a coarse fabric, characterized by a dark brown to dark gray optically inactive matrix (Pl. 4:f). The dark colors of the matrix and of the surface slip are indicative of firing in a reducing atmosphere. The main nonplastic components are elongate dark to almost black phyllites and a few quartzite and quartzite-schist fragments. The secondary nonplastic components consist of quartz and rare micrite. Frequent amorphous black concentrations, which seem to be clay pellets affected by the kiln atmosphere, are also present. This fabric is related to a low grade metamorphic environment, but the type of the phyllite, the texture, and the general appearance of the fabric make it stand out as different from the main phyllite fabric (Fabric Group 2B).

The sample represented is a jug of MM IB–IIA date in Barbotine Style. A metamorphic fabric with dark mudstones or textural concentration features in Barbotine Style has been encountered at Malia (Poursat and Knappett 2005, 21). A connection might exist between the two samples, but the color of the clay matrix is different (the Maliot is yellowish brown and optically active), and the provenance is also insecure. More samples of this ware are needed in order to investigate potential centers of production.

Sample HCH 04/47 (270)

This is a semicoarse fabric. The matrix has an orangish brown color, and it is optically active (Pl. 5:a). The main nonplastic components are fine-grained phyllite, quartzite, quartzite-schist, and very few fragments of quartz. Frequent clay pellets as well as clay striations indicate incomplete clay mixing. This feature indicates a different technique of manufacture from the main metamorphic group (Fabric Group 2), whereas the presence of quartzite-schist points also toward a different clay deposit for the raw material. There exists a parallel in the material under study from the Pediada survey, from Galatas Kephala, and although the provenance of the fabric remains open until more samples are examined, a potential source might be the alluvial deposits of the area of Galatas. The vessel represented from Hagios Charalambos is a possible cup of MM IA–IB date.

Sample HCH 04/43 (200)

This is a semicoarse fabric characterized by a yellowish-brown and highly micaceous matrix that is optically active (Pl. 5:b). The main nonplastic inclusions consist of quartz and muscovite mica schist. A few fragments of siltstone, phyllite, quartzite, and rare plagioclase feldspar are also present. The sample is from a tall carinated cup of MM IB date. This fabric bears no similarity to any of the other metamorphic fabrics due to the presence of the abundant mica. It is not possible to make any assumptions on the provenance of the raw material.

Fabric Group 3: Semicoarse with Carbonate Rock Fragments

Samples HCH 04/19 (3), 04/25 (1)

This is a semicoarse fabric characterized by a brown matrix that is moderately active (Pl. 5:c). The predominant nonplastic component is carbonate rock fragments. There are also some fragments of calcite, rare phyllite, siltstone, and plagioclase feldspar. It is a rather unusual fabric; the mineralogy is not distinctive of origin, but it seems that the raw material is derived from a marl deposit. Marly clays occur primarily in the Hagia Galini formation, which outcrops at the southern part of the Kastelli Plain. The color and texture of the fabric point toward a low firing temperature (probably below 750°C), which is explained by the early date of the samples. The vessels represented are a carinated bowl of LN II and a closed vessel of FN–EM I date, respectively. A possibility exists, therefore, that these vessels were imported into Hagios Charalambos from the broader area of Kastelli.

Fabric Group 4: Calcareous Grog Tempered

Samples HCH 04/31 (23), 04/33 (33)

This is a semifine fabric characterized by a brown matrix that is optically moderately active (Pl. 5:d). The main nonplastic component consists of small- and medium-sized carbonate rocks that are densely distributed in the clay matrix. The secondary nonplastic components are frequent fragments of quartz, rare white mica laths, and a few dark brown and optically inactive angular inclusions. The angularity and the straight contours of the latter, as well as the voids surrounding the grains caused by shrinkage due to the differential expansion of the clay matrix and the inclusions, led to the conclusion that it is grog added intentionally to the clay mix. Moreover, elongate voids indicate tempering with organic material. The presence of the carbonates along with fossils is indicative of the use of a calcareous raw material, most likely a Neogene marl. As with Fabric Group 3, the color and texture point toward a low firing temperature (below 750°C). The technology of manufacture is different, however, as indicated by the practice of tempering with grog and organic matter as well as by the use of a fossiliferous base clay. The vessels represented in this fabric group are burnished chalices of EM I date.

The practice of grog tempering appears in the FN and continues in the EM I–IIA period. It occurs mainly in East Crete (Day et al. 2005, 180; Nodarou 2012; for a detailed list of sites, see Papadatos et al., forthcoming), in the Pediada region, and particularly at the FN site of Modis Voni (E. Nodarou, pers. obs.; Papadatos and Tomkins 2011), but also in the south Greek Mainland and the island of Kythera (Broodbank and Kiriatzi 2007, 248).

Fabric Group 5: Red Fabric with Quartz and Clay Pellets

Samples HCH 04/10 (**197**), 04/44 (**196**)

This is a semicoarse fabric characterized by a red and optically active matrix (Pl. 5:e). The main nonplastic components are quartz, quartzite, and a few fragments of phyllite. Sample HCH 04/10 also contains rare fragments of quartz-mica schist and white mica laths. Bright red and optically active clay pellets are frequent in both samples. The vessels represented are tumblers of MM IB date.

In terms of provenance this fabric group does not seem to be local. The color of the matrix as well as the mineralogical composition point toward an alluvial deposit and find good parallels at Malia. A range of shapes in a similar red fabric has been described from Quartier Mu, and the source of the raw material is considered to be the coastal plain of Malia (Poursat and Knappett 2005, 17).

Fabric Group 6: Red Fabric with Quartz

Samples HCH 04/32 (**31**), 04/34 (**15**)

This is a semifine fabric characterized by an orangish-brown, noncalcareous matrix that is optically active (Pl. 5:f). The main nonplastic components are monocrystalline quartz, fine grained phyllite, quartzite-schist, and quartzite. There are also a few clay pellets as well as voids characteristic of tempering with organic matter. In terms of its composition, this fabric is similar to Fabric Group 5, and it seems to have been manufactured with an alluvial clay tempered with chaff. The color of the matrix is not as bright as in Fabric Group 5, but this is due to the lower firing temperature. The technological difference can be attributed to

the chronological and functional difference of the vessels included in the two fabric groups. The vessels represented in Fabric Group 6 are a chalice and an open bowl of FN–EM I date, and they might be connected with a Maliot center of production.

Fabric Group 7: Mirabello Fabric

Group 7A, Samples HCH 04/40 (**102**), 04/42 (**295**)

This fabric group is characterized by a very fine, brown, and optically inactive matrix in which the medium-sized nonplastic inclusions were added as temper (Pl. 6:a). The main nonplastic components are fragments of granodiorite composed of plagioclase feldspar with minor amounts of biotite and altered amphibole. In the matrix there are also abundant clay pellets. This fabric is well represented in Central and East Crete; it has been described macroscopically and microscopically at many instances ranging in date from EM to LM III (for overviews of the literature, cf. Betancourt 2008a, 30; Nodarou and Moody 2014). Its place of origin is the Mirabello region and, most specifically, the area of Gournia to Kalo Chorio. The vessels represented are a cup and a spouted two-handled cup of MM I–II date.

Group 7B, Sample HCH 04/2 (**79**)

This sample's mineralogical composition is similar to those in Group 7A, but its texture is totally different. The fabric is finer, the matrix has a very dark brown—almost black—color, and there are rare nonplastic inclusions consisting of granodiorite, plagioclase feldspar, and biotite (Pl. 6:b). The vessel represented in this fabric is a jug with mottled decoration (Vasiliki Ware) of EM IIB date.

Fine Fabrics

Fabric Group 8: Very Fine Fabric with Rare Nonplastics

Samples HCH 04/5 (**280**), 04/9 (**260**), 04/21 (**243**), 04/45 (**182**)

This is a very fine fabric characterized by a dark brown to greenish-brown matrix that is optically inactive (Pl. 6:c). It has very rare nonplastic inclusions consisting of small quartz fragments and biotite mica laths. Sandstone is very rare to absent.

Clay pellets are rare. The color of the matrix indicates the use of a calcareous raw material (a Neogene marl) that has been fired to a high temperature. The shapes represented are drinking and serving vessels (three jugs and a conical cup) and a two-handled cup dating from MM I to IIB. There is also a closed vessel dated to the EM II–MM II period.

The absence of nonplastic inclusions impedes any provenance assignment. The lack of analysis on EM and MM pottery from contemporary neighboring settlements limits the petrographic parallels from the area of Lasithi. Outside the plain, possible parallels were identified in assemblages from Malia and Pediada. The former are described by Poursat and Knappett (2005, 11), and the marl deposits at Chersonissos are suggested as possible sources of the raw material. The Pediada comparative material consists of MM I and LM III fine wares from the area of Voni-Galatas. With the present data one cannot be conclusive on the origin of this fabric group, but it is likely that this fine pottery is imported to the cave.

Fabric 9: Fine Calcareous Fabric with Pellets

Sample HCH 04/24 (132)

This is a very fine fabric characterized by a yellowish-brown matrix that is moderately active (Pl. 6:d). No nonplastic inclusions are present. The matrix is composed of small fragments of micrite with very little quartz and biotite mica laths. A few rounded clay concentrations ranging in color from reddish brown to dark brown are also present. There is nothing to link this fabric with any of the other fabrics encountered in the Hagios Charalambos assemblage. The vessel represented is a closed shape of EM II–MM II date.

Fabric 10: Fine Calcareous Fabric with Biotite Mica

Sample HCH 04/8 (236)

This is a fine fabric with a brown matrix that is optically moderately active (Pl. 6:e). The dominant components are biotite mica laths and small quartz fragments evenly distributed in the clay matrix. There are frequent clay pellets of reddish-brown color and a few microfossils (foraminifera).

The abundant biotite mica and the fossils are not present in any of the other fabrics in the Hagios Charalambos assemblage. The composition is not diagnostic of origin, but the color of the matrix and the presence of microfossils point to a raw material connected to Neogene marls. The vessel represented belongs to the Chamaizi type, and it is dated to the MM IIB period.

Fabric Group 11: Fine Fabric with Rounded Sandstones

Samples HCH 04/35 (41), 04/36 (39)

This is a very fine fabric characterized by a golden brown calcareous matrix that ranges from optically active to moderately active (Pl. 6:f). The nonplastic inclusions consist of very few rounded fragments of sandstone, which must have been added to the clay mix as temper. This composition is compatible with the ophiolites and the Asteroussia Nappe in the area of the Mesara (Wilson and Day 1994, 61, 68). The vessels represented in this fabric group belong to the fine Dark-on-Light Painted (Hagios Onouphrios) Style that is in accordance with a South-Central Cretan provenance.

Fabric 12: Very Fine Fabric with Pellets

Sample HCH 04/41 (293)

This is a very fine fabric characterized by a brown to greenish-brown matrix, which is high fired (almost vitrified) and optically inactive (Pl. 7:a). The large inclusions consist mainly of very few dark brown clay pellets. There are also rare fragments of quartz and biotite, and very rare fragments of rounded sandstone. The fineness of the fabric does not allow for a secure assignment of provenance. The rounded sandstones link this fabric with Fabric Group 11, and a connection with the area of the Mesara may be suggested. However, this fabric is also encountered in LM III pottery samples from the Pediada survey and more specifically from the area of Voni, whereas typological parallels from Malia may also link this sample with the marl deposits of the north coast. The sample represented is a white-on-dark spouted two-handled cup that belongs to the MM I–II period.

Fabric 13: Very Fine Fabric with Serpentinite

Sample HCH 04/48 (**287**)

This fabric seems connected to Fabric Group 12. It has the same brown, high-fired, and optically inactive matrix (Pl. 7:b). However, the mineralogy is different. The few nonplastic inclusions consist primarily of equant and elongate fragments of serpentinite along with some quartz, biotite, and rare phyllite. There are also a few dark brown clay concentrations. The mineralogy of the fabric points toward the environment of the ophiolite series. It could be a product of South-Central Crete (Mesara). The vessel represented is a closed vessel of MM I–II date.

Discussion of Raw Materials and Provenance

Analytical work carried out in the past decades in Crete has demonstrated the presence of a restricted number of pottery production centers during the Prepalatial period, and within this framework the Mesara has been the producer of a large portion of the fine painted pottery found at the sites of the northern coast (Wilson and Day 1994; Day, Wilson, and Kiriatzi 1997). The study of the Neopalatial kiln found at Kommos (Day and Kilikoglou 2001) has contributed substantially to this conclusion. A large part of southern Crete remains unknown in terms of pottery production, however, due to the dearth of published archaeological and analytical data. Recent work has demonstrated the complexity of the geology extending from Myrtos to the eastern Mesara, which consists of rocks such as ophiolites and flysch along with coastal sediments that are regularly repeated in the landscape (Poursat and Knappett 2005; Nodarou and Rathossi 2008). This repetitive appearance of compositionally homogeneous sediments impedes any secure provenance assignment.

Considering that the cave of Hagios Charalambos is located in this geological environment and that it was used for burial purposes throughout its history of use, pottery production in the area can only tentatively be hypothesized but not proved. This deficiency was partly counterbalanced by the publication of Quartier Mu at Malia (Poursat and Knappett 2005), which is dealing with the issue of pottery production on the southern coast as well as the ongoing study of the material from the Pediada survey (in collaboration with N. Panagiotakis and M. Panagiotaki), which is believed to shed new light on pottery production in this part of Crete.

The petrographic analysis of the Hagios Charalambos pottery assemblage resulted in the establishment of 13 fabric groups that reflect a multitude of raw materials used (for an overview of the fabric groups, see Table A.2). The predominance of the fabrics with metamorphic rocks is undeniable. Fabric Group 2 comprises 50% of the samples analyzed, and it is characterized by the presence of phyllite fragments in a red-firing matrix. This composition is compatible with a red clay deriving from an alluvial deposit. Deposits of this type are widely available in the immediate vicinity of the Hagios Charalambos Cave, in the alluvium of the Lasithi Plain that receives phyllitic material from nearby outcrops. Although no Prepalatial or Protopalatial settlement has been excavated so far on the Lasithi Plain, surface survey has identified several sites, two of which are very close to the cave. The earlier, Katsoucheiroi, was inhabited continuously from the LN until the EM II, but in MM I the inhabitants moved to a nearby lower site, called Plati (Watrous 1982, 267). Although no analysis has been carried out on pottery of these sites, it can be suggested that Fabric Group 2 represents local production of the Lasithi Plain (see also Langford-Verstegen 2008, 550). In this volume it is called the Lasithi Red Fabric Group. A future study of comparative material from settlement sites in the area would provide important insights into whether this production was taking place in the immediate vicinity of Hagios Charalambos or somewhere else on the plain. The broadly local manufacture of this pottery is reinforced by the representation of this fabric (a) in a large number of pots in the assemblage; (b) in a number of coarse and semi coarse shapes, mainly undecorated, such as cups, bowls, and jugs that would satisfy domestic as well as funerary needs; and (c) in the use of this fabric over a long period, from the FN to the MM II period.

With regard to the technology of manufacture, the change from the earlier (FN–EM II) to the later (EM III–MM II) phase is noteworthy. This change is reflected primarily in the domain of pyrotechnology, with the earlier samples being fired in lower temperatures than the later examples. Although Hagios Charalambos represents a consumption assemblage, it seems that this change in firing temperatures corresponds to actual technological advancements in the production technology of the local workshop(s).

Connected to the main metamorphic fabric are a series of small groups and loners that present certain characteristics pointing to a different source of raw material. The representation of these fabrics in the assemblage is very small, and the absence of comparative material makes any ascription of provenance impossible. The only exception is the case of the muscovite mica fabric (HCH 04/43 [**200**]). The Pediada survey provided a multitude of micaceous fabrics, and it can be suggested with a degree of certainty that the Hagios Charalambos sample comes from the Pediada.

With regard to the main fine fabric (Fabric Group 8), the absence of aplastic inclusions diagnostic of origin and the lack of analytical research on local assemblages do not allow a secure assignment of provenance. The only petrographic parallels available so far are from Galatas Pediada and Malia. In contrast to the local coarse Fabric Group 2, Fabric Group 8 is dated solely to the late Prepalatial–Protopalatial period (MM IA–IIB). Considering that important palatial centers begin to emerge at Malia in the Protopalatial period and at Galatas in the early Neopalatial period,

the possibility that the fine vessels of Fabric Group 8 may have been imported from these areas to the Lasithi Plain cannot be excluded.

The tempered Fabric Groups 1 and 4 constitute two well-known recipes of the Prepalatial period. The addition of inorganic (calcite and grog) as well as organic (chaff) materials in the clay mix constitute common practices in pottery production of the northern coast of Crete since the FN period.

Fabric Group 7 comprises samples that are knowingly imported to the cave. The granodioritic raw materials are found in the area of the Gulf of Mirabello, and pottery is produced in the area of Gournia to Kalo Chorio throughout the EM and MM periods. Indeed, the two sampled vessels are dated to the EM IIB and the MM I–II periods, suggesting that the Mirabello workshop(s) constituted a minor but persistent source of pottery imports for the community using the cave of Hagios Charalambos.

The last category of raw materials encountered in the Hagios Charalambos assemblage is connected with ophiolitic sources and represents fine wares, i.e., Fabrics 11 and 13 with sandstone and serpentine fragments, respectively. It is not certain whether the two fabrics derive from the same production center because they are of different composition and date. Fabric Group 11 represents a well-known category of Prepalatial painted pottery, which has been proven to be the product of Mesara workshop(s) (Wilson and Day 1994). For Fabric 13 the provenance is unclear. The presence of serpentine is not diagnostic of origin because it constitutes a secondary component of ophiolitic materials, outcrops of which are encountered across southeastern and Central Crete, from the area of Myrtos Ierapetra to the Mesara.

Conclusion

In pottery studies, petrography offers a valuable tool in order to characterize the assemblage and assess potential provenance. The analysis of the Hagios Charalambos pottery provided evidence of broadly local pottery production on the Lasithi Plain as indicated also from comparison with the material from Karphi. It has been possible to identify the local component of the assemblage and discuss changes in the firing technology over time.

The funerary character and use of the Hagios Charalambos Cave should not be overlooked in the assessment of the analytical results. The vessels deposited in the cave are not representative of a pottery assemblage encountered in a settlement, even if some of them were used in everyday activities before their deposition. Rather, they are vessels selected either as funerary offerings to the dead or to be used in ceremonies associated with the funerary rites. Until scientific analysis is performed on contemporary domestic assemblages, the results of the present analysis should be used with caution for the reconstruction of daily activities in the settlements of the Lasithi Plain.

The analytical study demonstrated that for their funerary rites and rituals the people of Hagios Charalambos utilized pottery made predominantly in the immediate vicinity or at least in the broader area of the Lasithi Plain. Although small and remote, the community (or communities) that used the cave for burials had the ability to participate in distant exchange networks that brought pottery into the area from centers as distant as the Mesara to the south or the Gulf of Mirabello to the east. The dating of these imports clearly indicate that this practice started in the EM I period and continued more intensively until the MM II period when the cave ceased to be used. Based on the number of samples, the connection with the Mesara and the Mirabello area seems to be rather limited and most probably indirect. On the other hand, the connection with the Pediada region and Malia seems to be more intensive and direct. The number of samples as well as the array of imports is definitely larger, including a variety of shapes, wares, and pottery fabrics. The connection with these areas has been attested already in the early Prepalatial period (FN–EM II), and it seems to become more intensive in the Protopalatial period when important palatial centers began to emerge at Malia and slightly later at Galatas. A broader analytical study of ceramics from the Lasithi Plain is required in order to enhance our knowledge on pottery production, use, and distribution in the area of the plain and discuss issues connected with potential political control by the palatial centers for the remote mountainous area of Lasithi.

Petrographic Descriptions of the Fabric Groups

The petrographic analysis was carried out at the INSTAP Study Center for East Crete using a Leica DMLP polarizing microscope. For the petrographic descriptions, the system used in Myer, McIntosh, and Betancourt (1995) was adopted.

Fabric Group 1: Calcite-Tempered

Samples HCH 04/22 (**37**), 04/27 (**24**), 04/28 (**2**)

I. MATRIX (GROUNDMASS)

A. Optical Properties and Color

1. Under plane-polarized light: orangish brown to brown, nonpleochroic.

2. Under cross-polarized light: reddish brown, the micromass is optically active.

B. Overall Grain Size and Modality of Inclusions and Voids

Coarse inclusions vary from 1.20 to 0.1 mm long dimension. The fine fraction is <0.1 mm and up to 60 μm long dimension. Inclusions: ca. 30%. Voids range in size from 1.60 to 0.15 mm long dimension; ca. 5%.

C. Overall Preferred Orientation of Inclusions and Voids

The nonplastics and vugs are randomly oriented. The planar voids display preferred orientation parallel to vessel margins. There is evidence for tempering with organic matter.

D. Silt-Sized Inclusions (under 62 μm)

1. Monocrystalline Quartz

Size range: up to ca. 62 μm.

Shape: equant, subangular to subrounded.

2. Calcite

Size range: up to ca. 62 μm.

Shape: equant to elongate, angular.

Comments: composed of microcrystalline limestone.

3. Textural Concentration Features (TCFs)

Size range: up to ca. 62 μm.

Shape: equant, subrounded.

Comments: they are dark red under cross-polarized light with no inclusions.

II. VOIDS

A. Size range: they vary in size from 1.60 to 0.15 mm long dimension.

B. Shape: there are meso and macro vugs, channels, and planar voids, double to open spaced. The vugs are irregular in shape, in most cases elongate; the planar voids are more regular in shape and elongate.

III. INCLUSIONS ABOVE SILT SIZE (ABOVE 62 M)

1. Calcite

    Size range: 1.20 to 0.1 mm long dimension.

    Shape: mainly equant but there are some elongate grains too, angular to subangular.

    Comments: the calcite constitutes the predominant nonplastic component; it occurs in large angular fragments and seems to have been added as temper in the clay paste. It is composed of microcrystalline limestone. In some cases it displays twinning.

2. Monocrystalline Quartz

    Size range: 0.32 to 0.1 mm long dimension.

    Shape: equant, subangular to subrounded.

    Comments: mainly merging boundaries.

3. Polycrystalline Quartz

    Size range: 0.25 to 0.1 mm long dimension.

    Shape: equant, subrounded.

4. Mica

    Size range: ca. 0.08 mm long dimension.

    Shape: laths.

    Comments: very rare biotite mica laths.

5. Epidote

    Size range: 0.8 mm long dimension.

    Shape: equant subangular.

    Comments: very rare fragments with merging boundaries.

6. Textural Concentration Features (TCFs)

    Size range: 0.24 to 0.1 mm long dimension.

    Shape: subrounded to rounded.

    Comments: they are dark red to dark brown under cross-polarized light and very compact in appearance without any nonplastic inclusions.

IV. MODALITY

Matrix (groundmass): 45%

Voids: 5%

Silt-sized inclusions

    Monocrystalline quartz under 1%

    Calcite 3%

    TCFs under 1%

Larger Inclusions

    Calcite 30%

    Monocrystalline quartz 10%

    Polycrystalline quartz under 1%

Mica under 2%

Epidote under 1%

TCFs under 1%

Fabric Group 2: Coarse Phyllite (Lasithi Red Fabric Group)

Group 2A, Samples HCH 04/1 (**69**), 04/4 (**303**), 04/12 (**204**), 04/26 (**11**), 04/29 (**12**), 04/37 (**63**), 04/38 (**60**), 04/39 (**70**)

Group 2B, Samples HCH 04/3 (**325**), 04/11 (**162**), 04/13 (**395**), 04/14 (**458**), 04/16 (**384**), 04/17 (Betancourt 2014b, 47, no. 22)

I. MATRIX (GROUNDMASS)

A. Optical Properties and Color

    1. Under plane-polarized light: orangish brown to dark brown, nonpleochroic.

    2. Under cross-polarized light: orangish brown and optically active for Group 2A, dark reddish brown to dark brown and optically inactive for Group 2B. In some cases the core is darker than the margins.

B. Overall Grain Size and Modality of Inclusions and Voids

    Coarse inclusions vary from 3.6 to 0.1 mm long dimension. The fine fraction is <0.1 mm and up to 60 μm long dimension. Inclusions ca. 45%–50%. Voids range in size from 3.2 to 0.3 mm long dimension; ca. 5%–7%.

C. Overall Preferred Orientation of Inclusions and Voids

    The nonplastics are randomly oriented. The planar voids display in some cases preferred orientation parallel to vessel margins.

D. Silt-sized Inclusions (under 62 μm)

    1. Monocrystalline Quartz

        Size range: up to ca. 62 μm.

        Shape: equant, subangular.

    2. Mica

        Size range: up to ca. 62 μm.

        Shape: laths.

        Comments: mainly biotite mica, very few fragments of white mica.

    3. Feldspar

        Size range: up to ca. 62 μm.

        Shape: equant, subangular.

        Comments: very few fragments of plagioclase and very rare fragments of alkali feldspar.

4. Phyllite

Size range: up to ca. 62 μm.

Shape: elongate, laths.

5. Quartzite

Size range: up to ca. 62 μm.

Shape: equant, subangular.

6. Textural Concentration Features (TCFs)

Size range: up to ca. 62 μm.

Shape: equant, subrounded.

Comments: there are dark red or dark brown to black under cross-polarized light, with no inclusions or internal texture.

II. VOIDS

A. Size range: from 3.2 to 0.3 mm long dimension.

B. Shape: they range from single to open spaced, mainly meso and macro planar voids and meso vugs.

III. INCLUSIONS ABOVE SILT SIZE (ABOVE 62 M)

1. Metamorphic Rock Fragments

(a) Phyllite

Size range: 3.6 to 0.16 mm long dimension.

Shape: elongate.

Comments: the inclusions are indicative of a low-grade metamorphic environment. The phyllites display a range of colors under cross-polarized light, from golden and silvery brown to orange and dark brown. Some fragments are finer grained than others; some seem to be metamorphosed siltstones. In a few cases folding of the schistosity is visible. They are composed mainly of biotite mica and quartz; in some cases there are also white mica laths.

(b) Quartzite-Schist

Size range: 2.0 to 0.24 mm long dimension.

Shape: equant to slightly elongate.

Comments: in some cases it grades into biotite-mica-schist.

2. Monocrystalline Quartz

Size range: 1.0 to 0.1 mm long dimension (mode = 0.24 mm long dimension).

Shape: equant, angular to subangular.

Comments: most fragments have straight extinction; a few have undulose extinction.

3. Polycrystalline Quartz

Size range: 0.8 to 0.2 mm long dimension.

Shape: equant or slightly elongate.

Comments: fine grained, in some cases grading into chert.

4. Chert

Size range: 0.8 to 0.4 mm long dimension.

Shape: equant, subangular.

Comments: composed of microcrystalline quartz.

5. Sandstone

Size range: 2.0 to 0.4 mm long dimension.

Shape: equant to slightly elongate, subrounded.

Comments: some fragments are composed of poorly sorted quartz grains set in a fine clay matrix (quartz arenite), while fewer display a larger percentage of fine-grained matrix (graywackes).

6. Micrite

Size range: 1.0 to 0.2 mm.

Shape: equant, subangular to subrounded.

Comments: very few fragments are composed of microcrystalline limestone.

7. Mica

Size range: ca. 0.24 mm long dimension.

Shape: laths.

Comments: mainly biotite mica and a few fragments of white mica.

8. Feldspar

Size range: 0.4 to 0.16 mm long dimension.

Shape: equant to slightly elongate.

Comments: very few fragments of alkali and plagioclase feldspar.

9. Opaques

Size range: 0.32 to 0.3 mm.

Shape: equant, angular.

Comments: very rare fragments, probably iron concentrations?

10. Textural Concentration Features (TCFs)

Size range: 0.8 to 0.1 mm long dimension.

Shape: equant, rounded to subrounded.

Comments: their color ranges from dark red to dark brown. Some fragments have a very compact appearance and contain no inclusions. Others contain a few quartz and biotite mica inclusions.

IV. MODALITY

Matrix (groundmass) 35%–40%

Voids 5%–7%

Silt-sized inclusions

Monocrystalline quartz ca. 2%

Mica under 1%

Feldspar under 1%

Phyllite under 2%

Quartzite under 1%

TCFs under 1%

Larger inclusions

Metamorphic rocks 30%–40%

Monocrystalline quartz 10%

Polycrystalline quartz under 2%

Chert under 1%

Sandstone 8%

Micrite under 1%

Mica under 2%

Feldspar under 1%

Opaques under 1%

TCFs under 2%

Fabric Group 3: Semicoarse with Carbonate Rock Fragments

Samples HCH 04/19 (**3**), 04/25 (**1**)

I. MATRIX (GROUNDMASS)

A. Optical Properties and Color

1. Under plane-polarized light: dark brown, nonpleochroic.

2. Under cross-polarized light: dark orangish brown at the margins, dark brown at the core. The micromass is moderately active.

B. Overall Grain Size and Modality of Inclusions and Voids

Coarse inclusions vary from 1.6 to 0.1 mm long dimension. The fine fraction is <0.1 mm and up to 60 μm long dimension. Inclusions ca. 30%. Voids range in size from 1.8 to 0.16 mm long dimension; ca. 10%.

C. Overall Preferred Orientation of Inclusions and Voids

The nonplastics are randomly oriented. The planar voids display preferred orientation parallel to vessel margins.

D. Silt-Sized Inclusions (under 62 μm)

1. Monocrystalline Quartz

Size range: up to ca. 62 μm.

Shape: equant, angular to subangular.

Comments: merging boundaries.

2. Carbonate Rocks

Size range: up to ca. 62 μm.

Shape: equant, subrounded to subangular, occasionally amorphous.

Comment: the most common nonplastic inclusion; merging boundaries.

3. Mica

Size range: up to ca. 62 μm.

Shape: laths.

Comments: very few fragments of biotite mica.

4. Phyllite

Size range: up to ca. 62 μm.

Shape: elongate, laths.

Comments: it is composed of quartz and biotite or white mica.

5. Epidote

Size range: up to ca. 62 μm.

Shape: equant.

6. Chert

Size range: up to ca. 62 μm.

Shape: equant, subrounded.

7. Textural Concentration Features (TCFs)

Size range: up to ca. 62 μm.

Shape: equant, subrounded to rounded.

Comments: they are dark red or dark brown to black under cross-polarized light, with no inclusions or internal texture.

II. VOIDS

A. Size range: from 1.8 to 0.16 mm long dimension

B. Shape: they are close to single spaced, mainly meso and macro planar voids and meso vugs.

III. INCLUSIONS ABOVE SILT SIZE (ABOVE 62 M)

1. Carbonate Rocks

Size range: 1.2 to 0.1 mm

Shape: equant to elongate, angular to subrounded

Comments: it is the predominant component of this fabric, it is composed of micritic limestone.

2. Siltstone

Size range: 1.6 to 0.32 mm long dimension.

Shape: elongate.

Comments: the inclusions have a dark brown color and are very fine grained without any inclusions.

3. Phyllite

Size range: 1.6 to 0.6 mm long dimension.

Shape: elongate.

Comments: there are very few fragments. They are similar to the mudstones in color and texture, the only difference being that they contain small inclusions of biotite mica and quartz. They are most likely metamorphosed siltstones.

4. Quartzite-Schist

Size range: 0.4 to 0.16 mm long dimension.

Shape: elongate.

Comments: very rare fragments.

5. Monocrystalline Quartz

Size range: 0.6 to 0.1 mm long dimension (mode = 0.16 mm long dimension).

Shape: equant, angular to subangular.

Comments: quartz and micrite constitute the predominant nonplastic component of this fabric.

6. Polycrystalline Quartz

Size range: 0.6 to 0.2 mm long dimension.

Shape: equant or slightly elongate.

Comments: fine grained, in some cases grading into chert.

7. Chert

Size range: 0.8 to 0.1 mm long dimension.

Shape: equant, subangular.

Comments: composed of microcrystalline quartz.

8. Sandstone

Size range: 1.4 to 0.8 mm long dimension.

Shape: equant, subrounded.

Comments: it is very fine grained, composed of small quartz fragments in a clay-rich matrix.

9. Plagioclase Feldspar

Size range: 1.2 mm long dimension.

Shape: elongate, subangular.

Comments: very rare fragments.

10. Textural Concentration Features (TCFs)

Size range: 1.0 to 0.1 mm long dimension.

Shape: equant to slightly elongate, subangular to subrounded.

Comments: their color and texture are very similar to the siltstones. They are dark red to dark brown without any inclusions.

MODALITY

Matrix (groundmass) 60%

Voids 10%

Silt-sized inclusions

Monocrystalline quartz ca. 1%

Carbonate rocks 2%

Mica under 1%

Phyllite under 1%

Epidote under 1%

Chert under 1%

TCFs under 1%

Larger inclusions

Carbonate rocks 30%

Siltstone 10%

Phyllite under 2%

Quartzite-schist under 1%

Monocrystalline quartz 10%

Polycrystalline quartz under 1%

Chert under 1%

Sandstone under 2%

Feldspar under 1%

TCFs under 2%

Fabric Group 4: Calcareous with Angular Inclusions

Samples HCH 04/31 (**23**), 04/33 (**33**)

I. MATRIX (GROUNDMASS)

A. Optical Properties and Color

1. Under plane-polarized light: dark brown, nonpleochroic

2. Under cross-polarized light: brown, the micromass is moderately active.

B. Overall Grain Size and Modality of Inclusions and Voids

Coarse inclusions vary from 1.2 to 0.1 mm long dimension. The fine fraction is <0.1 mm and up to 60 μm long dimension. Inclusions ca. 30%. Voids range in size from 1.5 to 0.2 mm long dimension; ca. 5%.

C. Overall Preferred Orientation of Inclusions and Voids

The nonplastics are randomly oriented, the planar voids display preferred orientation parallel to vessel margins. There is evidence for tempering with organic matter.

D. Silt-Sized Inclusions (under 62 μm)

1. Monocrystalline Quartz

Size range: up to ca. 62 μm.

Shape: equant, subangular.

2. Micrite

Size range: up to ca. 62 μm.

Shape: equant, subrounded.

3. Mica

Size range: up to ca. 62 μm.

Shape: laths.

Comments: mainly biotite and a few white mica fragments.

4. Dark Brown Angular Inclusions

Size range: up to ca. 62 μm.

Shape: equant, angular.

II. Voids

A. Size range: from 2.0 to 0.2 mm long dimension.

B. Shape: they range from single to open spaced, with mainly meso and macro planar voids.

III. Inclusions above Silt Size (above 62 μm)

1. Monocrystalline Quartz

Size range: mode = 0.1 mm long dimension.

Shape: equant, angular to subangular.

Comments: evenly distributed in the clay matrix.

2. Carbonate Rocks

Size range: 0.8 to 0.1 mm long dimension.

Shape: equant, subrounded.

Comments: composed of micritic limestone.

3. Dark Brown Angular Inclusions

Size range: 1.2 to 0.2 mm long dimension.

Shape: equant to slightly elongate, angular to subangular.

3. Chert

Size range: 0.2 mm long dimension.

Shape: equant, subangular.

4. Calcite

Size range: 0.16 mm long dimension.

Shape: equant, angular.

5. Mica

Size range: 0.2 mm long dimension.

Shape: laths.

Comments: mainly biotite and a few white mica fragments.

6. Fossils

Comments: foraminifera, occasionally identified by their casts in the clay matrix.

IV. Modality

Matrix (groundmass) 60%

Voids 5%

Silt-sized inclusions

Monocrystalline quartz 1%

Micrite   2%

Mica under 2%

Grog under 1%

Larger inclusions

Monocrystalline quartz 5%

Carbonate rocks ca. 10%

Brown angular inclusions 3%

Chert under 1%

Calcite under 1%

Mica under 1%

Fossils under 1%

Fabric Group 5: Red Fabric with Quartz and Clay Pellets

Samples HCH 04/10 (**197**), 04/44 (**196**)

I. Matrix (Groundmass)

A. Optical Properties and Color

1. Under plane-polarized light: dark orangish brown, nonpleochroic.

2. Under cross-polarized light: the red to reddish-brown micromass is moderately active at the core and optically active at the margins.

B. Overall Grain Size and Modality of Inclusions and Voids

Coarse inclusions vary from 3.6 to 0.1 mm long dimension. The fine fraction is <0.1 mm and up to 60 μm long dimension. Inclusions ca. 30%. Voids range in size from 1.5 to 0.2 mm long dimension; ca. 5%.

C. Overall Preferred Orientation of Inclusions and Voids

The nonplastics and voids are randomly oriented.

D. Silt-Sized Inclusions (under 62 μm)

1. Monocrystalline Quartz

Size range: up to ca. 62 μm.

Shape: equant to slightly elongate, subangular.

2. Mica

Size range: up to ca. 62 μm.

Shape: laths.

Comments: it is mainly biotite mica, with very few fragments of white mica.

3. Phyllite

Size range: up to ca. 62 μm.

Shape: elongate, laths.

4. Textural Concentration Features (TCFs)

Size range: up to ca. 62 μm.

Shape: equant, subrounded to rounded.

Comments: they are dark red under cross-polarized light, with no inclusions or internal texture.

II. Voids

A. Size range: from 1.5 to 0.2 mm long dimension.

B. Shape: they range from single to open spaced and are mainly meso and macro vugs.

III. Inclusions above Silt Size (above 62 M)

1. Metamorphic Rock Fragments

(a) Phyllite

Size range: 1.2 to 0.16 mm long dimension.

Shape: elongate.

Comments: the inclusions are indicative of a low-grade metamorphic environment. Under cross-polarized light the color of the phyllite ranges from dark brown to silvery. In the former it is composed of biotite mica and quartz, in the latter of white mica and quartz. The phyllite fragments are quite fine grained.

(b) Quartzite

Size range: 0.8 to 0.2 mm long dimension.

Shape: equant to slightly elongate.

(c) Quartz-Mica Schist

Size range: 0.9 to 0.2 mm long dimension.

Shape: elongate.

2. Monocrystalline Quartz

Size range: 0.48 to 0.1 mm long dimension.

Shape: equant, angular to subangular.

3. Polycrystalline Quartz

Size range: 0.8 to 0.2 mm long dimension.

Shape: equant.

Comments: very few fragments.

4. Mica

Size range: ca. 0.2 mm long dimension

Shape: laths.

Comments: mainly biotite mica and a few fragments of white mica.

5. Feldspar

Size range: 0.2 mm long dimension.

Shape: equant, subangular.

Comments: very rare fragments of alkali feldspar.

6. Textural Concentration Features (TCFs)

Size range: 0.56 to 0.1 mm long dimension.

Shape: equant, rounded to subrounded.

Comments: their color is dark red. They have a very compact appearance and contain no inclusions. They seem concordant with the micromass.

IV. Modality

Matrix (groundmass) 65%

Voids 5%

Silt-sized inclusions

Monocrystalline quartz ca. 2%

Mica 1%

Phyllite under 2%

TCFs under 1%

Larger inclusions

Metamorphic rocks 30%

Monocrystalline quartz 10%

Polycrystalline quartz under 1%

Mica under 2%

Feldspar under 1%

TCFs 2%

Fabric Group 6: Red Fabric with Quartz

Samples HCH 04/32 (**31**), 04/34 (**15**)

I. Matrix (Groundmass)

A. Optical Properties and Color

1. Under plane-polarized light: orangish brown, nonpleochroic.

2. Under cross-polarized light: orange to reddish brown, with the micromass optically active.

B. Overall Grain Size and Modality of Inclusions and Voids

Coarse inclusions vary from 1.2 to 0.1 mm long dimension. The fine fraction is <0.1 mm and up to 60 μm long dimension. Inclusions ca. 20%. Voids range in size from 1.0 to 0.2 mm long dimension; ca. 3%.

C. Overall Preferred Orientation of Inclusions and Voids

The nonplastics and voids are randomly oriented. There is evidence for tempering with organic matter.

D. Silt-Sized Inclusions (under 62 μm)

1. Monocrystalline Quartz

Size range: up to ca. 62 μm.

Shape: equant to slightly elongate, angular to subangular.

2. Mica

Size range: up to ca. 62 μm.

Shape: laths.

Comments: biotite and white mica.

3. Phyllite

Size range: up to ca. 62 μm.

Shape: elongate, laths.

4. Quartzite

Size range: up to ca. 62 μm.

Shape: equant.

5. Textural Concentration Features (TCFs)

Size range: up to ca. 62 μm.

Shape: equant, subrounded to rounded.

Comments: dark red under cross-polarized light, with no inclusions or internal texture.

II. Voids

A. Size range: from 1.0 to 0.2 mm long dimension.

B. Shape: they range from double to open spaced, and are mainly meso vugs and macro planar voids.

III. Inclusions above Silt Size (above 62 m)

1. Metamorphic Rock Fragments

(a) Phyllite

Size range: 1.0 to 0.2 mm long dimension.

Shape: elongate.

Comments: a few fine grained phyllite fragments. Under cross-polarized light their color ranges from dark brown to golden brown. They are composed of biotite mica and quartz.

(b) Quartzite

Size range: 0.7 to 0.1 mm long dimension.

Shape: equant to slightly elongate.

Comments: in some cases they grade into schist.

(c) White Mica Schist

Size range: 1.0 to 0.2 mm long dimension.

Shape: elongate.

Comments: rare fragments, mainly in sample HCH 04/32 (**31**).

2. Monocrystalline Quartz

Size range: 0.7 to 0.1 mm long dimension.

Shape: equant, angular to subangular.

3. Mica

Size range: ca. 0.2 mm long dimension.

Shape: laths.

Comments: biotite and white mica.

4. Feldspar

Size range: 0.2 mm long dimension.

Shape: equant, subangular.

Comments: very rare fragments of plagioclase feldspar.

5. Epidote

Size range: 0.2 mm long dimension.

Shape: equant, subangular.

6. Textural Concentration Features (TCFs)

Size range: 0.6 to 0.1 mm long dimension.

Shape: equant, rounded to subrounded.

Comments: their color is dark red. They have a very compact appearance and contain no inclusions. They are disconcordant with the micromass.

IV. Modality

Matrix (groundmass) 70%

Voids 3%

Silt-sized inclusions

Monocrystalline quartz ca. 2%

Mica 1%

Metamorphic rocks under 2%

TCFs under 1%

Larger inclusions

    Metamorphic rocks 15%

    Monocrystalline quartz 20%

    Mica under 2%

    Feldspar under 1%

    Epidote under 1%

    TCFs 2%

Fabric Group 7: Granodiorite Fabric (Mirabello Fabric)

Group 7A, Samples HCH 04/40 (**102**), 04/42 (**295**)

Group 7B, Sample HCH 04/2 (**79**)

I. MATRIX (GROUNDMASS)

A. Optical Properties and Color

  1. Under plane-polarized light: greenish brown to brown, nonpleochroic.

  2. Under cross-polarized light: dark greenish brown to dark reddish brown. The micromass is optically inactive.

B. Overall Grain Size and Modality of Inclusions and Voids

Coarse inclusions vary from 1.6 to 0.1 mm long dimension. The fine fraction is <0.1 mm and up to 60 μm long dimension. Inclusions ca. 20%. Voids range in size from 0.8 to 0.2 mm long dimension; ca. 2%.

C. Overall Preferred Orientation of Inclusions and Voids

The nonplastics and voids are randomly oriented.

D. Silt-Sized Inclusions (under 62 μm)

  1. Igneous Rock Fragments

    Size range: up to ca. 62 μm.

    Shape: equant to slightly elongate, subangular.

  2. Plagioclase Feldspar

    Size range: up to ca. 62 μm.

    Shape: equant, subangular.

  3. Biotite

    Size range: up to ca. 62 μm.

    Shape: elongate.

  4. Textural Concentration Features (TCFs)

    Size range: up to ca. 62 μm.

    Shape: equant, subrounded to rounded.

    Comments: they are dark red under cross-polarized light, with no inclusions or internal texture.

II. VOIDS

A. Size range: from 0.8 to 0.2 mm long dimension.

B. Shape: they range from open to double spaced, mainly meso vugs.

III. INCLUSIONS ABOVE SILT SIZE (ABOVE 62 μM)

  1. Igneous Rock Fragments

    Size range: 1.6 to 0.16 mm long dimension.

    Shape: equant to slightly elongate.

    Comments: the inclusions are composed of plagioclase feldspar, biotite, and altered amphibole. This rock and mineral suite is indicative of an environment with acid igneous rocks, namely the granodiorites that outcrop in the area of the Gulf of Mirabello.

  2. Plagioclase Feldspar

    Size range: 1.2 to 0.16 mm long dimension.

    Shape: equant, angular to subangular.

  3. Sandstone

    Size range: 1.6 to 0.2 mm long dimension.

    Shape: elongate.

    Comments: they are composed mainly of quartz fragments in a clay-rich matrix.

  4. Biotite

    Size range: 0.8 to 0.2 mm long dimension.

    Shape: equant to elongate.

  5. Textural Concentration Features (TCFs)

    Size range: 0.8 to 0.1 mm long dimension.

    Shape: equant, rounded to subrounded.

    Comments: their color is dark red to dark brown; they have a very compacted appearance and contain no inclusions.

IV. MODALITY

Matrix (groundmass) 65%

Voids 2%

Silt-sized inclusions

    Igneous rock fragments ca. 2%

    Plagioclase feldspar under 1%

    Biotite under 1%

    TCFs 2%

Larger inclusions

    Igneous rock fragments 15%

    Plagioclase feldspar 5%

    Sandstone under 1%

Biotite under 2%

TCFs 10%

Fabric Group 8: Very Fine Fabric with Rare Nonplastics

Samples HCH 04/5 (**280**), 04/9 (**260**), 04/21 (**243**), 04/45 (**182**)

I. MATRIX (GROUNDMASS)

A. Optical Properties and Color

1. Under plane-polarized light: orangish brown to brown, nonpleochroic.

2. Under cross-polarized light: dark orangish brown at the core to greenish brown at the margins. The micromass is optically active.

B. Overall Grain Size and Modality of Inclusions and Voids

Coarse inclusions vary from 1.4 to 0.08 mm long dimension. The fine fraction is <0.08 mm and up to 60 µm long dimension. Inclusions ca. 5%. Voids range in size from 0.8 to 0.08 mm long dimension; ca. 5%.

C. Overall Preferred Orientation of Inclusions and Voids

The nonplastics are randomly oriented. The fine planar voids display preferred orientation parallel to vessel margins.

D. Silt-Sized Inclusions (under 62 µm)

1. Monocrystalline Quartz

Size range: up to ca. 62 µm.

Shape: equant, angular to subangular.

2. Mica

Size range: up to ca. 62 µm.

Shape: laths.

3. Quartzite Schist

Size range: up to ca. 62 µm.

Shape: slightly elongate, subangular.

4. Polycrystalline Quartz

Size range: up to ca. 62 µm.

Shape: equant, subangular.

Comments: occasionally ranging into chert.

5. Textural Concentration Features (TCFs)

Size range: up to ca. 62 µm.

Shape: equant, subrounded to rounded.

Comments: they are orangish brown to dark reddish brown under cross-polarized light, with no inclusions or with very small inclusions.

II. VOIDS

A. Size range: from 0.8 to 0.08 mm long dimension.

B. Shape: they are single to open spaced, mainly meso planar voids and meso vugs.

III. INCLUSIONS ABOVE SILT SIZE (ABOVE 62 M)

1. Monocrystalline Quartz

Size range: 0.6 to 0.1 mm long dimension.

Shape: equant, angular to subangular.

Comments: quartz and micrite constitute the predominant nonplastic component of this fabric.

2. Textural Concentration Features (TCFs)

Size range: 0.4 to 0.08 mm long dimension.

Shape: equant, subrounded.

Comments: they are orangish brown to dark reddish brown under cross-polarized light, with no inclusions or with very small inclusions consisting of small quartz fragments and biotite mica.

MODALITY

Matrix (groundmass) 85%

Voids 5%

Silt-sized inclusions

Monocrystalline quartz 8%

Mica 2%

Quartzite-schist under 1%

Polycrystalline quartz under 1%

TCFs under 1%

Larger inclusions

Monocrystalline quartz 1 %

TCFs 2%

Fabric 9: Fine Calcareous Fabric with Pellets

Sample HCH 04/24 (**132**)

I. MATRIX (GROUNDMASS)

A. Optical Properties and Color

1. Under plane-polarized light: brown, nonpleochroic.

2. Under cross-polarized light: yellowish brown. The micromass is moderately active.

B. Overall Grain Size and Modality of Inclusions and Voids

Coarse inclusions vary from 0.8 to 0.08 mm long dimension. The fine fraction is <0.08 mm and up to 60 μm long dimension. Inclusions ca. 3%. Voids range in size from 0.4 to 0.08 mm long dimension; ca. 2%.

C. Overall Preferred Orientation of Inclusions and Voids

The voids and nonplastics are randomly oriented.

D. Silt-Sized Inclusions (under 62 μm)

1. Monocrystalline Quartz

Size range: up to ca. 62 μm.

Shape: equant, angular to subangular.

2. Micrite

Size range: up to ca. 62 μm.

Shape: equant, rounded.

Comments: composed of micritic limestone.

3. Mica

Size range: up to ca. 62 μm.

Shape: laths.

Comments: biotite mica.

4. Textural Concentration Features (TCFs)

Size range: up to ca. 62 μm.

Shape: equant, subrounded to rounded.

Comments: they are brown to dark brown under cross-polarized light, fine grained containing small inclusions of biotite mica and micrite.

II. Voids

A. Size range: from 0.4 to 0.08 mm long dimension

B. Shape: they are open to double spaced, mainly meso planar voids and meso vugs.

III. Inclusions above Silt Size (above 62 μm)

1. Micrite

Size range: 0.4 to 0.08 mm long dimension.

Shape: equant, rounded to subrounded.

Comments: micrite and TCFs constitute the predominant component of this fabric.

2. Textural Concentration Features (TCFs)

Size range: 0.8 to 0.08 mm long dimension.

Shape: equant, subrounded.

Comments: they are brown to dark brown under cross-polarized light, fine grained containing small inclusions of biotite mica and micrite.

IV. Modality

Matrix (groundmass) 80%

Voids 2%

Silt-sized inclusions

Monocrystalline quartz under 1%

Micrite 10%

Mica 1%

TCFs 1%

Larger inclusions

Micrite 10%

TCFs 5%

Fabric 10: Fine Calcareous with Biotite Mica

Sample HCH 04/8 (236)

I. Matrix (Groundmass)

A. Optical Properties and Color

1. Under plane-polarized light: brown, nonpleochroic.

2. Under cross-polarized light: golden brown. The micromass is optically moderately active.

B. Overall Grain Size and Modality of Inclusions and Voids

Coarse inclusions vary from 0.8 to 0.08 mm long dimension. The fine fraction is <0.08 mm and up to 60 μm long dimension. Inclusions ca. 15%. Voids range in size from 1.2 to 0.16 mm long dimension; ca. 3%.

C. Overall Preferred Orientation of Inclusions and Voids

The voids and nonplastics are randomly oriented.

D. Silt-Sized Inclusions (under 62 μm)

1. Monocrystalline Quartz

Size range: up to ca. 62 μm.

Shape: equant to slightly elongate, angular to subangular.

2. Biotite Mica

Size range: up to ca. 62 μm.

Shape: laths.

3. Micrite

Size range: up to ca. 62 μm.

Shape: equant, subangular.

4. Polycrystalline Quartz

Size range: up to ca. 62 μm.

Shape: equant, subangular.

Comments: it occasionally ranges into chert.

5. Textural Concentration Features (TCFs)

Size range: up to ca. 62 μm.

Shape: equant, subrounded to rounded.

Comments: they are dark reddish brown under cross-polarized light, with no inclusions.

II. Voids

A. Size range: from 1.2 to 0.16 mm long dimension.

B. Shape: they are open spaced, mainly meso vugs and a few planar voids.

III. Inclusions above Silt Size (above 62 m)

1. Monocrystalline Quartz

Size range: 0.16 to 0.08 mm long dimension.

Shape: equant, angular to subangular.

Comments: quartz and micrite constitute the predominant nonplastic component of this fabric.

2. Micrite

Size range: 0.12 to 0.08 mm long dimension.

Shape: equant to slightly elongate, angular to subangular.

Comments: it is composed of micritic limestone.

3. Textural Concentration Features (TCFs)

Size range: 0.2 to 0.08 mm long dimension.

Shape: equant to slightly elongate, subrounded.

Comments: they are dark reddish brown under cross-polarized light, with no inclusions.

Modality

Matrix (groundmass) 75%

Voids 3%

Silt-sized inclusions

Monocrystalline quartz 8%

Mica 8%

Micrite 10%

Polycrystalline quartz under 1%

TCFs under 1%

Larger inclusions

Monocrystalline quartz 3%

Micrite 3%

TCFs 1%

Fabric Group 11: Fine Fabric with Rounded Sandstones and Phyllite

Samples HCH 04/35 (**41**), 04/36 (**39**)

I. Matrix (Groundmass)

A. Optical Properties and Color

1. Under plane-polarized light: greenish brown to brown, nonpleochroic.

2. Under cross-polarized light: golden brown to brown. The micromass ranges from moderately active to active.

B. Overall Grain Size and Modality of Inclusions and Voids

Coarse inclusions vary from 1.2 to 0.1 mm long dimension. The fine fraction is <0.1 mm and up to 60 μm long dimension. Inclusions ca. 10%. Voids range in size from 0.4 to 0.1 mm long dimension; ca. 2%.

C. Overall Preferred Orientation of Inclusions and Voids

The voids and nonplastics are randomly oriented.

D. Silt-Sized Inclusions (under 62 μm)

1. Monocrystalline Quartz

Size range: up to ca. 62 μm.

Shape: equant to slightly elongate, angular to subangular.

2. Biotite Mica

Size range: up to ca. 62 μm.

Shape: laths.

3. Textural Concentration Features (TCFs)

Size range: up to ca. 62 μm.

Shape: equant, subrounded.

Comments: they are dark reddish brown under cross-polarized light, with no inclusions.

II. Voids

A. Size range: from 0.4 to 0.1 mm long dimension.

B. Shape: they are open spaced, mainly meso vugs.

III. Inclusions above Silt Size (above 62 m)

1. Monocrystalline Quartz

Size range: 0.16 to 0.08 mm long dimension.

Shape: equant, angular to subangular.

2. Sandstone

Size range: 1.2 to 0.16 mm long dimension.

Shape: equant to slightly elongate, rounded to subrounded.

Comments: they are composed of small quartz fragments and biotite mica laths.

3. Phyllite

Size range: 0.4 to 0.24 mm long dimension.

Shape: elongate.

Comments: composed of quartz and biotite mica laths.

4. Textural Concentration Features (TCFs)

Size range: ca 0.16 mm long dimension.

Shape: equant to slightly elongate, subrounded.

Comments: they are dark reddish brown under cross-polarized light, with no inclusions.

IV. MODALITY

Matrix (groundmass) 80%

Voids 2%

Silt-sized inclusions

Monocrystalline quartz under 2%

Mica 2%

Micrite 10%

TCFs under 1%

Larger inclusions

Monocrystalline quartz under 1%

Sandstone 3%

Phyllite 1%

TCFs under 1%

Fabric 12: Very Fine Fabric with Pellets

Sample HCH 04/41 (**293**)

I. MATRIX (GROUNDMASS)

A. Optical Properties and Color

1. Under plane-polarized light: dark grayish brown, nonpleochroic.

2. Under cross-polarized light: greenish brown to brown. The micromass is opticallly inactive.

B. Overall Grain Size and Modality of Inclusions and Voids

Coarse inclusions vary from 0.6 to 0.08 mm long dimension. The fine fraction is <0.08 mm and up to 60 μm long dimension. Inclusions ca. 15%. Voids range in size from 0.2 to 0.16 mm long dimension; ca. 1%.

C. Overall Preferred Orientation of Inclusions and Voids

The voids and nonplastics are randomly oriented.

D. Silt-Sized Inclusions (under 62 μm)

1. Monocrystalline Quartz

Size range: up to ca. 62 μm.

Shape: equant to slightly elongate, angular to subangular.

2. Mica

Size range: up to ca. 62 μm.

Shape: laths.

Comments: biotite.

3. Textural Concentration Features (TCFs)

Size range: up to ca. 62 μm.

Shape: equant, subrounded to rounded.

Comments: dark to light brown under cross-polarized light, with no inclusions.

II. VOIDS

A. Size range: from 0.2 to 0.16 mm long dimension

B. Shape: they are open spaced, mainly meso planar voids.

III. INCLUSIONS ABOVE SILT SIZE (ABOVE 62 M)

1. Monocrystalline Quartz

Size range: 0.12 to 0.08 mm long dimension.

Shape: equant, angular to subangular.

2. Sandstone

Size range: 0.2 to 0.1 mm long dimension.

Shape: equant to slightly elongate, subangular.

Comments: composed of fine quartz fragments.

3. Silty Pellets

Size range: 0.6 to 0.08 mm long dimension.

Shape: equant to elongate, subangular.

Comments: they are dark brown under cross-polarized light, very fine grained with no inclusions.

4. Textural Concentration Features (TCFs)

Size range: 0.3 to 0.08 mm long dimension.

Shape: equant, subrounded.

Comments: they are dark to light brown under cross-polarized light, with no inclusions.

IV. MODALITY

Matrix (groundmass) 75%

Voids 1%

Silt-sized inclusions

Monocrystalline quartz under 1%

Mica 1%

TCFs under 1%

Larger inclusions

Monocrystalline quartz under 1%

Sandstone 1%

Silty pellets 3

TCFs 2%

Fabric 13: Very Fine Fabric with Serpentinite

Sample HCH 04/48 (**287**)

I. MATRIX (GROUNDMASS)

A. Optical Properties and Color

1. Under plane-polarized light: dark brown, non-pleochroic

2. Under cross-polarized light: brown to greenish brown. The micromass is optically inactive.

B. Overall Grain Size and Modality of Inclusions and Voids

Coarse inclusions vary from 0.48 to 0.1 mm long dimension. The fine fraction is <0.1 mm and up to 60 μm long dimension. Inclusions ca. 15%. Voids range in size from 0.4 to 0.16 mm long dimension; ca. 2%.

C. Overall Preferred Orientation of Inclusions and Voids

The voids and nonplastics are randomly oriented.

D. Silt-Sized Inclusions (under 62 μm)

1. Monocrystalline Quartz
   Size range: up to ca. 62 μm.
   Shape: equant, angular to subangular.

2. Mica
   Size range: up to ca. 62 μm.
   Shape: laths.
   Comments: biotite.

3. Polycrystalline Quartz
   Size range: up to ca. 62 μm.
   Shape: equant, subangular.

4. Phyllite
   Size range: up to ca. 62 μm.
   Shape: elongate.

5. Textural Concentration Features (TCFs)
   Size range: up to ca. 62 μm.
   Shape: equant, subrounded to rounded.
   Comments: they are dark brown under cross-polarized light, with no inclusions.

II. VOIDS

A. Size range: from 0.4 to 0.16 mm long dimension.

B. Shape: open spaced, mainly meso vugs.

III. INCLUSIONS ABOVE SILT SIZE (ABOVE 62 M)

1. Monocrystalline Quartz
   Size range: 0.24 to 0.1 mm long dimension.
   Shape: equant to slightly elongate, angular to subangular.

2. Serpentinite
   Size range: 0.4 to 0.1 mm long dimension.
   Shape: equant to slightly elongate, subangular.

3. Sandstone
   Size range: 0.4 to 0.1 mm long dimension.
   Shape: equant, subrounded.
   Comments: the sandstone is composed of fine quartz fragments.

4. Phyllite
   Size range: 0.48 to 0.15 mm long dimension.
   Shape: elongate.
   Comments: it is golden brown to dark brown under cross-polarized light, very fine grained.

5. Textural Concentration Features (TCFs)
   Size range: 0.3 to 0.08 mm long dimension.
   Shape: equant, subrounded.
   Comments: they are dark brown under cross-polarized light. The coarser ones have quartz inclusions, and the finer ones contain no inclusions.

IV. MODALITY

Matrix (groundmass) 75%

Voids 2%

Silt-sized inclusions

Monocrystalline quartz under 1%

Mica under 1%

Polycrystalline quartz under 1%

Phyllite under 1%

TCFs under 1%

Larger inclusions

    Monocrystalline quartz 1%

Serpentinite 2%

Sandstone under 1%

Phyllite under 1%

TCFs 1%

Appendix B

The Pottery Statistics

Philip P. Betancourt, Susan C. Ferrence, and Louise C. Langford-Verstegen

After the sherds that joined together were combined to form vases and sections of vases, the entire assemblage of vessels and fragments was counted and recorded as statistics (Tables B.1–B.21). The pieces were strewn on tables and sorted by fabric and then by shapes and class. The identifications by class and position on the vase were recorded (complete vessel, largely restorable vessel, rim sherd, handle, base, spout, leg, or body sherd). The total of 18,064 entries is large enough for useful statistics, but this sample is by no means the entire deposit that was placed in the cave. It includes everything collected from the 1976 to 1983 campaign, a small number of sherds found in the dump from that early project, and all items found in the new excavations of 2002 and 2003. The recent project processed all the soil with a water separation machine (Peterson 2009), so the retrieved pottery approaches 100% of what was in the soil excavated in those two years. The largest amount of unrecovered pottery is what still survives in the cave beneath tons of loose rocks supporting a cracked and unstable ceiling at the east of Room 5. This dangerous part of the deposit would require sophisticated and expensive mining techniques to remove it.

The statistics, recorded in 19 tables, provide a good picture of the pottery. The regional ceramics from Lasithi (Tables B.7 [partial], B.8 [partial], B.19) is recognizable by its use of the Lasithi Red Fabric Group (Petrographic Fabric Group 2, see App. A). It accounts for 10,633 entries (59%) with a minimum of 1,228 vessels, which is also 59% of the total. The remaining percentage includes the early vases whose origin is unknown as well as the imports into the region. The origin for most of this imported material remains uncertain. The distinctive pieces include 135 sherds from MM IA North-Central Crete including Malia (Table B.10), an additional 214 sherds from MM IB–II Malia (Table B.17), and smaller amounts from the Mirabello region (Table B.8, 50 sherds of Vasiliki Ware), Knossos (Table B.12), and elsewhere. Future scholarship will hopefully be able to identify the sources of more of this material.

Table	Pottery Description	Source	Sherd Total/MNV	Date
B.2	Outside cave, coarse burnished	Unknown	6/6	FN–EM I
B.3	Incised decoration	Unknown	2/1	LN II
B.3	Monochrome Style, FN–EM I	Unknown	199/23	FN–EM I
B.3	Scored Style	Unknown	2/2	FN–EM I
B.3	Pyrgos Style	Unknown	35/6	EM I
B.3	Monochrome Style, EM I–IIA	Unknown	120/16	EM I–IIA
B.4	Hagios Onouphrios Style	Unknown	7/7	EM I
B.5	Koumasa Style	Unknown	7/6	EM IIA
B.6	Monochrome Style, pale fabrics	Unknown	24/13	EM IIA
B.7	Monochrome Style, red fabrics, dark surface, offering tables	Lasithi Region	37/13	EM IIA
B.8	Vasiliki Ware	Gulf of Mirabello	50/18	EM IIB
B.8	Vasiliki Style imitations, pale fabrics	Unknown	10/9	EM IIB
B.8	Vasiliki Style imitation, red fabric group	Lasithi Region	45/24	EM IIB
B.9	Light-on-dark pottery, pale fabrics	Unknown	159/57	EM III–MM II
B.10	Diagonal Line Style	Malia(?)	135/29	MM IA
B.10	Various small jugs	Unknown	209/62	MM I
B.11	Barbotine Style	North Central Crete	11/3	MM I
B.12	Knossian pottery	Knossos	6/6	EM II–MM I
B.13	Fine red pottery	Pediada including Kastelli	27/16	MM IB
B.14	Fine red pale fabric	Pediada	62/34	MM IB
B.15	Kamares Ware	Unknown	23/17	MM IB–IIB
B.16	Chamaizi Pots	Unknown	31/12	MM IIB
B.17	Malia pottery	Malia	214/52	MM I–IIB
B.18	Various nonlocal groups, pale fabrics	(Mostly unknown)	6,088/403	EM III–MM IIB
B.19	Lasithi Red Fabric Group	Lasithi region	10,551/1,153	EM II–MM II
B.21	Outside cave, LM III	Unknown	4/4	LM III

Table B.1. Summary of the 19 tables that record the pottery from the site.

The results are summarized in Table B.1. The total number of sherds (18,064) allows an estimate of the minimum number of vases (1,992) based on the examination of the surviving pieces (for cups, for example, the handles are counted). Assuming that the number of offerings is related to the number of burials, the charts show that the population gradually increased. The early periods have far fewer offerings in comparison with the large amount of pottery after EM IIB, but the pattern of about one pouring vessel for one or two drinking vessels seems correct. The chalices used before EM IIB were probably communal drinking vessels.

Looking at the shapes in categories leads to the recognition of some important patterns:

EM IIB Vasiliki Ware

Drinking vessels, 11 entries, minimum of 10 vessels
Pouring vessels, 94 entries, minimum of 40 vessels

Although these numbers are small, they establish several points in regard to the burial customs. Vasiliki Ware was used primarily as a pouring vessel.

Fewer goblets were used, and the teapots and other specialized shapes in this class were not used.

In examining the statistics of the later pottery by functional categories to provide additional information, the minimum numbers of vessels provide useful data:

EM III–MM IIB

Shallow bowls, for food, minimum of 25 vessels

Drinking vessels, minimum of 712 vessels

Medium-sized and large jugs (for pouring), minimum of 532 vessels

Other pouring vessels, minimum of 22 vessels

Small-sized jugs (height under 15 cm), for offerings, minimum of 91 vessels

Jars, small or medium-sized, for offerings, minimum of 70 vessels

Jars, large size (height over 20 cm), possibly for burials, minimum of 11 vessels

The 712 drinking vessels can be compared with the total number of burials in the cave. From a study of the human bones, the remains of almost 400 individuals are present in the remains that have been preserved (McGeorge 2008, 580). An estimate of over 1,000 burials is probably a good approximation based on an estimate of the number of discarded bones from the early excavations. The pottery suggests that most of them were from EM III and later. For the cups and other drinking vessels, the conclusion is that not every burial had offerings. The comparison between drinking and pouring vessels is not completely clear because a large number of the sherds from jugs could not be distinguished between small (under 15 cm high) and larger vases (over 15 cm high). The pattern of between one and two drinking vessels for each serving vessel rules out customs like the communal toasting ceremony recorded in the Mesara with conical cups that were "apparently mass-produced for ritual purposes" (Branigan 1970b), suggesting either offerings to the deceased or a modest ritual inside the original tomb involving drinking and/or the pouring of libations.

The pattern of depositing mostly drinking and pouring vessels plus small jugs containing liquid that had been established by EM IIA clearly continued throughout the history of the community. The implication for continuity of burial practices suggests no great population change during this long period. A few new customs, however, are clearly recognizable. The shallow bowls are mainly from EM III to MM I. Whatever they held (perhaps the meat offerings represented by the large number of animal bones in the deposit?), they were evidently replaced by some other container after MM I, perhaps one of wood. The large jars (minimum of 11 vessels) are all late in the sequence. Like the larnakes, they probably represent the well-known rise of burial in clay vessels that occurred in Crete during the Middle Bronze Age (Pini 1968)

A change in ceremony can be noted within EM IIA. Earlier burials used chalices, a drinking vessel whose large size suggests communal ceremonies, while the EM II substitution of individual goblets implies a more individualized function. The offering tables of EM I to IIA (Table B.7) may have been accessories that provided the communal food for these ceremonies. Periodic changes in the imported pottery classes are nicely represented in the statistics. The fashion for mottled Vasiliki Ware during EM IIB shows up as both imports and imitations (Table B.8). The distinctive Diagonal Line Style and the Barbotine Style jugs were imported in MM IA, probably from North Central Crete (Tables B.10–B.12). The trefoil-mouthed jugs (Table B.14) and the Fine Red Fabric (Table B.13) reflect the economic rise of Kastelli and then of Galatas and other centers in the Pediada during MM IB. The rise in popularity of polychrome pottery has at least a small impact on Lasithi's ceramic use (Table B.15). A closer relationship between the Lasithi Plain and Malia during MM IIB resulted in a substantial increase in pottery from this palace site (Table B.17).

Elite members of the community can be suggested by the presence of gold jewelry, stone beads and pendants, seals, and objects of ivory, but not a single metal vessel was found in the cave. If elite citizens wished to be distinguished by their serving and drinking sets, they must have chosen some of the imported alternatives to the local clay jugs and cups. The two-handled cups, the fine cups with dark red slip (Table B.13), the Kamares Style (Table B.15), and the trefoil jugs imported from various places in the Pediada including Galatas (Table B.14) could have served this segment of the society.

Pottery Outside the Mouth of the Cave

A few sherds of coarse, black to red burnished pottery were discovered near the front of the cave's mouth in a disturbed context that was below the dump from the early excavations in the cave (Table B.2; see also Ch. 3). These fragments, which were not associated with any human bones, suggest that visitors camped there near the cave at the end of the Final Neolithic and during EM I. The cave would have been a source of water then, and a nearby camp is understandable.

FN–EM I

Total number of sherds: 6 (**4**, **7**, **8**, **21**, **22**, **24**)

Shape and Cataloged Examples	C	NC	R	H	B	S	L	Body	Total	%	MNV
FN–EM I											
Monochrome Style, Dark Burnished Class											
Open vessel (**4**, **7**, **8**)								3	3	50%	3
EM I											
Coarse, Black or Red Monochrome Burnished Pottery											
Chalice (**24**)								1	1	16.6%	1
Cup or bowl (**21**, **22**)			2						2	33.3%	2

Table B.2. Exterior of the cave, FN–EM I. C = complete; NC = nearly complete; R = rim, H = handle; B = base; S = spout; L = leg; MNV = minimum number of vessels.

The Monochrome Style, Scored Style, Pyrgos Style, and Hagios Onouphrios Style from Neolithic, EM I, and EM I–IIA

The pottery from Neolithic to EM IIA spans a long time period that consisted of several hundred years (Tables B.3, B.4; see also Chs. 3, 4). In spite of this long period, the total number of sherds from these phases is relatively small (358 sherds, representing a minimum of 55 vessels). Because several of these pottery varieties continue from FN into EM I, and others were used both in EM I and EM IIA, close dating is not possible with several of these classes. Both the total number of vases and fragments and the minimum number of vessels (MNV) are given.

LN II

Incised Coarse Dark Burnished Class (2 sherds, MNV 1), **1**

FN–EM I

Monochrome Style, dark burnished fabric, chaff and calcite tempered (1 sherd, MNV 1), **2**

Monochrome Style, dark burnished fabric, calcite tempered (1 sherd, MNV 1), **3**

Monochrome Style, dark burnished fabric, with phyllite temper (197 sherds, MNV 21), **5**, **6**, **9–18**

Scored Style (2 sherds, MNV 2), **19**, **20**

EM I

Dark Gray Burnished Class, including the Pyrgos Style (35 sherds, MNV 6), **23**, **25–27**

Hagios Onouphrios Style (7 sherds, MNV 7), **38–44**

EM I or EM IIA

> Monochrome Style, coarse, burnished, red fabric, dark surface (117 sherds, MNV 13), **28–34**
>
> Monochrome Style, soft reddish-brown fabric (2 sherds, MNV 2), **35, 36**
>
> Monochrome Style, calcite tempered, variegated surface (1 sherd, MNV 1), **37**

The fragmentary nature of this pottery, with several examples of vessels represented by single sherds, indicates that the process that transported the material to the cave in MM IIB did not move many examples of complete or nearly complete vases from these early periods in the life of the community. These early fragments were mixed at random with the other pottery, as if those who moved the deposit gathered up occasional early material remaining in the original tomb or tombs. It is also possible to conclude that the community that used the primary burial place represented here was very small in these early years.

One important characteristic of this part of the deposit is the large number of different classes. In these early periods, Lasithi was obviously importing a substantial percentage of its ceramics.

Shape and Cataloged Examples	C	NC	R	H	B	S	L	Body	Total	%	MNV
LN II											
INCISED, COARSE DARK BURNISHED CLASS											
Carinated bowl (**1**)			1					1	2	0.5%	1
FN–EM I											
MONOCHROME STYLE, DARK BURNISHED CLASS, CALCITE TEMPERED											
Open vessel (**2**)								1	1	0.5%	1
Closed vessel (**3**)			1						1	0.5%	1
MONOCHROME STYLE, DARK BURNISHED CLASS WITH PHYLLITE											
Cup (**5, 6, 9–12**)			59		1				60	16.8%	10
Open vessel (**13–15**)			2		17			99	118	33.0%	6
Open vessel with lug (**16**)				1					1	0.5%	1
Jar (**17, 18**)			9		2			7	18	5.0%	4
Scored Style											
Open vessel (**19, 20**)								2	2	0.6%	2
EM I											
DARK GRAY BURNISHED CLASS (INCLUDING THE PYRGOS STYLE)											
Chalice (**23, 25–27**)			11					24	35	9.8%	6
EM I–IIA											
MONOCHROME STYLE, COARSE, BURNISHED, RED FABRIC, DARK SURFACE											
Chalice (**30–33**)		1						17	18	5.0%	3
Goblet (**34**)								2	2	0.6%	2
Open vessel (**28, 29**)			9	7	4			77	97	27.1%	8
MONOCHROME STYLE, SOFT REDDISH-BROWN FABRIC											
Open vessel (**35**)				1					1	0.5%	1
Miniature vessel (**36**)				1					1	0.5%	1
MONOCHROME STYLE, CALCITE TEMPERED, VARIEGATED SURFACE											
Jug with pointed spout (**37**)			1						1	0.5%	1

Table B.3. Monochrome Style, Scored Style, and Pyrgos Style, LN II and FN–EM I. For abbreviations, see caption of Table B.2.

Shape and Cataloged Examples	C	NC	R	H	B	S	L	Body	Total	%	MNV
EM I											
Hagios Onouphrios Style											
Open vessel (38)								1	1	14.3%	1
Jug (39–44)		1		2	1			2	6	85.7%	6

Table B.4. Hagios Onouphrios Style, EM I. For abbreviations, see caption of Table B.2.

The Koumasa Style, EM IIA

A total of seven sherds (6 MNV; **45–50**) can be assigned with certainty to the Koumasa Style of EM IIA (Table B.5; see also Ch. 4), which, together with several pieces from Tables B.3, B.6, and B.7, were surely also made during this period. The small number (under 1% of the total) represents a limited number of original burials.

Only seven sherds in the Koumasa Style come from the cave. All of the vessels are fragmentary,

and the remaining pieces of the vases were not found. The numbers are too few for internal percentages within the assemblage to be meaningful. The vases in this style form an almost uniform pattern of usage. All the sherds except for one body sherd are open shapes, but they are mostly fairly large, suitable for serving or communal drinking.

Shape and Cataloged Examples	C	NC	R	H	B	S	L	Body	Total	%	MNV
EM IIA											
Koumasa Style											
Cup (45)		1							1	14.3%	1
Wide-mouthed jug or cup (46)				1					1	14.3%	1
Open vessel with spout (47, 48, 50)			1			2			3	42.9%	3
Bridge-spouted open vessel (49)						1			1	14.3%	1
Closed vessel								1	1	14.3%	–

Table B.5. Koumasa Style, EM IIA. For abbreviations, see caption of Table B.2.

Monochrome Style, Pale Burnished Fabrics, EM IIA

A total of 24 sherds (MNV 13; **51–58**) of unpainted, pale colored EM IIA vessels come from the excavations (Table B.6; see also Ch. 4). The fragments represent a minimum of 13 vases. Nine of the 13 vases are small jugs of special and distinctive

shapes. These small jugs mark the beginning of the tradition of burying small closed vessels with narrow necks in tombs. In later times this class of vase regularly contained oil.

Shape and Cataloged Examples	C	NC	R	H	B	S	L	Body	Total	%	MNV
EM IIA											
MONOCHROME STYLE, PALE FABRICS											
Bowl with horizontal handle (**51**)			1						1	4.2%	1
Open vessel			1		2				3	12.5%	2
Cylindrical jug (**52–54**)		1	3					8	12	50%	5
Conical irregular jug (**55–57**)		1		4				2	7	29.2%	4
Pyxis (**58**)					1				1	4.2%	1

Table B.6. Monochrome Style, pale burnished fabrics, EM IIA. For abbreviations, see caption of Table B.2.

Dark Burnished Offering Tables in Monochrome Style with Coarse Phyllite Fabric, EM IIA

The selection of flat-bottomed offering tables in the ossuary (Table B.7; see also Ch. 5), the largest group of this artifact from any Minoan tomb, is perhaps related to the fact that the fabric analysis suggests that these vessels are products of the Lasithi region. The deposit has a total of 37 sherds representing a minimum of 13 offering tables (**59–71**).

Shape and Cataloged Examples	C	NC	R	H	B	S	L	Body	Total	%	MNV
EM IIA											
MONOCHROME STYLE, RED TO BROWN FABRICS, DARK SURFACE											
Offering table, Type B (**59, 60**)		2	3						5	13.5%	2
Offering table, Type D (**61**)	1								1	2.7%	1
Offering table, Type E (**62–64**)			3						3	8.1%	3
Offering table, Type I (**65**)	1								1	2.7%	1
Offering table, unknown type (**66–70**)			16	1	3			6	26	70.2%	5
Offering table, small (**71**)		1							1	2.7%	1

Table B.7. Monochrome Style, red to brown coarse fabrics with phyllite, dark surface, EM IIA. For abbreviations, see caption of Table B.2.

Vasiliki Ware and Its Imitations, EM IIB

Among the vases found in the excavations were three classes of Vasiliki Ware and its imitations (Table B.8; see also Ch. 6):

Vasiliki Ware from the region of the Gulf of Mirabello (50 sherds, MNV 18), **72–82**

Vasiliki Style Imitations in Pale Fabrics (10 sherds, MNV 9), **83–90**

Vasiliki Style Imitations in Lasithi Red Fabric Group (45 sherds, MNV 24), **91–100**

Vasiliki Ware is a mottled black to red to brown pottery found only in EM IIB contexts (Betancourt 1979). It is one of the ceramic classes in Crete that can be positively attributed to a specific locality because the fabric contains small grains of igneous rocks in the diorite to granodiorite series that is found only along the northeastern coast of Crete from Gournia to Priniatikos Pyrgos. This clay recipe, called Mirabello Fabric, has been described in print several times (Myer, McIntosh, and Betancourt 1995; Nodarou and Moody 2014, with added bibliography). Two different classes can be recognized among the imitations, one made from pale fabrics and the other using the Lasithi Red Fabric Group that is the local fabric in this cave (for the petrographic description, see Fabric Group 2 in App. A). Only this red fabric group was local to the region, so both the original Vasiliki Ware and its imitations in pale clay were imported into Lasithi.

The relative proportions among the fragmentary mottled material and the minimum numbers of vases demonstrate the very specialized selection used in the cave. Open shapes account for only 12 of the total number of sherds, while 91 pieces are fragments from jugs. In terms of minimum numbers, the excavations found evidence for 11 open vases (10 goblets and a bowl) and 40 jugs. The conclusion is that most of the pottery in this class was brought to the site as containers for liquids, with a smaller number used as drinking vessels. Both the large and small Vasiliki Ware jugs have cylindrical necks that would allow them to be used as transport vases by inserting a cylindrical stopper (made easily from wooden limbs) into the neck.

This use contrasts strongly with the presence of Vasiliki Ware in the region nearer to where it was made originally. This pottery was beautifully decorated with a lustrous slip that received interesting mottled ornament, and it was used both for ordinary vases and for some elaborate display and serving pieces in the area around the Gulf of Mirabello. The finest ornament was placed on the serving vessels: teapots, bridge-spouted jars, and jugs with very tall vertical spouts. Some of the Vasiliki Ware shapes used in the Gulf of Mirabello region are missing from this cave deposit, and among the missing shapes are the most elaborate ones: teapots with exaggerated spouts and the very largest jugs that were given extremely tall spouts. Also completely absent from the cave are shallow, open bowls, conical bowls without spouts, and bridge-spouted jars, all of which were common in the region of the Gulf of Mirabello. The most likely explanation for this pattern is that the Lasithi community had a specialized and limited exchange pattern in regard to this class. It either imported mostly jugs used for liquids, or it only used the other vessels for nonfunerary purposes. The region used a few of the very small drinking vessels, but not the large serving pieces, and the region's potters only copied this restricted inventory.

Shape and Cataloged Examples	C	NC	R	H	B	S	L	Body	Total	%	MNV
EM IIB											
VASILIKI WARE											
Spouted Goblet (**72**)					1			1	2	1.9%	2
Jug, small (**73**, **74**)	2		1		3				6	5.7%	5
Jug, medium or large (**75–82**)				5	7			12	24	22.9%	9
Jug, unknown size					2			16	18	17.1%	2
VASILIKI STYLE IMITATIONS IN PALE FABRICS											
Goblet (**83–86**)			1		4			1	6	5.7%	5
Jug, small (**87**)					1				1	1.0%	1
Jug, carinated (**88**, **89**)		2							2	1.9%	2
Jug, globular with incised dashes (**90**)	1								1	1.0%	1

Table B.8. Vasiliki Ware and its imitations, EM IIB. For abbreviations, see caption of Table B.2.

Shape and Cataloged Examples	C	NC	R	H	B	S	L	Body	Total	%	MNV
EM IIB, cont.											
VASILIKI STYLE IMITATIONS IN LASITHI RED FABRIC GROUP											
Spouted conical bowl (**91**)	1							1	2	1.9%	1
Goblet (**92–94**)				3					3	2.9%	3
Jug, small		4		3	10				17	16.2%	7
Jug, medium or large (**95–100**)		2	12	6		2		1	23	21.9%	13

Table B.8, cont. Vasiliki Ware and its imitations, EM IIB. For abbreviations, see caption of Table B.2.

Light-on-Dark Pottery, Pale Fabrics, EM III–MM II

By EM III, the local pottery production making vases in the Lasithi Red Fabric Group was well established (see Table B.19). Although the local output made a substantial quantity of vessels, specialized imports were used to augment the local production. Among the imports were vases in the light-on-dark style. For EM III to MM II, this class includes 23 whole and restorable vases plus enough additional sherds to bring the minimum number of vessels to 57 (Table B.9; see also Ch. 7). The increased number of whole and restorable vases in comparison with the earlier periods indicates that the movement of burials to the cave had access to a greater number of better preserved burials in comparison with material moved from the earlier history of the community. The following categories can be recognized:

Drinking vessels (109 entries, 19 MNV), **102–110**

Small jugs (18 entries, 15 MNV), **111–123**

Medium or large jugs (23 entries, 14 MNV), **124–131, 133, 134**

Shallow conical bowl (1 vessel), **101**

Other shapes, including the jar with vertical handles, spouted bowl, teapot, spouted jar, jar, and pyxis (9 entries, 8 MNV), **135–142**

Along with the jugs and pouring vessels that form the main part of the ceramic assemblage, the imports in light-on-dark pottery included a number of special shapes that were uncommon in the highland plain. The eight vessels in this category represent 12% of the total, which is a significant percentage.

Shape and Cataloged Examples	C	NC	R	H	B	S	L	Body	Total	%	MNV
EM III–MM I											
PALE FABRICS, LIGHT-ON-DARK											
Shallow conical bowl (**101**)	1								1	0.6%	1
MM I–II											
PALE FABRICS, LIGHT-ON-DARK											
Cup (**102–108**)	2	1	77	7	2			17	106	66.2%	16
Tumbler (**109, 110**)		1			2				3	1.9%	3
Jug, small (**111–123**)	2	8		2	3			3	18	11.25%	15
Jug, medium or large (**124–131**)				5	5	7		3	20	12.5%	12
Closed vessel (**132**)								1	1	0.9%	1
Jar with vertical handles (**133**)			2						2	1.2%	1

Table B.9. Light-on-dark pottery, pale fabrics, EM III–MM II. For abbreviations, see caption of Table B.2.

Shape and Cataloged Examples	C	NC	R	H	B	S	L	Body	Total	%	MNV
MM I–II, cont.											
PALE FABRICS, LIGHT-ON-DARK, cont.											
Jug without spout (**134**)		1							1	0.6%	1
Wide-mouthed jug (**135**)		1							1	0.6%	1
Spouted bowl (**136**)		1							1	0.6%	1
Teapot (**137**)		1							1	0.6%	1
Spouted jar or teapot (**138**)			1						1	0.6%	1
Jar (**139–141**)	3								3	1.9%	3
Pyxis (**142**)		1							1	0.9%	1

Table B.9, cont. Light-on-dark pottery, pale fabrics, EM III–MM II. For abbreviations, see caption of Table B.2.

Small Jugs in Pale Fabrics, MM I

The largest group of this class is decorated with diagonal lines in either dark-on-light or light-on-dark format (Table B.10; see also Ch. 8). Other groups include jugs with barbotine decoration, examples with large bosses on the body, and several other classes (Table B.11; see also Ch. 9). Find spots for the Diagonal Line Style and the Barbotine Style indicate that most of these juglets probably come from North-Central Crete.

Jugs in the Diagonal Line Style

The Diagonal Line Style is a decorative system used on a series of small- to medium-sized jugs. The vessels were made by hand and decorated with simple linear motifs painted diagonally across the body. Both dark-on-light and light-on-dark examples exist, but the dark-on-light ones are more common. Of the 135 sherds and vases, a minimum of 29 juglets in this class come from the cave (**143–162**). The jugs have constricted necks, which would cause the liquid inside to pour out slowly.

Jugs with Barbotine Decoration

The Barbotine Style from this site consists only of jugs with horizontal rows of small knobs. Too

few pieces are present for internal statistics. Two fabrics are present, a fine one used for small jugs and a slightly coarser version for the less common larger variety. A total of 11 sherds (MNV 3; **180–187**) come from the cave.

Small Jugs with Bosses on the Body

A small group of juglets has large bosses on the body. As with the other specialized vessels, the distinctive form is probably intended to make the class more recognizable as a label for its contents. The 22 sherds and vases represent a minimum number of six vessels with four cataloged pieces (**163–166**).

Various Classes of Small Jugs

About two-thirds of the imported small MM I jugs represent a series of disparate traditions. The pale fabrics of these vessels are not local to Lasithi or nearby parts of the Pediada, and several different sources are certain. Like the Diagonal Line Style, some of them surely come from somewhere near the northern coast of Crete. The 187 sherds and vases represent a minimum number of 56 vessels with less cataloged pieces (**167–179**).

Shape and Cataloged Examples	C	NC	R	H	B	S	L	Body	Total	%	MNV
MM IA											
DIAGONAL LINE STYLE, DARK-ON-LIGHT											
Small jug (**143–157**)	3			26	19			81	129	37.5%	24
Medium-sized jug (**159–161**)	1			1				2	4	1.2%	3
DIAGONAL LINE STYLE, LIGHT-ON-DARK											
Small jug (**158**)	1								1	0.3%	1
Medium-sized jug (**162**)	1								1	0.3%	1
MM IA–IB											
VARIOUS SMALL JUGS											
Jugs with bosses (**163–166**)	1			4	1			16	22	6.4%	6
Misc. small jugs (**167–179**)	7	2		74	5	11		88	187	54.5%	56

Table B.10. Diagonal Line Style and other small jugs, MM IA–IB. For abbreviations, see caption of Table B.2.

Shape and Cataloged Examples	C	NC	R	H	B	S	L	Body	Total	%	MNV
MM I											
BARBOTINE STYLE											
Jugs, fine fabric (**180–186**)		1			1			7	9	81.8%	2
Jug, coarse fabric (**187**)								2	2	18.2%	1

Table B.11. Barbotine Style, MM I. For abbreviations, see caption of Table B.2.

Knossian Pottery

The six sherds from Knossos are too few to present meaningful statistics in terms of the relative number of classes or shapes (Table B.12; see also Ch. 10). With a minimum number of six vessels, two closed vessels and four goblets are present, although only five are cataloged (**188–192**). The main conclusion from this data is that very few Knossian vessels reached Hagios Charalambos before the end of MM IIB.

Shape and Cataloged Examples	C	NC	R	H	B	S	L	Body	Total	%	MNV
EM II–MM II											
KNOSSIAN FABRIC											
Goblet (**188–190**)					4				4	66.7%	4
Open-spouted jar (**191**)			1						1	16.7%	1
Bridge-spouted vessel (**192**)						1			1	16.7%	1

Table B.12. Knossian pottery, EM II–MM II. For abbreviations, see caption of Table B.2.

Fine Red Pottery with Dark Red or Black Slip

The vessel shapes in this category are unusual, and many of them are unique to the deposit (Table B.13; see also Ch. 11).

Tumbler (6 sherds, MNV 5), **193–197**

Goblet with conical bowl (1 sherd), **198**

Two-handled cup (1 sherd), **199**

Tall carinated cup (13 sherds, MNV 5), **200**

Carinated cup (1 sherd), **201**

Miniature pithos (1 sherd), **202**

Small jug(?) (3 sherds, MNV 1), **203**

Jug with incised band (1 sherd), **204**

These pieces make up a very specialized class of import. A minimum of 16 vessels are present. All of them are small and delicate, with hard fabrics and lustrous red or black surfaces. They are made on the potter's wheel, and the quality of both manufacturing and firing is excellent. These vessels are all small skillfully made products imported for their high quality. Many of them are unique items for this deposit, and they are very different from the local vessels and from most other imported vases. They are elite products.

Shape and Cataloged Examples	C	NC	R	H	B	S	L	Body	Total	%	MNV
MM IB											
Fɪɴᴇ Rᴇᴅ Pᴏᴛᴛᴇʀʏ											
Tumbler (**193–197**)				5				1	6	22.2%	5
Goblet with conical bowl (**198**)					1				1	3.7%	1
Kantharos (**199**)			1						1	3.7%	1
Tall carinated cup (**200**)				3	4			6	13	48.1%	5
Carinated cup (**201**)								1	1	3.7%	1
Miniature pithos (**202**)								1	1	3.7%	1
Small jug(?) (**203**)								3	3	11.1%	1
Jug with incised band (**204**)		1							1	3.7%	1

Table B.13. Fine red pottery with dark red or black slip, MM IB. For abbreviations, see caption of Table B.2.

Vases in Fine Red Fabric with Parallels from Galatas

Vessels with parallels from Galatas are a distinctive class of import (Table B.14; see also Ch. 12). The most common shape is a tall closed vessel with a trefoil mouth.

Straight-sided cup (3 sherds, MNV 3), **205**

Jug with trefoil mouth (54 sherds and vases, MNV 26), **206–217**

Side-spouted jug or cup (5 vases, MNV 5), **218–222**

Shape and Cataloged Examples	C	NC	R	H	B	S	L	Body	Total	%	MNV
MM IB											
Pᴏᴛᴛᴇʀʏ ᴡɪᴛʜ Pᴀʀᴀʟʟᴇʟs ғʀᴏᴍ Gᴀʟᴀᴛᴀs											
Straight-sided cup (**205**)					3				3	4.8%	3
Jug with trefoil mouth (**206–217**)	5	7	12	6	17	1		6	54	87.1%	26
Side-spouted jug or cup (**218–222**)	4		1						5	8.1%	5

Table B.14. Pottery with parallels from Galatas, MM IB. For abbreviations, see caption of Table B.2.

Kamares Ware, MM IB–IIB

A small selection of pottery has polychrome decoration (Table B.15; see also Ch. 13). Of the 23 sherds and vases that make up the minimum of 17 vessels, six were cups, six were jugs, and five were jars (**223–232**). All of them were small, attractive elite products.

Shape and Cataloged Examples	C	NC	R	H	B	S	L	Body	Total	%	MNV
MM IB–II											
KAMARES WARE											
Straight sided-cup (**223**)				2				2	4	17.4%	1
Semiglobular cup (**224**)			1						1	4.3%	1
Carinated cup (**225**)			1						1	4.3%	1
Cup, unknown shape				1	1			3	5	21.7%	3
Jugs (**226–231**)	1			1		1		3	6	26.1%	6
Jar (**232**)			1	1					2	8.7%	1
Jars with horizontal handles				4					4	17.4%	4

Table B.15. Kamares Ware, MM IB–IIB. For abbreviations, see caption of Table B.2.

Chamaizi Pots, MM IIB

The Chamaizi pot is a small closed vessel with a rounded body, a vertical neck without any spout, and a small handle. The shape is not common at the cave, but it is very distinctive. Two sizes occur in the assemblage (Table B.16; see also Ch. 14)— small (30 sherds, MNV 11, **233–238**) and medium (1 sherd, MNV 1, **239**). A total of 97% of the sherds are in the smaller size.

Shape and Cataloged Examples	C	NC	R	H	B	S	L	Body	Total	%	MNV
MM IB											
PALE FABRIC											
Chamaizi pot, small (**233–238**)	1	1	7	9				12	30	96.8%	11
Chamaizi pot, medium size (**239**)			1						1	3.2%	1

Table B.16. Chamaizi pots, MM IIB. For abbreviations, see caption of Table B.2.

Pottery with Parallels from Malia, MM I–IIB

A total of 214 sherds (52 MNV) of the pottery in the cave has parallels from MM IB to MM II Malia (Table B.17; see also Ch. 15). This number is large in comparison with the other imported vases, but it represents only 1.2% of the sherds and vases in the ossuary. Fragments from carinated cups form the largest group (155 sherds, 38 MNV), but many different shapes are present:

Carinated cup (155 sherds, MNV 38), **248, 250–260**

Spouted two-handled cup, incised (4 sherds, MNV 4), **240–244**

Tall conical cup with grooves (1 sherd), **261**

Semiglobular cup with grooves (1 sherd), **249**

Cups with grooves (47 sherds, MNV 3), none cataloged

Jug with incised lines (4 sherds, MNV 3), **245–247**

Basket-shaped vessel (1 sherd), **263**

Scoop (1 sherd), **262**

The statistics make the pattern of this pottery extremely clear. Of the minimum total of 52 vases, 88% (46 individuals) are fancy drinking vessels, 8% (4 individuals) are unusual pouring vessels, and the remaining pieces are small pots of various shapes. In other words, 96% of this pottery was imported as elite drinking and serving vessels. This statistic is very interesting, and it contributes good information on the relationship between the community in Lasithi and the palace at Malia.

Shape and Cataloged Examples	C	NC	R	H	B	S	L	Body	Total	%	MNV
MM IB–IIB											
Pottery with Parallels from Malia, Pale Fabrics											
Spouted two-handled cup, incised (**240–244**)		2						2	4	1.9%	4
Jugs, incised (**245–247**)			1					3	4	1.9%	3
Carinated jug with wide mouth (**248**)		1							1	0.5%	1
Semiglobular cup with grooves (**249**)								1	1	0.5%	1
Carinated cup without grooves (**250**)			1						1	0.5%	1
Carinated cup with grooves, straight rim (**251–253**)			11	16				1	28	13.1%	15
Carinated cup with undulating rim (**254–256**)			6	2				1	9	4.2%	6
Carinated cup with grooves, unknown rim (**257–260**)			2	8	40			66	116	54.2%	15
Straight-sided cup with grooves (**261**)					1			1	0.5%		1
Cup with grooves, unknown shape			4	3				40	47	22.0%	3
Scoop (**262**)				1					1	0.5%	1
Basket-shaped vessel (**263**)				1					1	0.5%	1

Table B.17. Pottery with parallels from Malia, pale fabrics, MM I–IIB. For abbreviations, see caption of Table B.2.

Various Nonlocal Pale Colored Fabrics, EM III–MM IIB

Many of the vases in these miscellaneous classes of imported pottery cannot be readily identified in regard to shape (Table B.18; see also Ch. 16). The chart records 6,088 sherds, of which 2,351 (39%) are unidentifiable as to shape and an additional 328 (5%) are cups of unknown shape. They are still important because they allow a better computation of the amount of nonlocal ceramics. The minimum number of vessels from the entire chart is 403. By far the largest group that can be identified is the selection of jugs, which account for 3,255 sherds (54%). The other shapes represent a smaller percentage of the miscellaneous group.

Shape and Cataloged Examples	C	NC	R	H	B	S	L	Body	Total	%	MNV
EM III–MM IIB											
VARIOUS NONLOCAL FABRICS											
Shallow bowl (**264, 265**)			2					2	4	0.07%	2
Rounded cup (**266**)				1					1	0.02%	1
Conical cup (**267**)	1								1	0.02%	1
Cup, unknown shape (**268–271**)			27	75	29			197	328	5.4%	50
Cup, unknown shape, pronounced base					16				16	0.3%	6
Bowl (**272**)		1							1	0.02%	1
Miniature tripod vessel (**273**)							1		1	0.02%	1
Jugs (**274–285**)				366	538	172		2,179	3,255	53.5%	295
Jug or jar or cup, unknown shape			5	267	89			1,990	2,351	38.6%	—
Side-spouted vessel (**286**)			2	1	1				4	0.07%	3
Closed vessel (**287**)								1	1	0.02%	1
Bridge-spouted jar (**288, 289**)			5	1		1		1	8	0.13%	5
Side-spouted jug (**290**)						4			4	0.07%	4
Open-spouted vessel						4		1	5	0.02%	3
Spouted two-handled cup (**291–299**)			15	4	14	6		42	81	1.3%	20
Misc. jar (**300, 301**)			10	9		2		6	27	0.4%	10

Table B.18. Various nonlocal fabric groups, pale fabrics, EM III–MM IIB. For abbreviations, see caption of Table B.2.

The Lasithi Red Fabric Group, EM III–MM IIB

The regional production made in a coarse version of the Lasithi Red Fabric Group accounts for the largest percentage of pottery in the cave: 10,551 sherds and vases, or 1,153 as the minimum number of vessels (Table B.19; see also Ch. 17). This production can be distinguished from a finer textured and somewhat paler fabric with parallels from Galatas (Table B.14), even though both classes have geologically similar inclusions. Cups and jugs make up a majority of the production deposited at this site.

Within the total of 10,551 sherds and vases in this category, 5,073 entries for cups were present in the assemblage, representing a minimum of 585 vessels (**312–416, 418, 419, 421**). They were divided into the typological categories shown in Table B.20 for the sake of statistical examination. The division is based on the type of handle (circular cross section or elliptical cross section) and whether the handle is attached on the outside of the rim

or across the rim as well as whether the interior is left plain or is painted. The use of unpainted interiors is a chronological factor because it was abandoned early in MM I. The other characteristics are workshop traditions in manufacturing the cups, and they may represent different potters or different production traditions.

The next largest group after the cups consisted of jars. Few of these vessels were complete enough to recognize their shape. A total of 1,669 sherds were present, representing a minimum of 55 vessels (if they had only two handles) or fewer examples if they had multiple handles. The estimate of 33 MNV assumes an average of three handles per vessel. This category includes small jars that could have contained offerings as well as larger examples that may have been burial containers for the deceased.

The next largest class consists of jugs. Most of these jugs were large enough to be regarded as pouring vessels for beverages, but in many cases it was not possible to estimate the overall size based on small fragments. A total of 890 sherds of ordinary jugs were present, representing a minimum of 197 pouring vessels (**431–445**).

The only other large group consisted of shallow bowls. The 128 sherds came from a minimum of 22 bowls (**302–311**). These bowls were the local or regional version of the imported shallow bowls recorded on Tables B.9 and B.18. The shape was common in EM III to MM IA, but individual vessels can seldom be assigned a specific date if they come from mixed contexts like this ossuary. They represent a continuation of the contact with the Gulf of Mirabello region that began with Vasiliki Ware in EM IIB.

Shape and Cataloged Examples	C	NC	R	H	B	S	L	Body	Total	%	MNV
EM III–MM II											
LASITHI RED FABRIC GROUP											
Shallow conical bowl (**302–308**)			88	1				33	122	1.2%	20
Shallow conical bowl with tab handles (**309–311**)			3	2				1	6	0.06%	2
Rounded cup with lugs low on body (**312**)	1								1	0.01%	1
Cup Type 6 (**333**)		1	82	4	4			10	101	1.0%	5
Cup Type 7 (**314, 316**)			1	7					8	0.08%	8
Cup Type 8 (**319**)	1			2					3	0.03%	3
Cup Type 9 (**315, 317, 320, 312**)		2		8					10	0.1%	10
Cup Type 10 (**322, 328, 346, 401**)				4					4	0.04%	4
Cup Type 11 (**317, 324, 325, 399**)		1	81	2	8			1	93	0.9%	6
Cup Type 12			2					2	4	0.04%	2
Cup Type 16 (**313, 327**)		1	34	1	4			2	42	0.4%	4
Cup Type 17 (**402, 403**)			7	1	6			18	32	0.3%	4
Cup Type 18 (**397**)		3		27					30	0.3%	30
Cup Type 19 (**330, 349, 353–356, 386, 394**)	3	6		11					20	0.2%	20
Cup Type 20				4					4	0.04%	4
Cup Type 21			96	36	84			51	267	2.5%	36
Cup Type 23				9					9	0.1%	9
Cup Type 24 (**396**)		1		4					5	0.05%	5
Cup Type 25 (**408–410**)		3		6					9	0.1%	9
Cup Type 26 (**400**)	1			2					3	0.03%	3
Cup Type 27 (**395, 398**)	1		56	4	35			6	102	1.0%	6
Cup Type 28				2					2	0.02%	2
Cup Type 28 with spout (**414**)	1			1					2	0.02%	2
Cup Type 29				2					2	0.02%	2
Cup Type 30				1					1	0.01%	1
Cup Type 32			31					6	37	0.3%	—
Cup Type 33 (**404**)			11	1	44			25	81	0.8%	10
Cup Type 34 (**358, 360, 363–369, 375, 376, 378**)	8	1		38					47	0.4%	47
Cup Type 34 with spout (**377**)		1							1	0.01%	1
Cup Type 35 (**380, 382, 415**)	1	3		33					37	0.3%	37
Cup Type 36				4					4	0.04%	4

Table B.19. Lasithi Red Fabric Group, EM III–MM II. For abbreviations, see caption of Table B.2

Shape and Cataloged Examples	C	NC	R	H	B	S	L	Body	Total	%	MNV
EM III–MM II, cont.											
LASITHI RED FABRIC GROUP, CONT.											
Cup Type 37 (**370–373**, **385**)		2	357	37	391			249	1,036	10.1%	39
Cup Type 38 with spout		1	1			5			7	0.07%	6
Cup Type 39				7					7	0.07%	7
Cup Type 40		1		5					6	0.06%	6
Cup Type 41				8					8	0.08%	8
Cup Type 42				2					2	0.02%	2
Cup Type 43			241	14	133			29	417	4.0%	14
Cup Type 43 with spout						1			1	0.01%	1
Cup Type 44				1					1	0.01%	1
Cup Type 46				6					6	0.06%	6
Cup Type 48			148	2	1			6	157	1.5%	2
Cup Type 49			310		303			1,170	1,783	16.9%	—
Cup Type 50				226					226	2.1%	75
Cup Type 51				415					415	3.9%	138
Miniature tumbler (**416**)				1					1	0.01%	1
Brazier (**417**)				1					1	0.01%	1
Wide-mouthed cup (**418**)				1					1	0.01%	1
Cup without handle (**419**)		1							1	0.01%	1
Vessel with loop handles (**420**)				1					1	0.01%	1
Miniature cup (**421**)	1								1	0.01%	1
Tripod vessels (**422–427**)			4	6	2		29		41	0.4%	12
Cooking pot without handles (**428**)		1							1	0.01%	1
Small tripod vessel (**429**)					1				1	0.01%	1
Cooking dish (**458**)					1				1	0.01%	1
Jug or Kantharos (**430**)			2	2				7	11	0.1%	3
Jug (**431–445**)	5	13	72	179	199	26		396	890	8.5%	197
Vessel with knob on handle				1					1	0.01%	1
Tube-spouted and bridge-spouted jugs (**446–449**)	2		19	7		6			34	0.3%	12
Closed vessel (**450**)			1						1	0.01%	1
Jar with flat ledge rim (**451**)			8						8	0.08%	3
Jar with ropework decoration (**452**)				3	2	1		12	18	0.2%	6
Various jars (**453–457**)	1		20	109	259	37		1,243	1,669	15.7%	33
Vessels with uncertain shape (mostly cup and jugs)				798	50			1,145	1,993	18.8%	266
Various closed shapes								642	642	6.1%	—
Jar or cup				15				55	70	0.7%	6
Lid (**459**)	2			1					3	0.03%	2

Table B.19, cont. Lasithi Red Fabric Group, EM III–MM II. For abbreviations, see caption of Table B.2

Cup typology by date	Rounded handle		Elliptical/strap handle		No handle preserved	Fragments
	Outside rim	Across rim	Ouside rim	Across rim		
EM III–MM IA						
NO HANDLE, WITH BOSS						
Rounded shape	—	—	—	—	—	Type 1
PAINT INSIDE RIM						
Rounded shape	Type 2	Type 3	Type 4	Type 5	Type 6	—
Straight-sided with conical shape	Type 7	Type 8	Type 9	Type 10	Type 11	—
Flaring rim shape	Type 12	Type 13	Type 14	Type 15	Type 16	—
Unknown shape	—	—	—	—	—	Type 17
MM I–II						
FULLY PAINTED INTERIOR						
Rounded shape	Type 18	Type 19	Type 20	Type 21	Type 22	—
Straight-sided with conical shape	Type 23	Type 24	Type 25	Type 26	Type 27	—
Flaring rim shape	Type 28	Type 29	Type 30	Type 31	Type 32	—
Unknown shape	—	—	—	—	—	Type 33
EM II–MM II						
PAINT MISSING						
Rounded shape	Type 34	Type 35	Type 36	Type 37	Type 38	—
Straight-sided with conical shape	Type 39	Type 40	Type 41	Type 42	Type 43	—
Flaring rim shape	Type 44	Type 45	Type 46	Type 47	Type 48	—
Unknown shape	—	—	—	—	—	Type 49
PAINTED						
Handles only, cup shape unknown	—	—	—	—	—	Type 50
PAINT MISSING						
Handles only, cup shape unknown	—	—	—	—	—	Type 51

Table B.20. Typology of local cups in the Lasithi Red Fabric Group. Many are painted with White-on-Dark decoration.

LM III Pottery from Outside the Cave

A few LM III sherds were found by Costis Davaras when he excavated several trenches in the vicinity of the cave in 1976 in attempts to find the original entrance (Table B.21; for discussion, see Davaras 1976; Betancourt and Poulou-Papadimitriou 2014, 39). The entrance was not found at that time (it was revealed in 1982). The sherds demonstrate that people visited the vicinity of the cave after it was no longer being used for burial. Too few sherds are present for meaningful statistics. These pieces can be excluded from the statistics used for pottery deposited inside the cave.

Shape and Excavation Numbers	w	R	H	B	S	L	Body	Total	%	MNV
LM III Pottery										
Kylix (HCH 46, HCH 87)				2				2	50%	2
Flask (HCH 04-392)				1				1	25%	1
Stirrup jar (HCH 140)	1							1	25%	1

Table B.21. LM III pottery from outside the cave. For abbreviations, see Table B.2.

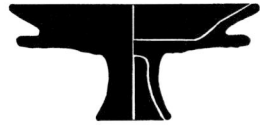

Appendix C

The Elusive Site of the Primary Burials of the Hagios Charalambos Cave: A Speculative Scenario

Costis Davaras

Quite naturally, the site of the primary inhumations deposited as secondary burials in the Hagios Charalambos Cave is considered to have been an early circular or tholos tomb in the area of the highland plain of Lasithi and its hilly border, today undiscovered. However, in hindsight, this very sensible idea seems to me to be far from certain.

First of all, this area, rather intensively surveyed in the recent past (Watrous 1982; cf. already Pendlebury 1936–1937), especially the area near the cave, has not yielded any evidence about the existence of such a big tholos tomb, which should have been of considerable dimensions if we judge from the great amount of pottery, as published in this volume, and mainly of that of the skulls and long bones selected to be transferred to the cave and deposited with care and respect. Such a big tholos tomb, as implicated, obviously not smaller than the largest ones in the Mesara area, could hardly have disappeared, if it ever existed at some "reasonable" distance from the cave, without leaving some visible architectural traces.

Secondly, we are here far away from the Mesara plain and the Asterousia Mountains, home of the Prepalatial and later tholos tombs; the nearest one is the isolated tholos tomb at Krasi Pediados, at the foothills leading up to the upland plateau of Lasithi, indeed not very far away despite mountainous barriers in between, a not typical but rather idiosyncratic tomb, even thought earlier to have been used by a Cycladic colony on Crete. Certainly, albeit the recent discovery of tholos tombs as far away as the Siteia district and their exemplary publication in a short time (Papadatos and Sofianou 2013) have demonstrated that distance from the Mesara area is not an absolutely prohibitive factor for their distribution, we surely must accept that the probability of discovering a tomb of this category diminishes with distance, especially when no sea routes were available.

Thirdly—and most importantly—we should consider the possibility as to whether such a quite unusual—indeed unheard of—funerary practice for tholos tombs ever existed, namely to remove the bones of the primary burials and their accompanying grave goods from inside the tholos to a remote cave instead of building a proper ossuary, a rectangular house tomb adjacent to it, following the architectural rule concerning these tombs.

According to the usual burial custom observed in the much later rock-cut chamber tombs regarding the earlier depositions, along with their *kterismata*, being carelessly pushed or swept aside, toward the tomb's walls, to make room for new corpses, it is generally accepted that the same should have happened in the earlier Mesara-type tholoi (cf., e.g., Peatfield 1987, 90; Murphy 2011, 38–40). Nevertheless, this is far from certain, due to the fact that these circular, large structures were used for several centuries, receiving the mortal remains of a great number of departed, thus making it difficult, if not impossible, to distinguish between primary and "secondary" burials. In fact, we cannot even be sure whether any primary burials took place at all inside the tholoi: the osteological material found during the excavations gives the strong impression that it belongs to a filled up ossuary, exactly like the house tombs. Even if some burials show an intact, articulated skeleton, as very rarely happens, this could perhaps be explained as a latest, exceptional inhumation, rather as an "intruder," an exception contradicting the rule. One has to ponder on the unambiguous statement that "[t]o day no Prepalatial or Protopalatial primary burial has been securely identified" (Hatzaki 2012, 308). Indeed, "ethnographic studies ...have established that secondary manipulation is a transcultural form of reintegration rite for the mourners" (Flouda 2014). The situation being so complicated, in order to be relatively sure, we should expect the discovery of a new Prepalatial tholos tomb to be excavated, unlike those explored in the twentieth century, by modern archaeologists applying extremely sophisticated excavation methods for the osteological material like those recently used by a Belgian team at Sissi, Central-Eastern Crete (Schoep, Crevecoeur, and Schmitt 2015). Here we should remark that certainly not the totality of the grave goods belonging to the primary burials was accompanying the selected bones to their final resting place. Scholarship concurs that in such cases the most precious or useful items, especially the ones of metal, were withheld or looted by the living.

On the other hand, as has been stressed, the vessels deposited in our cavern are not representative of the pottery assemblage found in settlements but were either selected as a funerary offering to the dead or even used in ceremonies associated with the funerary rites (Nodarou, this volume), obviously accompanying the primary inhumations.

Probably the extent of the secondary burial at a final resting place—evidently associated with the realm of the ancestors—as a funerary custom in Prepalatial Crete has not been sufficiently appreciated in its totality. The situation of the large EM I–II cemetery at Hagia Photia where, contrary to my first impressions, exclusively secondary burials were practiced, persuaded me that the extent of this burial custom has been somehow underestimated. But that is quite beyond the scope of this appendix.

What is here in question is, of course, why in the case of the Hagios Charalambos Cave ossuary was it felt necessary to evacuate the place of the primary inhumations: could that be that this was a distant Mesara-type tholos tomb? I feel that this was not the case. In short, I posit that an alternative, indeed a speculative scenario could be considered instead: that this place was not a built tomb but a natural receptacle for inhumations, traditional in Neolithic and Prepalatial Crete, a cave, exactly like the contemporary Trapeza Cave, just at the opposite side of the Lasithi *altiplano*. Anyway, caves were, by their own nature, liminal places par excellence (cf. Shelmerdine 2008, 86).

If that were really the case, then this transfer of the primary burials should be examined more carefully. In any case, I am persuaded that the grave goods must belong to the primary deposits, accompanying their proprietors in the inhumation or even used ritually during it, and certainly not, as sometimes thought, after the decomposition of the body and the disarticulation of the skeleton and the later removal of the bones, in a secondary burial.

Let us say, parenthetically, that especially regarding the six sistra, found in unheard of abundance here and discussed in the previous volume (cf. Betancourt 2011, 3–4), I believe that they were not used as musical instruments accompanying funerary rites. On the contrary, they should have belonged to the departed as objects of their status and professional equipment in life, particularly associated with them, not as musicians but rather as important religious persons. The sistra would have been used to perform some ritual action where its rhythmic sound should be very appropriate for the purpose of inducing with their rhythmic sonic drive an ASC (Altered State of Consciousness; cf. Tyree 2001, 41 n. 19; Morris and Peatfield 2002; 2006; Hayden 2003; Herva 2006, 591, 594; cf. Rehak and Younger 1998, 165; about gourd-rattles as magical devices, see Harrison 1912, 78). The sistron players would have performed as shamans who are thought to have being performing even within Minoan caves (Tully 2015), a trenchant hypothesis. Shamans have been thought to exist not only in Minoan Crete (Schefold 1960, 3) but also in the early Cyclades (Renfrew 1984, 27).

The only extant example in Minoan iconography of a sistron is, of course, the famous Harvester Vase from Hagia Triada, originally covered with a gold foil (Bucholz 1999, 574), seen functioning in the hand of a prominent man with open mouth, "chanting or shouting," who was followed by a group of chanting persons, giving them the rhythm. Surely this explanation is more plausible than making of the sistron-shaking man a mere secular *Kapellmeister* or a "rustic coryphaeus" from tragedy, according to Evans (cf. Marinatos 2010, 21). On the contrary, I should rather prefer to see here two independent—albeit interdependent—ritual actions: on the one hand the choir chanting a hymn, evidently relative to the harvest in process, and on the other the sistron-player functioning in some magical, shamanic act connected here with harvest and fertility (cf. Forsdyke 1954, 7: ". . . not a musical instrument in a civilized sense . . . an essential feature of fertility rites . . ."), such as communicating in an ASC with the divinity of fertility. Perhaps an epiphany was evoked. As we know, in Egypt, the country of origin for this instrument (cf. Watrous 1995, 397), idiophones like sistra "were particularly associated with religious worship" (Shaw and Nicholson 2008, 214). Their origin might have been a bundle of papyroi with a rushing sound (Larkes 1980, 75). An early scholar, A. Erman (Erman 1905, 71), went so far as to believe that the Harvester Vase was an Egyptian import, not a Minoan masterpiece. On the other side, Xanthoudides saw Cretans armed with tridents (Xanthoudides 1904, 130).

Be that as it may, to return to our subject, if our proposition about some cave and not a tholos tomb being the place of the primary burials could be considered as an acceptable working hypothesis, which cavern could be seen as a likely candidate? With a leap of imagination, I dare to suggest the venerable Psychro Cave itself. In favor of this, one could enumerate the following arguments:

1. The distance between the two caves is actually a short one, about 1 km, bringing the transfer of the bones and their paraphernalia in one or rather several ritual processions into the realm of possibility, if not probability.

2. The chronological concordance between the two caves coincides rather well, although not exactly: The Psychro Cave started to function as a sacred cave about MM I (Tyree 2001, 40 n. 11; pers. comm. 2015), whereas the Hagios Charalambos pottery covers the period from FN to MM IIB.

3. There are no inhumations in the Psychro Cave, as one should normally expect following the example of other caverns, e.g., that of the nearby Trapeza Cave, except for a few isolated EM I inhumations found at some spots in the deepest layers (Faure 1964, 68). Tyree herself is very skeptical about the existence at all of human bones at Psychro (pers. comm. 2015).

4. It is conceivable that the Psychro Cave was originally—as expected—used as a burial cave, but the inhumations (selected bones and grave goods) were transferred to another cavern. The Hagios Charalambos Cave is actually both the nearest one and the best suited in order to "purify" it so that it could function properly as a sanctuary. Probably this exhumation process and subsequent transfer took some considerable span of time, thus explaining the discrepance between the latest deposits at Hagios Charalambos and the earliest ones of the Psychro sanctuary.

The establishment of both of the most characteristic worship places of Minoan religion, the peak sanctuaries and the sacred caves, is a collateral phenomenon with a major event in Cretan history which occurred in MM IB, the emergence of the palaces, coincident with the establishment of a system of centralized economy. It is clear that at this time, historically so important, people must have been looking for the most appropriate places to found the aforementioned new type of sanctuaries. The Psychro Cave was a first class candidate for this purpose, indeed the best one on the island, as far as size, grandeur, depth, and stalactitic/sta-lagmitic decoration. It had copious amounts of water inside, both dripping and in pools. It met all the criteria set by the Cretan caves expert, Loeta Tyree, for choosing a cavern to be turned into a sacred cave, including that of a steep downward sloping passage (Tyree 2013, 177–179).

One can then presume that after the founding of the Psychro sanctuary and the starting of bringing and deposing offerings, the old, traditional custom of inhumations there did not continue for a long time. It was eventually interrupted at some point during MM IIB, and a decision was made by some high religious authority, probably residing in the first "palace" of Malia where the highland plain belonged (Knappett 1999 625, fig. 6; Betancourt 2007) and whence much of the earlier pottery comes (Watrous 2004). A decision was made to clean the grotto from the burials in order to "purify" it from the "miasma" of human death and leave it free for the worship of the divine. In this process, a few, if any, earlier burials in the deepest layers and recesses of the cavern could perhaps have escaped removal.

If our speculative scenario comes near the truth, one could ask if there is some historic parallel, which warrants emphasis. Indeed, about a millennium later, we see the "purification" of the island of Delos, birthplace of the divine twins Apollo and Artemis, in an attempt to render Delos suitable for a proper worship of them: the existing burials had to be removed and reburied in the nearby island of Rheneia. These secondary burials were excavated by the École française d'Athènes (EFA) in 1898–1899: "Aussi était-il defendu de naître et mourir à Délos: c'est donc dans l'île d'en face, à Rhénée qu'on voyait le jour et rendait l'âme" (Bruneau and Ducat 1983, 33). This purification took place in two phases: a partial one was executed in the 6th century B.C., during the rule of Peisistratos, who ordered that all tombs within sight around the temple, as incompatible to it, should be dug up and the contents should be transferred and reburied. A complete purge of the entire island of all dead was done about one century later, in 426 B.C. during the Peloponnesian war under instruction of the Delphic Oracle (Thuc. 3.104). On the thus totally purged Delos, it was characteristically ordered that no one any more should be permitted to either die or be born.

By the by, Delos has some strange connections with Crete in other ways: Among its numerous outstanding ruins is a monumental public fountain, the "Minoan Fountain," as it was enigmatically called. According to the official guide of the EFA, the name does not implicate any particular reference to the pre-Hellenic era (Bruneau and Ducat 1983, 142–144: "Fontaine Minoé"), but, prudently, no explanation is offered. Surely, we know that "there is no Bronze Age attestation at all for the name of Minos" (Burkert 2004, 14). However, an explanation is needed for that name referring to the glorious past of Crete, besides the additional fact that an inscription found here mentions three "Nymphes Minoïdes." Obviously the name does not implicate any direct connection with Bronze Age Crete, but, at least, some link to the island must exist, not the contemporary one to Delos but certainly an earlier link, at least some Mycenaean connection (Bruneau and Ducat 1983, 18).

Less cautiously Evans, after narrating how Theseus arrived at Delos after his victory over the Minotaur, and after setting up the image of Aphrodite and the horned Keraton altar and dancing the Geranos Dance with the other young men who had survived the labyrinth (Plut. *Vit. Thes.*; cf. Lawler 1946), and subsequently marrying Ariadne, speaks about "the interesting fact that there was a fountain called Minoê in Delos and that 'Minoid nymphs' were here adored" (Evans 1930, 74 n. 6). Very sensibly, he pushed the issue no further, certainly realizing that one cannot know and explain everything about ancient religion. This fountain is of a "current type" (Bruneau and Ducat 1983, 144, fig. 29), although with its descending flight of stairs, here broad and straight, it somehow resembles a Minoan lustral basin, probably quite immaterially as nobody on Delos had ever seen such a basin, unless, of course, an idea of it had somehow survived. However, at the "palace" of Zakros we encounter idiosyncratic built pools known as "spring chamber, built well, and fountain," with straight, rather broad flights of stairs descending likewise to the water surface, unattested in the other Minoan "palaces."

In this connection, the four Hyperborean Maidens should also be mentioned, whose *theke* here has been compared to the antechambers of the Minoan tholos tombs as well as other contact points with Crete (Long 1961), beside the fact that this enigmatic Hyperborean country is considered to be a doublet of Crete (Picard 1927, 352, following R. Vallois).

Be that as it may, if no case for a closer connection with Crete can be made, a subject examined already several decades ago (Picard 1929, 2; Gallet de Santerre 1948; Long 1961; cf. Davis 1992, 727), one could not avoid connecting Delos with Daskalio (Italian *scoglio*, "precipitous;" therefore, the established misspelling *Daskaleio* is a clear case of *Volksetymologie*) Cavos at nearby Keros, an EC I and later place of pilgrimage. It is "the oldest maritime sanctuary" (Renfrew, Boyd, and Ramsey 2012), an *Ur-Delos* as it were, preceding it by some 2,000 years but roughly contemporary with the Hagia Photia cemetery on Crete where the Cycladic presence is strongly felt. At Keros marble vessels and figurines were found in abundance, ritually broken and deliberately deposited, not unlike the votive offerings at the later Minoan peak sanctuaries. Consequently, an evasive religious past of hoary antiquity leaving visible traces here and there in the Aegean cannot be excluded.

Another link of Delos with early Crete could be in the instrumental role played by Eileithyia—a divinity especially worshipped in Crete—in the very *raison d'être* as a sanctuary of the island, who suddenly [appeared?] [emerged?] from the sea, namely the birth of Apollo and Artemis. Their birth had a miraculous character for the ancients (Bruneau and Ducat 1983, 32), something like our Christmas we might say: in this issue, we all can agree that "the links between Eileithyia, an earlier Minoan goddess, and a still earlier Neolithic prototype are, relatively, firm" (Willetts 1958, 221; cf. Picard 1927, 352; see also Lawler 1954, 104, for possible association with the bee).

Another parallel between Delos and the Dictaean Cave is the prohibition of dying at the place of the divine birth. The identification of the Dictaean Cave is uncertain, and opinions differ about whether it can be identified with the Psychro Cave. According to Antoninus Liberalis (*Met.* 19: "ἐν Κρήτη μέγεται εἶναι ἱερὸν ἄντρον μελισσῶν. . . κατέχουσι δὲ τό ἄντρον ἱεραὶ μέλισσαι, τροφοὶ τοῦ Διὸς,)" four rascals, Laios, Kerberos, Keleos, and Aigolios, disastrously invaded the sacred cave of Zeus to steal the honey on which the infant god was nurtured" (Crowther 1988, 40), putting on bronze armor to protect themselves from the sacred bees. But as none could die in the cave

of Zeus' birth, Themis and the Fates turned them into various kinds of birds.

Treading on very thin ice, one might here wonder whether the famous Protopalatial golden Bee Pendant from the Chrysolakkos at Malia could possibly belong as an *insignium dignitatis* to some high "Priestess of the Sacred Bees" of the Dictaean Cave, a sanctuary that perhaps belonged to the religious authority of the Malian "palace." It was a fortunate stroke of serendipity for an object like this to escape looting by the excessive zeal of the active locals, who after having "fort ruiné cette construction . . . y trouvèrent quantité d'objets d'or qu'ils durent porter à Candie pour les faire fonder . . ." (Demargne 1930, 405). This correlation can only be true, of course if this Greek tradition about sacred bees could go so far back in time, and if we can finally agree with Sinclair Hood (Hood 1976), most adequately argued, and recently with Janice Crowley (Crowley 2014, 133), that this masterpiece of jewelry shows really the beneficent and blessed honey bees, "signifying immortality and resurrection" (cf. also Laffineur 1985, 255; Dietrich 1990, 14), and not the obnoxious carnivorous wasps, as opinions continue to differ. Generally, one even today insistently avoids putting this item in the sacred or secular sphere, albeit a lengthy paper has been written about it (Bloedow and Björk 1989; cf. Richards-Mantzoulinou 1979).

Let us note, that the tiny golden sphere inside a "cage" with golden wires as "bars" moving freely inside it and producing thus a "whisper," could well be considered as a miniscule, symbolic sistron above the pendant (cf. Laffineur 1985, 260 and n. 109), thus completing its "bee" symbolism, and highlighting obviously one of the functions of the person carrying it.

Therefore, this said, if we can readily agree with Harissis and Harissis (Harissis and Harissis 2009), and especially with Crowley (Crowley 2014), that certain dots on some Minoan golden signet rings are a "shorthand" rendition of bees, of course in the appropriate proportions in the scene, "as seems quite reasonable," in the frame of ritual scenes on them, along with beehives, as an icon, a memorable unit beside other, top Minoan religious icons ("kneeling to boulder, pulling the tree, the bird as flying messenger, serving at the shrine, and greeting the deity appearing on high;" Crowley 2014, 137), this telling, frequent presence of bees acquires a deeper, religiously more meaningful, "holistic" background.

Certainly, entomologists may be scientifically right, that the anatomy of these insects corresponds rather to that of "social wasp *polistes*" (LaFleur et al. 1979, with earlier interpretations), or "queen-bees" (Matz 1962, 436, 441, pl. 33), in this "supreme achievement of EM goldworking" (K. Branigan); but the excavator points out aptly that "il nous semble possible de lui trouver un sens dans la vie des abeilles" (Demargne 1930, 414), referring also to the Egyptian and Cretan hieroglyphic system (Evans 1909, 212, no. 86; Crowley 2014, 133; cf. Grumach 1964 about a honey sign). Of course, the Bee Pendant is of an Egyptian inspiration (Poursat 1985, 54–55: "abeilles"; cf. Somville 1978, 130, fig. 1; Watrous 1995, 397) as well as the corresponding Minoan syllabogram (cf. Woud Ruisen 1997). Demargne, interpreting this jewel in the frame of the Minoan spirit, reasonably believes that it was "destinée à quelque personne de grande condition" (Demargne 1930, 416).

The question of the sacred bees and the religious meaning of honey has often been examined in the past (e.g., Elderkin 1939; Somville 1978; Davaras 1984; 2003, 148–149, s.v. Honig; Giuman 2009). At Delphi, Pythia herself was tellingly called *Melissa*, like the priestesses of Artemis at Ephesos, the "Delphic Bee" ("μελίσσα Δελφίς;" see Pind. *Pyth.* 4.60; cf. *Ol.* 6.47; cf. Somville 1978, 132; cf. Dietrich 1990, 7; cf. Elderkin 1939, 203; cf. Lowler 1954, 103) clearly a pre-Greek word (LaFleur, Matthews, and McCorkle 1979, 208 n. 10).

Now, it should be of the greatest importance if one could identify with some assurance the Dictaean Cave with that of Psychro; unfortunately, scholarly opinions vary considerably. Here is hardly the place to discuss at length this moot problem or make detailed references to the eminent scholars who are *pro* or *contra*. Assuredly, it is not a question of a majority of opinions or eminence in scholarship but of value of arguments, which cannot be discussed here. An examination has been done some decades ago (Willetts 1962, 215–216).

Hogarth, the main excavator of Psychro, followed by Cretologists, to name just a few, like Xanthoudides (1904, passim), Evans (1921, 159, 163), Guarducci, Boardman (1961), Alexiou (1964, 70, with some hesitation), Kenna (1966, 33); Warren (1990, 67), Rehak and Younger (1998, 143),

and Hitchcock (2007, fig. 10.1), made a good case for his time: "The Lassithi *massif*, with its enclosed lacustine plain, is the ancient Dicte, placed by Strabo 1,000 stades east of Ida" (Hogarth 1899–1900, 45), and not far from Lyttos, of course. Following Dionysius of Halikarnassus, who pointed out a deep and scarcely accessible cavern, he concluded that "the Dictaean was the prolific Psychro Cave . . . For no other natural grot exists in the Lassithi *massif*, certainly none to be compared for size and beauty" where Minos went down into the ἱερὸν ἄντρον and reappeared with the Law."

Here, let us note that obviously this massif, the third highest of Crete, was named "Dicte"— the official name today—after this identification of the Psychro Cave with the Dictaean Cave, a name established in archaeological papers needing a name for the mountain (e.g., Rehak and Younger 1998, 942; Shelmerdine 2008, 78: "Lasithi (Dictaean) Mountains . . . a formidable barrier between central and east Crete . . ."). Early travelers called it just the "Lasethe Mountains" (Spratt 1865, passim).

Other scholars, early or recent, do not accept at all this identification (among them Cook 1914–1925, 926; Nilsson 1925; 367; S. Marinatos, in favor of the Arkalochori Cave instead; Long 1958, 36, n. 2; Faure 1964, passim, *in extenso*; Burkert 1985, 25 and n. 8; Crowther 1988; Rutkowski and Nowicki 1996; Mazarakis Ainian 1997, 334; L. Tyree, pers. comm. 2015). Very skeptical about the cave is Thorne (2000, 160 n. 50), who places firmly the sanctuary of Dictaean Zeus (*di-ka-ka-jo di-we* in the Knossian Linear B tablet Fp1) at Palaikastro. Of course, many scholars do not commit themselves. The subject has also been examined by others (Bosanquet and Hutchinson 1939–1940). Anyway, a recent monograph about the Psychro Cave, in fact the only one, has the eloquent title *The Cave Sanctuary of Zeus at Psychro* (Watrous 1996). On her part, Tyree believes that it was a Zeus cave but not that of Dictaean Zeus (L. Tyree, pers. comm. 2015), followed by Casson (1938, 368); a suggestion that has not been retained in a somewhat parallel way with Watrous, who thinks that the Psychro Cave is not in the Dictaean Cave of the historical era, but "however, remains the best candidate for the shrine of Diktaian Zeus mentioned in the Linear B tablets (Fp1: *di-ka-ta-jo di-we*) and Hesiod [*Theog.* 481–484]"

(Watrous 1982, 62). In any case, it has been maintained that the cult of Zeus in Crete was earlier than previously thought, and connected to the Minoan peak sanctuaries (Bloedow 1991). This controversy need not concern us here any longer.

The fact remains that we are speaking about a big cave, an ἄντρον, which does not exist east of Psychro, unless we could take in consideration the nearby Trapeza Cave, much smaller, simpler and shallow, which the locals of Tzermiado, in touristic rivalry with the villagers of Psychro, named in road signs "Kronion," the "sacred cave of Kronos"! The Archaeological Service has been unable to remove these road signs, which fake Greek mythology and perplex the visitors.

The identification of the Dictaean Cave is a matter of considerable scholarly debate that bristles with difficulties, beyond our discussion here. At any rate, perhaps just one thought-provoking point, somehow overlooked, if at all relevant, could be mentioned, for what it is worth: that the Harpies, pursued in the air by the Argonauts Boreads, sons of Boreas, sword in hand, were saved in the last moment through the intervention of Iris, sent by Zeus, over the Strophades islet near Zakynthos (Ap. Rhod. 2.291–300). Then these horrible birds, "the dogs of great Zeus," αἱ μὲν ἔδυσαν κευθμῶνα Κρήτης Μινωίδος, "they entered their den in Minoan Crete" (trans. R.S. Seaton, Cambridge, MA, 1912). Let us note that the word ἅρπυια is of uncertain etymology, perhaps of pre-Greek origin according to a recent and competent opinion (Beekes 2010, 139). Already Spyridon Marinatos, almost one century ago, has been able to distinguish in Minoan iconography some winged creatures, and he maintained that they could be the precursors of the Greek Harpies (Marinatos 1927–28, 23).

Actually this Harpies' grotto has been duly mentioned by Faure as unindentifiable. However, the most likely candidate that the Ancients could have in mind should be the Psychro Cave, being at the same time the most impressive, deepest and difficult to approach of all Cretan caves, beyond featuring a really liminal entrance. Interestingly, Evans noted about other birds, doves, that, "In Crete, indeed . . . swarms of these birds still haunt the inmost clefts of such old centres of Minoan cult such as the Dictaean and Kamares Caves or the great vault of Skoteino . . ." (Evans 1936, 411). Of them, to meet the above criteria, Kamares is too shallow, and Skoteino is not on a mountain, too easily approachable by working farmers. Even the Idaian Cave is rather a big rock-shelter than a real cave, as Tyree comments, and should not be considered. Surely, the Harpies' den should not necessarily coincide with the Dictaean Cave, even if that could be the Psychro Cave. Perhaps it was just an "ideal," mythological Cretan cave in the mind of the Ancients. Perhaps future research in the Psychro Cave will bring some welcome evidence on the matter. For the time being, lacking solid evidence, suffice it to say that "the Diktaian Cave regrettably eludes discovery" (Tyree 2001, 43).

However, if our scenario comes to be near the truth, that a burial cave was eventually converted into a major sanctuary, perhaps as the place of a divine birth, then we can quote a dictum of Lurker: "the cavern was closely associated with the 'great mother' archetype. The female image of the hollow space was the seat of birth and death" (Lurker 1980, 39).

References

Alexiou, S. 1964. Μινωικὸς πολιτισμός, Herakleion.

Beekes, R.S.P. 2010. *Etymological Dictionary of Greek* (*Leiden Indo-European Etymological Dictionary Series* 10 [1–2]), Leiden.

Betancourt, P.P. 2007. "Lasithi and the Malia-Lasithi State," in *Krinoi kai Limenes. Studies in Honor of Joseph and Maria Shaw* (*Prehistory Monographs* 22), P.P. Betancourt, M.C. Nelson, and H. Williams, eds., Philadelphia, pp. 209–220.

———. 2011. "Newly Excavated Artifacts from Hagios Charalambos, Crete, with Egyptian Connections," *Journal of Ancient Egyptian Interconnections* 3 (2), pp. 1–5.

Bloedow, E.F. 1991. "Evidence for an Early Date for the Cult of Cretan Zeus," *Kernos* 4, pp. 139–177.

Bloedow, E.F., and C. Björk. 1989. "The Mallia Pendant: A Study in Iconography and Minoan Religion," *SMEA* 27, pp. 9–62.

Boardman, J. 1961. *The Cretan Collection in Oxford: The Dictaean Cave and Iron Age Crete*, Oxford.

Bosanquet, R., and R.W. Hutchinson. 1939–1940. "Dicte and the Temple of Dictaean Zeus," *BSA* 40, pp. 60–77.

Bruneau, P., and J. Ducat. 1983. *Guide de Délos* (*Sites et monuments* 1), Paris.

Bucholz, H.G., 1999. *Ugarit, Zypern und Ägäis. Kulturbeziehungen im zweiten Jahrtausend v. Chr.*, Münster.

Burkert, W. 2004. "Epiphanies and Signs of Power: Minoan Suggestions and Comparative Evidence," *Illinois Classical Studies* 29, pp. 1–23.

———. 1985. *Greek Religion*, Cambridge, MA.

Casson, S. 1938. "Archaeological Discovery 1936–1937," *The American Scholar* 7, pp. 368–373.

Cook, A.B. 1914–1925. *Zeus: A Study in Ancient Greek Religion* I–III, Cambridge.

Crowley, J.L. 2014. "Beehives and Bees in Gold Signet Ring Designs," in *KE-RA-ME-JA. Studies Presented to Cynthia W. Shelmerdine* (*Prehistory Monographs* 46), D. Nakassis, J. Gulizio, and S. James, eds., Philadelphia, pp. 129–139.

Crowther, C. 1988. "A Note on Minoan Dikta," *BSA* 83, pp. 37–44.

Davaras, C. 2003. *Führer zu den Altertümern Kretas*, Athens.

———. 1984. "Μινωικὸ κηριοφόρο πλοιάριο τῆς Συλλογῆς Μητσοτάκη," *ArchEph* 123, pp. 55–95.

Davis, J.L. 1992. "Review of Aegean Prehistory I: The Islands of the Aegean," *AJA* 96, pp. 699–756.

Demargne, P. 1930. "Bijoux minoens de Mallia," *BCH* 54, pp. 401–421.

Dietrich, B.C. 1990. "Early Oracular Practice: Inspiration by Divination," *Mediterranean Studies* 2, pp. 7–20.

Elderkin, G.W. 1939. "The Bee of Artemis," *AJP* 60, pp. 203–213.

Erman, A. 1905. *Die ägyptische Religion*, Berlin.

Evans, A.J. 1909. *Scripta Minoa: The Written Documents of Minoan Crete.* I: *The Hieroglyphic and Primitive Linear Classes*, Oxford.

———. 1921. *The Palace of Minos at Knossos* I, London.

———. 1930. *The Palace of Minos at Knossos* III, London.

———. 1936. *The Palace of Minos at Knossos* IV, London.

Faure, P. 1964. *Fonctions des caverns crétoises* (*Traveux et memoires* 14), Athens.

Flouda, G. 2014. "Minoan Communities and Commemorative Practices: The Late Prepalatial to Protopalatial Tholos Tomb A at Apesokari, Crete," *Centre for Hellenic Studies Research Bulletin* 3 (1), http://nrs.harvard.edu/urn-3:hlnc.jissue:CHS_Research_Bulletin.Vol_03.Issue_01.2014.

Forsdyke, J. 1954. "The 'Harvester' Vase of Hagia Triada," *Journal of the Warburg and Courtauld Institutes* 17, pp. 1–9.

Gallet de Santerre, H. 1948. "Délos, la Crète et le continent mycénien au second millénaire," *RA* 29–30, pp. 387–400.

Giuman, M. 2009. *Melissa: Archeologia delle api e del miele nella Grecia antica* (*Archeologica* 148), Rome.

Grumach, E. 1964. "Ein kretisches Honigzeichen?" *CretChron* 18, pp. 7–14.

Harissis, H.V., and A.V. Harissis, 2009. *Apiculture in the Prehistoric Aegean: Minoan and Mycenaean Symbols Revisited* (*BAR-IS* 1958), Oxford.

Harrison, J.E. 1912. *Themis: A Study of the Social Origins of Greek Religion* (*Meridian Books* M145), Cambridge.

Hatzaki, E. 2012. "Mortuary Practices and Ideology in Bronze Age and Early Iron Age Crete," in *Parallel Lives: Ancient Island Societies in Crete and Cyprus. Papers Arising from the Conference in Nicosia Organised by the British School at Athens, the University of Crete and the University of Cyprus, in November–December 2006* (*BSA Studies* 20), G. Cadogan, M. Iakovou, K. Kopaka, and J. Whitley, eds., London, pp. 307–313.

Hayden, B. 2003. *Shamans, Sorcerers, and Saints: The Prehistory of Religion*, Washington DC.

Herva, V.-P. 2006. "Flower Lovers after All? Rethinking Religion and Human Environment Relations in Minoan Crete," *WorldArch* 38, pp. 586–598.

Hitchcock, L.A. 2007. "Naturalising the Cultural: Architectonised Landscape as Ideology in Minoan Crete," in *Building Communities: House, Settlement, and Society in the Aegean and Beyond* (*BSA Studies* 5), R. Westgate, N. Fisher, and J. Whitley, eds., pp. 91–97.

Hogarth, D.G. 1899–1900. "The Dictaean Cave," *BSA* 6, pp. 94–116.

Hood, S. 1976. "The Mallia Gold Pendant: Wasps or Bees?" in *Tribute to an Antiquary. Essays Presented to Marc Fitch by Some of His Friends*, London, pp. 59–72.

Kenna, V.E.G. 1966. "Seal Use in Ancient Crete after the Destruction of the Palaces," *BICS* 13, pp. 68–75.

Knappett, C. 1999. "Assessing a Polity in Protopalatial Crete: The Malia-Lasithi State," *AJA* 103, pp. 615–639.

Laffineur, R. 1985. "Iconographie minoenne et iconographie mycénienne à l'epoque des tombes à fosse," in *L'iconographie minoenne. Actes de la Table ronde d'Athènes (21–22 avril 1983) (BCH Suppl. 11)*, Athens, pp. 245–266.

LaFleur, R.A., R.W. Matthews, and D.B. McCorcle. 1979. "A Re-examination of the Mallia Insect Pendant," *AJA* 83, pp. 208–212.

Lawler, L.B. 1946. "The Geranos Dance—A New Interpretation," *TAPA* 77, pp. 112–130.

———. 1954. "Bee Dances and the 'Sacred Bees,'" *Classical Weekly* 47 (7), pp. 103–106.

Long, C.R. 1958. "The Lasithi Dagger," *AJA* 82, pp. 35–46.

———. 1961. "Reply to M. Gallet de Santerre," *AJA* 65, pp. 65–66.

Lurker, M. 1980. *The Gods and Symbols of Ancient Egypt: An Illustrated Dictionary*, London.

Marinatos, N. 2010. *Minoan Kingship and the Solar Goddess: A Near Eastern Koine*, Urbana.

Marinatos, S. 1927–1928. "Γοργόνες καὶ Γοργόνεια," *ArchEph* 66–67, pp. 7–41.

Matz, F. 1962. *The Art of Crete and Early Greece: The Prelude to Greek Art*, New York.

Mazarakis Ainian, A. 1997. *From Rulers' Dwellings to Temples. Archaeology, Religion, and Society in Early Iron Age Greece (1100–700 B.C.) (SIMA 121)*, Jonsered.

Morris, C., and A. Peatfield. 2002. "Feeling through the Body: Gesture in Cretan Bronze Age Religion," in *Thinking through the Body: Archaeologies of Corporeality*, Y. Hamilakis, M. Pluciennik, and S. Tarlow, eds., New York and London, pp. 105–120.

———. 2006. "Experiencing Ritual: Shamanic Elements in Minoan Religion," in *Celebrations: Sanctuaries and Vestiges of Cult Activity. Selected Papers and Discussions from the Tenth Anniversary Symposium of the Norwegian Institute at Athens, 12–16 May 1999 (Papers from the Norwegian Institute at Athens 6)*, M. Wedde, ed., Athens, pp. 35–59.

Murphy, J.M.A. 2011. "Landscape and Social Narratives: A Study of the Regional Social Structures in Prepalatial Crete," in *Prehistoric Crete: Regional and Diachronic Studies in Mortuary Systems*, J.M.A. Murphy, ed., Philadelphia, pp. 23–47.

Nilsson, M. 1925. "Minoisch-mykenische kunst," in *Lehrbuch der Religionsgeschichte*, vol. 2, A. Bertholet and E. Lehmann, eds., Freiburg, pp. 306–312.

Papadatos, Y., and C. Sofianou. 2013. "Prepalatial Tholos Tomb at Mesorrachi Skopi, near Siteia, East Crete," *Aegean Archaeology* 10 [2009–2010], pp. 7–31.

Peatfield, A.A.P. 1987. "Palace and Peak: The Political and Religious Relationship between Palaces and Peak Sanctuaries," in *The Function of the Minoan Palaces. Proceedings of the Fourth International Symposium at the Swedish Institute in Athens, 10–16 June, 1984 (ActaAth 4°, 35)*, R. Hagg and N. Marinatos, eds., Stockholm, pp. 89–93.

Pendlebury, J.D.S. 1936–1937. "Lasithi in Ancient Times," *BSA* 37, pp. 194–200.

Picard, C. 1927. "La Crète et les légendes Hyperboréennes," *RA* 25, pp. 349–360.

———. 1929. "Chronique de la religion minoenne," *RHR* 99, pp. 1–16.

Poursat, J.-C. 1985. "Iconographie minoenne: Continuités et ruptures," in *L'iconographie minoenne. Actes de la Table ronde d'Athènes (21–22 avril 1983) (BCH Suppl. 11)*, Athens, pp. 51–57.

Rehak, P., and J.G. Younger. 1998. "Review of Aegean Prehistory VII: Neopalatial, Final Palatial, and Postpalatial Crete," *AJA* 102, pp. 91–173.

Renfrew, C. 1984. "Speculations on the Use of Early Cycladic Sculpture," in *Cycladica. Studies in Memory of N.P. Goulandris. Proceedings of the Seventh British Museum Classical Colloquium, June 1983*, J.L. Fitton, ed., London, pp. 24–30.

Renfrew, C., M. Boyd, and C.B. Ramsey. 2012. "The Oldest Maritime Sanctuary? Dating the Sanctuary at Keros and the Cycladic Bronze Age," *Antiquity* 86, pp. 144–160.

Richards-Mantzoulinou, E. 1979. "Μέλισσα πότνια," *AAA* 12, pp. 72–92.

Rutkowski, B., and K. Nowicki. 1996. *The Psychro Cave and Other Sacred Grottoes in Crete (Studies and Monographs in Mediterranean Archaeology and Civilization 2°, 1)*, Warshaw.

Schefold, K. 1960. *Meisterwerke griechischer Kunst*, Basel.

Schoep, I., I. Crevecoeur, and A. Schmitt. 2015. "An Archaeothanatological Approach to the Study of Minoan Funerary Practices: Case-Studies in the Early and Middle Minoan Cemetery at Sissi, Crete," *JFA* 40, pp. 1–7.

Shaw, I., and P. Nicholson. 2008. *The Princeton Dictionary of Ancient Egypt*, Princeton.

Shelmerdine, C.W., ed., 2008. *Cambridge Companion to the Aegean Bronze Age*, Cambridge.

Somville, P. 1978. "L'abeille et le taureau (ou la vie et la mort dans la Crète minoenne)," *Revue de l'histoire des religions* 194, pp. 129–146.

Spratt, T.A.B. 1865. *Travels and Researches in Crete* I–II, London.

Thorne, S. 2000. "Diktaian Zeus in Later Greek Tradition," in *The Palaikastro Kouros: A Minoan Chryselephantine Statuette and Its Aegean Bronze Age Context* (*BSA Studies* 6), J.A. MacGillivray, J.M. Driessen, and L.H. Sackett, eds., London, pp. 149–162.

Tully, C. 2015. "Trans-former/Performer: Shamanistic Elements in Late Bronze Age Minoan Cults." Paper read at the 2015 Annual Meeting of the Archaeological Institute of America, 8–11 January, New Orleans, LA.

Tyree, L. 2001. "Diachronic Changes in Minoan Cave Cult," in *POTNIA: Deities and Religion in the Aegean Bronze Age. Proceedings of the 8th International Aegean Conference, Goteborg University 12–15 April 2000* (*Aegaeum* 22), R. Laffineur and R. Hagg, eds., Liège, pp. 39–50.

———. 2013. "Defining Bronze Age Ritual Caves in Crete," in *Stable Places and Changing Perceptions: Cave Archaeology in Greece* (*BAR-IS* 2558), F. Mavridis and J.T. Jensen, eds., Oxford, pp. 176–187.

van Leuven, J.C. 1980. "Economic Determinism and Bronze Age Greece," *Historia* 29, pp. 129–141.

Warren, P.M. 1990. "Of Baetyls," *OpAth* 18, pp. 193–206.

Watrous, L.V. 1982. *Lasithi: A History of Settlement on a Highland Plain in Crete* (*Hesperia Suppl.* 18), Princeton.

———. 1995. "Some Observations on Minoan Peak Sanctuaries," in *POLITEIA: Society and State in the Aegean Bronze Age. Proceedings of the 5th International Aegean Conference, University of Heidelberg, Archäologisches Institut, 10–13 April 1994* (*Aegaeum* 12), R. Laffineur and W.-D. Niemeier, eds., Liège, pp. 394–403.

———. 1996. *The Cave Sanctuary of Zeus at Psychro: A Study of Extra-Urban Sanctuaries in Minoan and Early Iron Age Crete* (*Aegaeum* 15), Liège.

———. 2004. "New Pottery from the Psychro Cave and Its Implications for Minoan Crete," *BSA* 99, pp. 129–147.

Willetts, R.F. 1958. "Cretan Eileithyia," *CQ* 8, pp. 221–223.

———. 1962. *Cretan Cults and Festivals*, London.

Woudhuizen, F.C., 1997. "The Bee Sign (Evans No. 86): An Instance of Egyptian Influence on Cretan Hieroglyphic," *Kadmos* 36, pp. 97–110.

Xanthoudides, S. 1904. *Ο κρητικός πολιτισμός*, Athens.

Concordance of Excavation and Museum Numbers with Catalog Numbers

HCH 1	**65**	HCH 28	**433**	HCH 47	**271**
HCH 2	**304**	HCH 29	**95**	HCH 49	**9**
HCH 3	**305**	HCH 30	**413**	HCH 52	**415**
HCH 4	**379**	HCH 31	**128**	HCH 53	**321**
HCH 5	**205**	HCH 33	**212**	HCH 54	**442**
HCH 12	**452**	HCH 34	**301**	HCH 55	**198**
HCH 13	**93**	HCH 35	**456**	HCH 56	**122**
HCH 14	**101**	HCH 36	**451**	HCH 57	**148**
HCH 16	**258**	HCH 37	**423**	HCH 58	**179**
HCH 18	**281**	HCH 38	**117**	HCH 59	**152**
HCH 19	**165**	HCH 39	**285**	HCH 60	**289**
HCH 22	**200**	HCH 40	**231**	HCH 61	**38**
HCH 23	**10**	HCH 41	**261**	HCH 62	**72**
HCH 24	**66**	HCH 42	**236**	HCH 63	**328**
HCH 25	**306**	HCH 43	**420**	HCH 64	**276**
HCH 26	**358**	HCH 44	**188**	HCH 66	**251**
HCH 27	**353**	HCH 45	**189**	HCH 67	**193**

HCH 68	**194**	HCH 115	**147**	HCH 164	**154**
HCH 69	**105**	HCH 116	**126**	HCH 165	**225**
HCH 70	**297**	HCH 118	**300**	HCH 167	**123**
HCH 71	**224**	HCH 119	**125**	HCH 168	**292**
HCH 72	**138**	HCH 120	**303**	HCH 169	**293**
HCH 73	**102**	HCH 121	**296**	HCH 170	**418**
HCH 74	**48**	HCH 122	**245**	HCH 171	**56**
HCH 75	**41**	HCH 123	**35**	HCH 172	**310**
HCH 76	**302**	HCH 124	**395**	HCH 173	**322**
HCH 77	**404**	HCH 125	**398**	HCH 174	**299**
HCH 78	**399**	HCH 126	**67**	HCH 176	**182**
HCH 79	**444**	HCH 127	**448**	HCH 177	**172**
HCH 80	**454**	HCH 128	**171**	HCH 179	**191**
HCH 81	**273**	HCH 129	**87**	HCH 181	**155**
HCH 82	**58**	HCH 130	**146**	HCH 182	**107**
HCH 83	**459**	HCH 131	**75**	HCH 183	**124**
HCH 84	**246**	HCH 132	**380**	HCH 184	**187**
HCH 85	**127**	HCH 135	**237**	HCH 186	**232**
HCH 86	**25**	HCH 136	**238**	HCH 187	**185**
HCH 88	**393**	HCH 137	**181**	HCH 188	**184**
HCH 89	**403**	HCH 138	**183**	HCH 190	**94**
HCH 90	**277**	HCH 139	**108**	HCH 192	**455**
HCH 91	**153**	HCH 141	**84**	HCH 193	**263**
HCH 92	**149**	HCH 142	**85**	HCH 194	**133**
HCH 93	**307**	HCH 144	**77**	HCH 195	**262**
HCH 94	**317**	HCH 147	**361**	HCH 196	**16**
HCH 97	**270**	HCH 149	**34**	HCH 02-14	**412**
HCH 98	**268**	HCH 150	**402**	HCH 02-15	**254**
HCH 99	**269**	HCH 151	**346**	HCH 02-20	**178**
HCH 101	**288**	HCH 153	**135**	HCH 02-25	**202**
HCH 102	**78**	HCH 155	**422**	HCH 02-30	**324**
HCH 103	**82**	HCH 156	**57**	HCH 02-37	**267**
HCH 106	**309**	HCH 157	**129**	HCH 02-41	**52**
HCH 107	**76**	HCH 158	**80**	HCH 02-44	**421**
HCH 110	**265**	HCH 159	**81**	HCH 02-46	**214**
HCH 112	**150**	HCH 160	**1**	HCH 02-49	**362**
HCH 113	**143**	HCH 161	**427**	HCH 02-51	**244**
HCH 114	**151**	HCH 163	**230**	HCH 02-53	**458**

HCH 02-54	**176**	HCH 02-138	**43**	HCH 03-214	**253**
HCH 02-55	**216**	HCH 02-139	**392**	HCH 03-215	**2**
HCH 02-56	**162**	HCH 02-140	**44**	HCH 03-216	**164**
HCH 02-57	**335**	HCH 02-142	**298**	HCH 03-217	**257**
HCH 02-60	**377**	HCH 02-143	**453**	HCH 03-218	**166**
HCH 02-61	**440**	HCH 02-144	**406**	HCH 03-219	**318**
HCH 02-62	**382**	HCH 02-145	**313**	HCH 03-220	**416**
HCH 02-64	**213**	HCH 02-146	**53**	HCH 03-221	**287**
HCH 02-67	**401**	HCH 02-148	**314**	HCH 03-224	**195**
HCH 02-68	**316**	HCH 02-149	**315**	HCH 03-225	**54**
HCH 02-69	**282**	HCH 02-150	**320**	HCH 03-226	**327**
HCH 02-70	**326**	HCH 02-151	**378**	HCH 03-227	**337**
HCH 02-71	**435**	HCH 03-153	**373**	HCH 04-271	**14**
HCH 02-75	**96**	HCH 03-154	**381**	HCH 04-272	**32**
HCH 02-77	**384**	HCH 03-156	**391**	HCH 04-273	**27**
HCH 02-79	**445**	HCH 03-157	**357**	HCH 04-274	**30**
HCH 02-80	**97**	HCH 03-163	**390**	HCH 04-275	**26**
HCH 02-81	**100**	HCH 03-164	**4**	HCH 04-276+277	**79**
HCH 02-83	**385**	HCH 03-165	**334**	HCH 04-278	**39**
HCH 02-90	**386**	HCH 03-166	**192**	HCH 04-279	**50**
HCH 02-98	**411**	HCH 03-169	**7**	HCH 04-280	**294**
HCH 02-99	**177**	HCH 03-170	**352**	HCH 04-281	**295**
HCH 02-106	**274**	HCH 03-174	**252**	HCH 04-285	**49**
HCH 02-110	**186**	HCH 03-175	**250**	HCH 04-286	**47**
HCH 02-113	**329**	HCH 03-185	**130**	HCH 04-287	**17**
HCH 02-114	**170**	HCH 03-186	**330**	HCH 04-288	**241**
HCH 02-116	**429**	HCH 03-189	**8**	HCH 04-289	**242**
HCH 02-120	**3**	HCH 03-202	**24**	HCH 04-290	**247**
HCH 02-122	**204**	HCH 03-204	**190**	HCH 04-303	**45**
HCH 02-123	**259**	HCH 03-205	**278**	HCH 04-304	**169**
HCH 02-126	**450**	HCH 03-206	**279**	HCH 04-305	**86**
HCH 02-127	**449**	HCH 03-207	**275**	HCH 04-306	**219**
HCH 02-128	**264**	HCH 03-208	**359**	HCH 04-320	**180**
HCH 02-129	**256**	HCH 03-209	**255**	HCH 04-321	**40**
HCH 02-130	**424**	HCH 03-210	**426**	HCH 04-345	**290**
HCH 02-131	**425**	HCH 03-211	**249**	HCH 04-347	**272**
HCH 02-132	**215**	HCH 03-212	**201**	HCH 04-348	**134**
HCH 02-136	**331**	HCH 03-213	**372**	HCH 04-349	**248**

HCH 04-350	**159**	HCH 05-435	**71**	HNM 12,409	**111**
HCH 04-351	**136**	HCH 05-436	**109**	HNM 12,410	**120**
HCH 04-353	**46**	HCH 05-437	**19**	HNM 12,411	**405**
HCH 04-354	**286**	HCH 05-438	**20**	HNM 12,412	**98**
HCH 04-355	**31**	HCH 05-439	**70**	HNM 12,413	**168**
HCH 04-356	**51**	HCH 05-440	**15**	HNM 12,414	**432**
HCH 04-357	**239**	HCH 05-441	**11**	HNM 12,415	**284**
HCH 04-358	**161**	HCH 05-442	**12**	HNM 12,416	**431**
HCH 04-370	**227**	HCH 05-443	**33**	HNM 12,417	**88**
HCH 04-371	**28**	HCH 05-444	**266**	HNM 12,418	**90**
HCH 04-372	**131**	HCH 05-445	**104**	HNM 12,419	**371**
HCH 04-373	**36**	HCH 05-447	**223**	HNM 12,420	**363**
HCH 04-376	**291**	HCH 05-448	**6**	HNM 12,421	**360**
HCH 04-377	**29**	HCH 05-449	**5**	HNM 12,422	**73**
HCH 04-380	**203**	HCH 05-450	**18**	HNM 12,425	**89**
HCH 04-381	**196**	HCH 05-452	**160**	HNM 12,426	**113**
HCH 04-388	**228**	HCH 06-455	**446**	HNM 12,427	**140**
HCH 04-389	**62**	HCH 06-456	**430**	HNM 12,428	**349**
HCH 04-390	**63**	HCH 06-457	**42**	HNM 12,429	**364**
HCH 04-391	**60**	HCH 06-458	**428**	HNM 12,430	**340**
HCH 04-395	**457**	HCH 06-459	**417**	HNM 12,431	**394**
HCH 04-403	**308**	HCH 06-460	**199**	HNM 12,432	**414**
HCH 04-404	**37**	HNM 6,828	**137**	HNM 12,433	**312**
HCH 04-407	**21**	HNM 12,389	**400**	HNM 12,434	**319**
HCH 04-408	**22**	HNM 12,390	**233**	HNM 12,435	**222**
HCH 05-409	**69**	HNM 12,391	**437**	HNM 12,436	**115**
HCH 05-410	**132**	HNM 12,392	**441**	HNM 12,438	**139**
HCH 05-411	**325**	HNM 12,393	**436**	HNM 12,440	**209**
HCH 05-412	**311**	HNM 12,394	**370**	HNM 12,442	**220**
HCH 05-413	**280**	HNM 12,396	**374**	HNM 12,443	**221**
HCH 05-414	**217**	HNM 12,399	**397**	HNM 12,445	**103**
HCH 05-416	**235**	HNM 12,401	**116**	HNM 12,446	**351**
HCH 05-417	**260**	HNM 12,402	**118**	HNM 12,447	**389**
HCH 05-418	**197**	HNM 12,403	**336**	HNM 12,448	**350**
HCH 05-419	**243**	HNM 12,404	**338**	HNM 12,449	**91**
HCH 05-431	**106**	HNM 12,405	**354**	HNM 12,450	**348**
HCH 05-432	**342**	HNM 12,407	**339**	HNM 12,451	**419**
HCH 05-434	**283**	HNM 12,408	**173**	HNM 12,452	**388**

HNM 12,453	**332**	HNM 13,813	**167**	HNM 13,837	**410**
HNM 12,454	**434**	HNM 13,815	**175**	HNM 13,838	**121**
HNM 12,455	**345**	HNM 13,816	**387**	HNM 13,839	**408**
HNM 12,456	**383**	HNM 13,817	**333**	HNM 13,840	**368**
HNM 12,457	**323**	HNM 13,818	**226**	HNM 13,841	**92**
HNM 12,458	**114**	HNM 13,819	**163**	HNM 13,842	**55**
HNM 12,459	**438**	HNM 13,820	**447**	HNM 13,843	**83**
HNM 12,460	**74**	HNM 13,821	**347**	HNM 13,844	**369**
HNM 12,461	**174**	HNM 13,822	**341**	HNM 13,845	**210**
HNM 12,568	**64**	HNM 13,823	**110**	HNM 13,846	**443**
HNM 12,569	**59**	HNM 13,824	**343**	HNM 13,847	**375**
HNM 12,570	**61**	HNM 13,825	**119**	HNM 13,848	**409**
HNM 13,789	**13**	HNM 13,826	**145**	HNM 13,849	**356**
HNM 13,801	**229**	HNM 13,827	**141**	HNM 13,850	**207**
HNM 13,803	**407**	HNM 13,828	**158**	HNM 13,852	**208**
HNM 13,805	**355**	HNM 13,829	**144**	HNM 13,853	**211**
HNM 13,806	**99**	HNM 13,830	**365**	HNM 13,854	**376**
HNM 13,807	**157**	HNM 13,831	**366**	HNM 13,856	**23**
HNM 13,808	**112**	HNM 13,832	**218**	HNM 13,885	**68**
HNM 13,809	**156**	HNM 13,833	**206**	HNM 13,982	**240**
HNM 13,810	**396**	HNM 13,834	**344**	HNM 14,066	**142**
HNM 13,811	**234**	HNM 13,835	**367**		

References

Alexiou, S., and P. Warren. 2004. *The Early Minoan Tombs of Lebena, Southern Crete* (*SIMA* 30), Sävedalen.

Apostolakou, V., P.P. Betancourt, and T.M. Brogan. 2011. "The Alatzomouri Rock Shelter: Defining EM III in Eastern Crete," *Aegean Archaeology* 9, pp. 35–48.

Banti, L. 1930–1931. "La grande tomba a tholos de Haghia Triada," *ASAtene* 13–14, pp. 155–251.

Betancourt, P.P. 1979. *Vasilike Ware: An Early Bronze Age Pottery Style in Crete* (*SIMA* 56), Göteborg.

———. 1980. *Cooking Vessels from Minoan Kommos: A Preliminary Report* (*UCLAPap* 7), Los Angeles.

———. 1983. *The Cretan Collection in the University Museum, University of Pennsylvania* I: *Minoan Objects Excavated from Vasiliki, Pseira, Sphoungaras, Priniatikos Pyrgos, and Other Sites* (*University Museum Monograph* 47), Philadelphia.

———. 1984. *East Cretan White-on-Dark Ware: Studies on a Handmade Pottery of the Early to Middle Minoan Periods* (*University Museum Monograph* 51), Philadelphia.

———. 1985. *The History of Minoan Pottery*, Princeton.

———. 2002. "The 2002 Excavations at Hagios Charalambos," *Kentro: The Newsletter of the INSTAP Study Center for East Crete* 5, p. 1.

———. 2003. "The Impact of Cycladic Settlers on Early Minoan Crete," *Mediterranean Archaeology and Archaeometry* 3, pp. 3–12.

———. 2005. "Egyptian Connections at Hagios Charalambos," in EMPORIA: *Aegeans in the Central and Eastern Mediterranean. Proceedings of the 10th International Aegean Conference, Italian School of Archaeology in Athens, 14–18 April 2004* (*Aegaeum* 25), R. Laffineur and E. Greco, eds., Liège, pp. 449–453.

———. 2006. *The Chrysokamino Metallurgy Workshop and Its Territory* (*Hesperia Suppl.* 36), Princeton.

———. 2007. "Lasithi and the Malia-Lasithi State," in *Krinoi kai Limenes. Studies in Honor of Joseph and Maria Shaw* (*Prehistory Monographs* 22), P.P. Betancourt, M.C. Nelson, and H. Williams, eds., Philadelphia, pp. 209–219.

————. 2008a. *The Bronze Age Begins: The Ceramics Revolution of Early Minoan I and the New Forms of Wealth That Transformed Prehistoric Society*, Philadelphia.

————. 2008b. "Discussion and Conclusions," in Betancourt et al. 2008a, pp. 594–596.

————. 2011a. "The Diagonal Line Juglets: New Evidence from Hagios Charalambos," in *Our Cups Are Full: Pottery and Society in the Aegean Bronze Age. Papers Presented to Jeremy B. Rutter on the Occasion of His 65th Birthday*, W. Gauss, M. Lindblom, R.A.K. Smith, and J.C. Wright, eds., Oxford, pp. 25–30.

————. 2011b. "A Marine Style Gold Ring from the Hagios Charalambos Ossuary: Symbolic Use of Cockle Shells in Minoan Crete," in *Metallurgy: Understanding How, Learning Why. Studies in Honor of James D. Muhly* (*Prehistory Monographs* 29), P.P. Betancourt and S.C. Ferrence, eds., Philadelphia, pp. 117–123.

————. 2014a. *Hagios Charalambos: A Minoan Burial Cave in Crete*. I: *Excavation and Portable Objects* (*Prehistory Monographs* 47), Philadelphia.

————. 2014b. "Larnakes," in Betancourt 2014a, pp. 45–48.

Betancourt, P.P., S. Chlouveraki, C. Davaras, H.M. Dierckx, S.C. Ferrence, J. Hickman, P. Karkanas, P.J.P. McGeorge, J.D. Muhly, D.S. Reese, E. Stavropodi, L. Langford-Verstegen. 2008a. "Excavations in the Hagios Charalambos Cave: A Preliminary Report," *Hesperia* 77, pp. 539–605.

Betancourt, P.P., and C. Davaras, eds., 2003. *Pseira VII: The Pseira Cemetery. Part 2: Excavation of the Tombs* (*Prehistory Monographs* 6), Philadelphia.

Betancourt, P.P., C. Davaras, S.C. Ferrence, L.C. Langford-Vertegen, A. Stamos, M. Tsiboukaki, and G.M. Weng. 2014. "Excavations inside the Cave," in Betancourt 2014a, pp. 21–29.

Betancourt, P.P., H.M.C. Dierckx, and D.S. Reese. 2003. "Tomb 1: Catalog of Objects," in Betancourt and Davaras, eds., 2003, pp. 7–16.

Betancourt, P.P., and S.C. Ferrence. 2014a. "Figurines," in Betancourt 2014a, pp. 49–53.

————. 2014b. "Seals," in Betancourt 2014a, pp. 63–67.

Betancourt, P.P., V. Karageorghis, R. Laffineur, and W.-D. Niemeier, eds. 1999. *Meletemata. Studies in Aegean Archaeology Presented to Malcolm H. Wiener as He Enters His 65th Year* (*Aegaeum* 20), Liège.

Betancourt, P.P., and J.D. Muhly. 2008. "The Sistra," in Betancourt et al. 2008a, p. 577.

Betancourt, P.P., and N. Poulou-Papadimitriou. 2014. "Later Occupation of the Area," in Betancourt 2014a, pp. 37–41.

Betancourt, P.P., D.S. Reese, L. Langford-Verstegen, and S.C. Ferrence. 2008b. "Feasts for the Dead: Evidence from the Ossuary at Hagios Charalambos," in Hitchock, Laffineur, and Crowley, eds., 2008, pp. 161–166.

Betancourt, P.P., and J.S. Silverman. 1991. *Pottery from Gournia: The Cretan Collection in the University Museum, University of Pennsylvania*, vol. 2 (*University Museum Monograph* 52), Philadelphia.

Blackman, D., and K. Branigan. 1982. "The Excavation of an Early Tholos Tomb at Ayia Kyriaki, Ayiofarango, Southern Crete," *BSA* 77, pp. 1–57.

Boyd, H.A. 1905. "Gournia: Report of the American Exploration Society's Excavations at Gournia, Crete, 1904," *Transactions of the Department of Archaeology, Free Museum of Science and Art* 1, pp. 177–189.

Branigan, K. 1970a. *The Foundations of Palatial Crete: A Survey of Crete in the Early Bronze Age*, New York.

————. 1970b. *The Tombs of Mesara: A Study of Funerary Architecture and Ritual in Southern Crete, 2800–1700 B.C.*, London.

————. 1988. *Pre-Palatial: The Foundations of Palatial Crete. A Survey of Crete in the Early Bronze Age* (*States and Cities of Ancient Greece*), Amsterdam.

————. 1991. "Funerary Ritual and Social Cohesion in EBA Crete," *Journal of Mediterranean Studies* 1, pp. 9–13.

————. 1993. *Dancing with Death: Life and Death in Southern Crete, c. 3000–2000 B.C.*, Amsterdam.

————. 2008. "Communal Ceremonies in an Early Minoan Tholos Cemetery," in *Dioskouroi. Studies Presented to W.G. Cavanaugh and C.B. Mee on the Anniversary of Their 30-Year Joint Contribution to Aegean Archaeology* (*BAR-IS* 1889), C. Gallou, M. Georgiadis, and G.M. Muskett, eds., Oxford, pp. 15–22.

Broodbank, C., and E. Kiriatzi. 2007. "The First 'Minoans' of Kythera Revisited: Technology, Demography, and Landscape in the Prepalatial Aegean," *AJA* 111, pp. 241–274.

Cadogan, G. 1990. "Lasithi in the Old Palace Period," *BICS* 37, pp. 172–174.

————. 1995. "Mallia and Lasithi: A Palace State," in *Πεπραγμένα Ζ´ Διεθνούς Κρητολογικού Συνεδρίου* A´ (1), Rethymnon, pp. 97–104.

Chapouthier, F., and J. Charbonneaux. 1928. *Fouilles exécutées à Mallia: Premier rapport (1922–1924)* (*ÉtCrét* 1), Paris.

Chapouthier, F., and P. Demargne. 1942. *Fouilles exécutées à Mallia: Troisième rapport. Exploration du palais, bordures orientale et septentrionale (1927, 1928, 1931, 1932)* (*ÉtCrét* 6), Paris.

Christakis, K. 2013. "The Symi Sanctuary at the Transition from the Protopalatial to the Early Neopalatial Periods: The Evidence of the Pottery," in I*ntermezzo: Intermediacy and Regeneration in the Middle Minoan III Palatial Crete* (*BSA Studies* 21), C.F. Macdonald and C. Knappett, eds., London, pp. 169–177.

Coldstream, J.N., and G.L. Huxley. 1972. *Kythera: Excavations and Studies Conducted by the University of Pennsylvania Musuem and the British School at Athens*, London.

Davaras, C. 1976. "Ἅγιος Χαράλαμπος," *ArchDelt* 31 (Β'), p. 379.

———. 1982. "Σπήλαιο Ἁγίου Χαραλάμπους," *ArchDelt* 37 (Β', 2), pp. 387–388.

———. 1983. "Σπήλαιο Γεροντομουρί," *ArchDelt* 38 (Β', 2), p. 375.

———. 1986. "Πρώιμες μινωικές σφραγίδες καὶ σφραγιστικοί δακτύλιοι ἀπό τὸ σπήλαιο Γεροντομουρί Λασιθίου," *ArchEph* 125, pp. 9–48.

———. 1988. "Ἕνα παλαιοανακτορικὸ πρίσμα ἀπὸ τὸ σπήλαιο Γεροντομουρὶ Λασιθίου," *Cretan Studies* 1, pp. 49–55.

Davaras, C., and P.P. Betancourt. 2004. *The Hagia Photia Cemetery* I: *The Tomb Groups and Architecture* (*Prehistory Monographs* 14), Philadelphia.

Day, P.M., A. Hein, L. Joyner, V. Kilikoglou, E. Kiriatzi, A. Tsolakidou, and D.E. Wilson. 2012. "Appendix A: Petrographic and Chemical Analysis of the Pottery," in *The Hagia Photia Cemetery* II: *The Pottery* (*Prehistory Monographs* 34), C. Davaras and P.P. Betancourt, Philadelphia, pp. 115–138.

Day, P.M., L. Joyner, E. Kiriatzi, and M. Relaki. 2005. "Petrographic Analysis of Some Final Neolithic–Early Minoan II Pottery from the Kavousi Area," in *Kavousi* I: *The Archaeological Survey of the Kavousi Region* (*Prehistory Monographs* 16), D.C. Haggis, Philadelphia, pp. 178–195.

Day, P.M., L. Joyner, and M. Relaki. 2003. "A Petrographic Analysis of the Neopalatial Pottery," in *Mochlos* IB: *Period III. Neopalatial Settlement on the Coast: The Artisans' Quarter and the Farmhouse at Chalinomouri. The Neopalatial Pottery* (*Prehistory Monographs* 8), K.A. Barnard and T.M. Brogan, eds., Philadelphia, pp. 13–32.

Day, P.M., and V. Kilikoglou. 2001. "Analysis of Ceramics from the Kiln," in *A LM IA Ceramic Kiln in South-Central Crete* (*Hesperia Suppl.* 30), J.W. Shaw, A. Van de Moortel, P.M. Day, and V. Kilikoglou, Princeton, pp. 111–133.

Day, P.M., D.E. Wilson, and E. Kiriatzi. 1997. "Reassessing Specialisation in Prepalatial Cretan Ceramic Production," in Τεχνη: *Craftsmen, Craftswomen and Craftsmanship in the Aegean Bronze Age. Proceedings of the 6th International Conference, Philadelphia, Temple University, 18–21 April 1996* (*Aegaeum* 16), R. Laffineur and P.P. Betancourt, eds., Liège, pp. 275–290.

———. 1998. "Pots, Labels and People: Burying Ethnicity in the EM I Cemetery at Aghia Photia, Siteias," in *Cemetery and Society in the Aegean Bronze Age* (*Sheffield Studies in Aegean Archaeology* 1), K. Branigan, ed., Sheffield, pp. 133–149.

Demargne, P. 1945. *Fouilles exécutées à Mallia: Exploration des nécropoles (1921–1933), premíer fascicule* (*ÉtCrét* 7), Paris.

Demargne, P., and H. Gallet de Santerre. 1953. *Fouilles exécutées à Mallia: Exploration des maisons et quartiers d'habitation (1921–1948), premier fascicule* (*ÉtCrét* 9), Paris.

Deshayes, J., and A. Dessenne. 1959. *Fouilles exécutées à Mallia: Exploration des maisons et quartiers d'habitation (1948–1954), deuxième fascicule* (*ÉtCrét* 11), Paris.

Dierckx, H.M.C. 2008. "The Stone Vessels," in Betancourt et al. 2008a, pp. 562–569.

Dierckx, H.M.C., and B. Tsikouras. 2007. "Petrographic Characterization of Rocks from the Mirabello Bay Region, Crete, and Its Application to Minoan Archaeology: The Provenance of Stone Implements from Minoan Sites," in "Proceedings of the 11th International Congress, Athens, May 2007," special issue, *Bulletin of the Geological Society of Greece* 40, Athens, pp. 1768–1779.

Durkheim, É. 1912. *The Elementary Forms of the Religious Life*, London.

Evans, A.J. 1921–1935. *The Palace of Minos at Knossos* I–IV, London.

Faure, P. 1964. *Fonctions des cavernes crétoises* (*Travaux et mémoires des anciens membres étrangers de l'École et de divers savants* 14), Paris.

Ferrence, S.C. 2007. "Hippopotamus Ivory in EM–MM I Lasithi and the Implications for Eastern Mediterranean Trade: New Evidence from Hagios Charalambos," in *Krinoi kai Limenes. Studies in Honor of Joseph and Maria Shaw* (*Prehistory Monographs*

22), P.P. Betancourt, M.C. Nelson, and H. Williams, eds., Philadelphia, pp. 167–175.

————. 2008. "The Human Figurines," in Betancourt et al. 2008a, pp. 570–576.

Filippa-Touchais, A. 2000. "Η τριποδική χύτρα στον αιγαιακό χώρο κατά τη Μέση Χαλκοκρατία: Διάδοση και σημασία," in Πεπραγμένα Η′ Διεθνούς Κρητολογικού Συνεδρίου Α′ (3), Herakleion, pp. 421–436.

Floyd, C.R. 1999. "The Minoan Scoop," in Betancourt et al., eds., 1999, pp. 249–257.

Foster, K.P. 1982. Minoan Ceramic Relief, Göteborg.

Galanaki, K. 2006. "Πρωτομινωικό ταφικό σύνολο στην πρώην Αμερικανική Βάση Γουρνών Πεδιάδος," in Πεπραγμένα Θ′ Διεθνούς Κρητολογικού Συνεδρίου Α′ (2), Herakleion, pp. 227–242.

Galli, E. 2014. "Where the Past Lies: The Prepalatial Tholos Tomb at Krasi and Its Stratigraphic Sequence," Kritika Chronika 34, pp. 231–248.

Gerontakou, E. 2003. "Δύο Μεσομινωικοί Αποθέτες του Πλατάνου," in Αργοναύτης. Τιμητικός τόμος για τον καθηγητή Χρίστου Γ. Ντούμα από τους μαθητές του στο Πανεπιστήμιο Αθηνών (1980–2000), edited by A. Vlachopoulos and K. Birtacha, Athens, pp. 303–330.

Goodman, F.D. 1988. Ecstacy, Ritual, and Alternate Reality: Religion in a Pluralistic World, Bloomington.

Goody, J.R. 1961. "Religion and Ritual: The Definitional Problem," British Journal of Sociology 12, pp. 142–164.

Haggis, D.C. 1996. "Excavations at Kalo Khorio, East Crete," AJA 100, pp. 645–681.

————. 1997. "The Typology of the Early Minoan I Chalice and the Cultural Implications of Forms and Style in Early Bronze Age Ceramics," in TEXNH: Craftsmen, Craftswomen and Craftsmanship in the Aegean Bronze Age. Proceedings of the 6th International Conference, Philadelphia, Temple University, 18–21 April 1996 (Aegaeum 16), R. Laffineur and P.P. Betancourt, eds., Liège, pp. 291–299.

Haggis, D.C., M. Mook, T. Carter, and L.M. Snyder. 2007. "Excavations at Azoria, 2003–2004, Part 2. The Final Neolithic, Late Prepalatial, and Early Iron Age Occupation," Hesperia 76, pp. 665–716.

Hall, E.H. 1912. Excavations in Eastern Crete: Sphoungaras (University of Pennsylvania, the Museum Anthropological Publications 3 [2]), Philadelphia.

————. 1914. Excavations in Eastern Crete: Vrokastro (University of Pennsylvania, the Museum Anthropological Publications 3 [3]), Philadelphia.

Halstead, P., and J.C. Barrett, eds. 2004. Food, Cuisine and Society in Prehistoric Greece (Sheffield Studies in Aegean Archaeology 5), Sheffield.

Hamilakis, Y. 1998. "Eating the Dead: Mortuary Feasting and the Politics of Memory in the Aegean Bronze Age Societies," in Cemetery and Society in the Aegean Bronze Age (Sheffield Studies in Aegean Archaeology 1), K. Brannigan, ed., Sheffield, pp. 115–132.

Hawes, H.B., B.E. Williams, R.B. Seager, and E.H. Hall. 1908. Gournia, Vasiliki and Other Prehistoric Sites on the Isthmus of Hierapetra, Crete, Philadelphia.

Hayden, B.J. 2003. "Final Neolithic–Early Minoan I/IIA Settlement in the Vrokastro Area, Eastern Crete," AJA 107, pp. 363–412.

Hazzidakis, J. 1934. Les villas minoennes de Tylissos (ÉtCrét 3), Paris.

Hickman, J. 2008. "The Gold Strips," in Betancourt et al. 2008a, pp. 561–562.

Hitchcock, L.A., R. Laffineur, and J. Crowley, eds. 2008. DAIS: The Aegean Feast. Proceedings of the 12th Aegean Conference, University of Melbourne, Centre for Classics and Archaeology, 25–29 March 2008 (Aegaeum 29), Liège.

Hogarth, D.G. 1899–1900. "The Dictaean Cave," BSA 6, pp. 94–116.

Hood, M.S.F. 1990. "Autochthons or Settlers? Evidence for Immigration at the Beginning of the Early Bronze Age in Crete," in Πεπραγμένα του ΣΤ′ Διεθνούς Κρητολογικού Συνεδρίου Α′ (1), Chania, pp. 367–375.

IGME. 1989. Geological Map of Greece, Mochos Sheet, 1:50,000.

Jones, D.W. 1999. Peak Sanctuaries and Sacred Caves in Minoan Crete: A Comparison of Artifacts, Jonsered.

Kaiser, B. 1976. Untersuchungen zum minoischen Relief, Bonn.

Karantzali, E. 2008. "The Transition from EB I to EB II in the Cyclades and Crete: Historical and Cultural Repercussions for Aegean Communities," in Horizon. A Colloquium on the Prehistory of the Cyclades (McDonald Institute Monographs), N. Brodie, J. Doole, G. Gavalas, and C. Renfrew, eds., Cambridge, pp. 241–260.

Karkanas, P. 2008. "Local and Regional Geomorphology," in Betancourt et al. 2008a, pp. 547–549.

————. 2014. "Geology, Geomorphology, and Micromorphology Studies," in Betancourt 2014a, pp. 13–16.

Knappett, C. 1999. "Assessing a Polity in Protopalatial Crete: The Malia-Lasithi State," *AJA* 103, pp. 615–639.

Koehl, R.B. 2006. *Aegean Bronze Age Rhyta* (*Prehistory Monographs* 19), Philadelphia.

Langford-Verstegen, L. 2008. "Early and Middle Minoan Pottery," in Betancourt et al. 2008a, pp. 549–556.

Lawson, E.T., and R.N. McCauley. 1993. *Rethinking Religion: Connecting Cognition and Culture*, Cambridge.

Leach, E.R. 2002. "Ritual," in *Ritual and Belief: Readings in the Anthropology of Religion*, D. Hicks, ed., 2nd ed., Boston, pp. 114–121.

Levi, D. 1976. *Festòs e la civiltà minoica* (*Incunabula Graeca* 60), Rome.

Liard, F. 2012. "Petrographic Analysis of Three Neopalatial and Postpalatial Conical Cup Assemblages," in *Excavations at Sissi* III: *Preliminary Report on the 2011 Campaign* (*Aegus* 6), J. Driessen, I. Schoep, M. Anastasiadou, F. Carpentier, I. Crevecoeur, S. Déderix, M. Devolder, F. Gaignerot-Driessen, S. Jusseret, C. Langohr, Q. Letesson, F. Liard, A. Schmitt, C. Tsoraki, and R. Veropoulidou, Louvain, pp. 169–184.

Macdonald, C.F., and C. Knappett. 2007. *Knossos: Protopalatial Deposits in Early Magazine A and the South-West Houses* (*BSA Suppl.* 41), London.

MacGillivray, J.A. 1998. *Knossos: Pottery Groups of the Old Palace Period* (*BSA Studies* 5), London.

———. 2007. "Protopalatial (MM IB–MM IIA): Early Chamber beneath the West Court, Royal Pottery Stores, the Trial KV, and the West and South Polychrome Deposits Groups," in Momigliano, ed., 2007, pp. 105–149.

Marinatos, N. 1993. *Minoan Religion: Ritual, Image, and Symbol*, Columbia, SC.

Marinatos, S. 1929. "Πρωτομινωϊκὸς θολωτὸς τάφος παρὰ τὸ χωρίον Κράσι Πεδιάδος," *ArchDelt* 12 [1932], pp. 102–141.

Martlew, H. 1988. "Domestic Coarse Pottery in Bronze Age Crete," in *Problems in Greek Prehistory. Papers Presented at the Centenary Conference of the British School of Archaeology at Athens, Manchester, April 1986*, E.B. French and K.A. Wardle, eds., Bristol, pp. 421–424.

McGeorge, P.J.P. 2008. "The Human Remains," in Betancourt et al. 2008a, pp. 578–594.

Molloy, B.P.C., J. Day, S. Bridgford, V. Isaakidou, E. Nodarou, G. Kotzamani, M. Milić, T. Carter, P. Westlake, V. Klontza-Jaklova, E. Larsson, and B. Hayden. 2014. "Life and Death of a Bronze Age House: Excavation of Early Minoan I Levels at Priniatikos Pyrgos," *AJA* 118, pp. 307–358.

Momigliano, N. 1990. "The Development of the Footed Goblet ('Egg Cup') from EM II to MMIII at Knossos," in Πεπραγμένα του ΣΤ´ Διεθνούς Κρητολογικού Συνεδρίου Α΄ (1), pp. 477–487.

———. 1991. "MM IA Pottery from Evans' Excavations at Knossos: A Reassessment," *BSA* 86, pp. 149–271.

———. 2000. "Knossos 1902, 1905: The Prepalatial and Protopalatial Deposits from the Room of the Jars in the Royal Pottery Stores," *BSA* 95, pp. 65–105.

Muhly, J.D. 2008. "The Metal Artifacts," in Betancourt et al. 2008a, pp. 557–560.

———. 2014a. "Objects of Copper and Bronze," in Betancourt 2014a, pp. 55–56.

———. 2014b. "Objects of Gold, Silver, and Lead," in Betancourt 2014a, pp. 57–59.

———. 2014c. "Seal Rings," in Betancourt 2014a, pp. 61–62.

Myer, G.H., K.G. McIntosh, and P.P. Betancourt. 1995. "Definition of Pottery Fabrics by Ceramic Petrography," in *Pseira* I: *The Minoan Buildings on the West Side of Area A* (*University Museum Monographs* 94), P.P. Betancourt and C. Davaras, eds., Philadelphia, pp. 143–153.

Nodarou, E. 2003. *Pottery Production, Distribution and Consumption in Early Minoan West Crete: An Analytical Perspective*, Ph.D. diss., University of Sheffield.

———. 2011. *Pottery Production, Distribution and Consumption in Early Minoan West Crete: An Analytical Perspective* (*BAR-IS* 2210), Oxford.

———. 2012. "Pottery Fabrics and Recipes in the Final Neolithic and Early Minoan I Period: The Analytical Evidence from the Settlement and the Rock Shelter of Kephala Petras," in *Petras, Siteia: 25 Years of Excavations and Studies. Acts of a Two-Day Conference Held at the Danish Institute at Athens, 9–10 October 2010* (*Monographs of the Danish Institute at Athens* 16), M. Tsipopoulou, ed., Athens, pp. 81–88.

———. 2013. "Petrographic Analysis of the Pottery," in *Aphrodite's Kephali: An Early Minoan I Defensive Site in Eastern Crete* (*Prehistory Monographs* 41), P.P. Betancourt, Philadelphia, pp. 151–169.

Nodarou, E., and I. Iliopoulos. 2011. "Analysis of Postpalatial Pottery from Karphi," in *The Pottery from Karphi: A Re-Examination* (*BSA Studies* 19), L.P. Day, Athens, pp. 337–348.

Nodarou, E., and J. Moody. 2014. "Mirabello Fabric(s) Forever: An Analytical Study of the Granodiorite Pottery of the Vrokastro Area from the Final Neolithic Period to Modern Times," in *A Cretan Landscape through Time: Priniatikos Pyrgos and Environs* (*BAR-IS* 2634), B.P.C. Molloy and C.N. Duckworth, eds., Oxford, pp. 91–98.

Nodarou, E., and C. Rathossi. 2008. "Petrographic Analysis of Selected Animal Figurines from Syme Viannou," in *The Sanctuary of Hermes and Aphrodite at Syme Viannou* IV: *Animal Images of Clay* (Βιβλιοθήκη τῆς ἐν Ἀθήναις Ἀρχαιολογικῆς Ἑταιρείας 256), P. Muhly, ed., Athens, pp. 165–182.

Nowicki, K. 1996. "Lasithi (Crete): One Hundred Years of Archaeological Research," *Aegean Archaeology* 3, pp. 27–47.

Panagiotakis, N., and M. Panagiotaki. 2011. "Architectural Remains of the Bronze Age Pediada in Their Landscape," in Πεπραγμένα Ι' Διεθνοῦς Κρητολογικοῦ Συνεδρίου Α' (2), Chania, pp. 727–746.

Panagiotakis N., M. Panagiotaki, M. Guzowska, and K. Paschalidis. 2011. "Μελετώντας τη Μέση και την Ύστερη Εποχή του Χαλκού στην Επαρχία Πεδιάδας της Κεντρικής Κρήτης," in Πεπραγμένα Ι' Διεθνοῦς Κρητολογικοῦ Συνεδρίου Α' (2), Chania, pp. 747–773.

Papadatos, Y. 2005. *Tholos Tomb Gamma: A Prepalatial Tholos Tomb at Phourni, Archanes* (*Prehistory Monographs* 17), Philadelphia.

———. 2008. "The Neolithic-Early Bronze Age Transition in Crete: New Evidence from the Settlement at Petras Kephala, Siteia," in *Escaping the Labyrinth: The Cretan Neolithic in Context* (*Sheffield Studies in Aegean Archaeology* 8), V. Issakidou and P. Tomkins, eds., Oxford, pp. 261–275.

Papadatos, Y., and P. Tomkins. 2011. "Final Neolithic and Early Minoan Pottery from the Pediada Survey," in Πεπραγμένα Ι' Διεθνοῦς Κρητολογικοῦ Συνεδρίου Α' (2), Chania, pp. 713–726.

Papadatos, Y., P. Tomkins, E. Nodarou, and I. Iliopoulos. Forthcoming. "The Beginning of Early Bronze Age in Crete: Continuities and Discontinuities in the Ceramic Assemblage at Petras Kephala, Siteia," in *Proceedings of the International Conference "The Early Bronze Age in the Aegean: New Evidence" Held in Athens, 11–14 April 2008*, Athens.

Peatfield, A. 1987. "Palace and Peak: The Political and Religious Relationship Between Palaces and Peak Sanctuaries," in *The Function of the Minoan Palaces* (*SkrAth* 4°, 35), R. Hägg and N. Marinatos, eds., Stockholm, pp. 89–93.

Pelon, O. 1970. *Fouilles exécutées à Mallia: Exploration des maisons et quartiers d'habitation (1963–1966). Le quartier E* (*ÉtCrét* 16), Paris.

Pendlebury, J.D.S. 1936–1937. "Lasithi in Ancient Times," *BSA* 37, pp. 194–200.

———. 1939. *The Archaeology of Crete: An Introduction*, London.

Pendlebury, J.D.S., H.W. Pendlebury, and M.B. Money-Coutts. 1935–1936. "Excavations in the Plain of Lasithi. I: The Cave of Trapeza," *BSA* 36, pp. 1–131.

Peterson, S.E. 2009. *Retrieval of Materials with Water Separation Machines* (*INSTAP Archaeological Excavation Manual* 1), Philadelphia.

Pini, I. 1968. *Beiträge zur minoischen Gräberkunde*, Wiesbaden.

Poursat, J.-C. 1996. *Fouilles exécutées à Malia: Le Quartier Mu* III. *Artisans Minoens: Les maisons-ateliers du Quartier Mu* (*ÉtCrét* 32), Athens.

Poursat, J.-C., and C. Knappett. 2005. *Fouilles exécutées à Malia: Le Quartier Mu* IV. *La poterie du Minoen Moyen II: Production et utilisation.* (*ÉtCrét* 33), Paris.

Rethemiotakis, G., and K.S. Christakis. 2004. "Contacts between Knossos and the Pediada Region in Central Crete," in *Knossos: Palace, City, State. Proceedings of the Conference in Herakleion Organised by the British School at Athens and the 23rd Ephoreia of Prehistoric and Classical Antiquities of Herakleion, in November 2000, for the Centenary of Sir Arthur Evans's Excavations at Knossos* (*BSA Studies* 12), G. Cadogan, E. Hatzaki, and A. Vasilakis, eds., London, pp. 169–175.

Rutkowski, B. 1986. *The Cult Places in the Aegean World*, Warsaw.

Rutkowski, B., and K. Nowicki. 1996. *The Psychro Cave and Other Sacred Grottoes in Crete* (*Studies and Monographs in Mediterranean Archaeology and Civilization* 2°, 1), Warsaw.

Sakellarakis, Y., and E. Sapouna-Sakellaraki. 1997. *Archanes: Minoan Crete in a New Light*, Athens.

Seager, R.B. 1906–1907. "Report of Excavations at Vasiliki, Crete, in 1906," *University of Pennsylvania Transactions of the Department of Archaeology, Free Museum of Science and Art* 2, pp. 111–132.

Seager, R. 1912. *Explorations on the Island of Mochlos*, Boston.

Serpetsidaki, I. 2006. "Προανακτορικός σπηλαιώδης τάφος στο Κυπαρίσσι Τεμένους," in Πεπραγμένα

Θ΄ Διεθνούς Κρητολογικού Συνεδρίου Α΄ (2), Herakleion, pp. 243–258.

Shank, E. 2005. "New Evidence for Anatolian Relations with Crete in EM I–IIA," in *Emporia: Aegeans in the Central and Eastern Mediterranean. Proceedings of the 10th International Aegean Conference, Italian School of Archaeology in Athens, 14–18 April 2004* (*Aegaeum* 25), R. Laffineur and E. Greco, eds., Liège, pp. 103–106.

Soles, J. 1992. *The Prepalatial Cemeteries at Mochlos and Gournia and the House Tombs of Bronze Age Crete* (*Hesperia Suppl.* 24), Princeton.

Todaro, S. 2001. "Nuove prospettive sulla produzione in stile Pyrgos nella Creta meridionale in caso della pisside e della coppa su base ad anello," *Creta Antica* 2, pp. 11–28.

Tomkins, P. 2007. "Neolithic: Strata IX–VIII, VII–VIB, VIA–V, IV, IIIB, IIIA, IIB, IIA and IC Groups," in Momigliano, ed., 2007, pp. 9–48.

Tsipopoulou, M., and M. Wedde. 2000. "Διαβάζοντας ένα χωμάτινο παλίμψηστο: Στρωματογραφικές τομές στο ανακτορικό κτήριο του Πετρά Σητείας," in *Πεπραγμένα Η΄ Διεθνούς Κρητολογικού Συνεδρίου* Α΄ (3), Herakleion, pp. 359–377.

Tyree, E.L. 1974. *Cretan Sacred Caves: Archaeological Evidence*, Ph.D. diss., University of Missouri, Columbia.

Tyree, L., A. Kanta, and H.L. Robinson. 2008. "Evidence for Ritual Eating and Drinking: A View from Skoteino Cave," in Hitchcock, Laffineur, and Crowley, eds., 2008, pp. 179–187.

Tzedakis, Y., and H. Martlew, eds. 1999. *Minoans and Mycenaeans: Flavours of Their Time*, Athens.

Tzedakis, Y., H. Martlew, and M.K. Jones. eds. 2008. *Archaeology Meets Science: Biomolecular Investigations in Bronze Age Greece. The Primary Evidence, 1997–2003*, Oxford.

van Gennep, A. 1961. *The Rites of Passage*, M.B. Vizedom and G.L. Caffee, trans, London.

Vasilakis, A., and K. Branigan. 2010. *Moni Odigitria: A Prepalatial Cemetery and Its Environs in the Asterousia, Southern Crete* (*Prehistory Monographs* 30), Philadelphia.

Walberg, G. 1976. *Kamares: A Study of Palatial Middle Minoan Pottery* (*Uppsala Studies in Ancient Mediterranean and Near Eastern Civilizations* 8), Uppsala.

———. 1983. *Provincial Middle Minoan Pottery*, Mainz am Rhein.

Watrous, L.V. 1982. *Lasithi: A History of a Settlement on a Highland Plain in Crete* (*Hesperia Suppl.* 18), Princeton.

Wilson, D.E. 1985. "The Pottery and the Architecture of the EM IIA West Court House at Knossos," *BSA* 80, pp. 281–364.

———. 2007. "Early Prepalatial (EM I–EM II): EM I Well, West Court House, North-East Magazines and South Front Groups," in Momigliano, ed., 2007, pp. 49–77.

Wilson, D.E., and P. Day. 1994. "Ceramic Regionalism in Prepalatial Central Crete: The Mesara Imports at EM I to EM IIA Knossos," *BSA* 89, pp. 1–87.

———. 2000. "EM I Chronology and Social Practice: Pottery from the Early Palace Tests at Knossos," *BSA* 95, pp. 21–63.

Wright, J.C., ed. 2004. *The Mycenaean Feast* (*Hesperia* 73 [2]), Princeton.

Xanthoudides, S. 1906. "Προϊστορικὴ οἰκία εἰς Χαμαίζι Σητείας," *ArchEph* 45, pp. 117–154.

———. 1918. "Μέγας πρωτομινωϊκὸς τάφος Πύργου," *ArchDelt* 4 [1921], pp. 136–170.

———. 1924. *The Vaulted Tombs of Mesara*, London.

Zois, A. 1969. *Προβλήματα χρονολογίας τῆς μινωϊκῆς κεραμεικῆς: Γούρνες—Τύλισος—Μάλια*, Athens.

Index

Figures

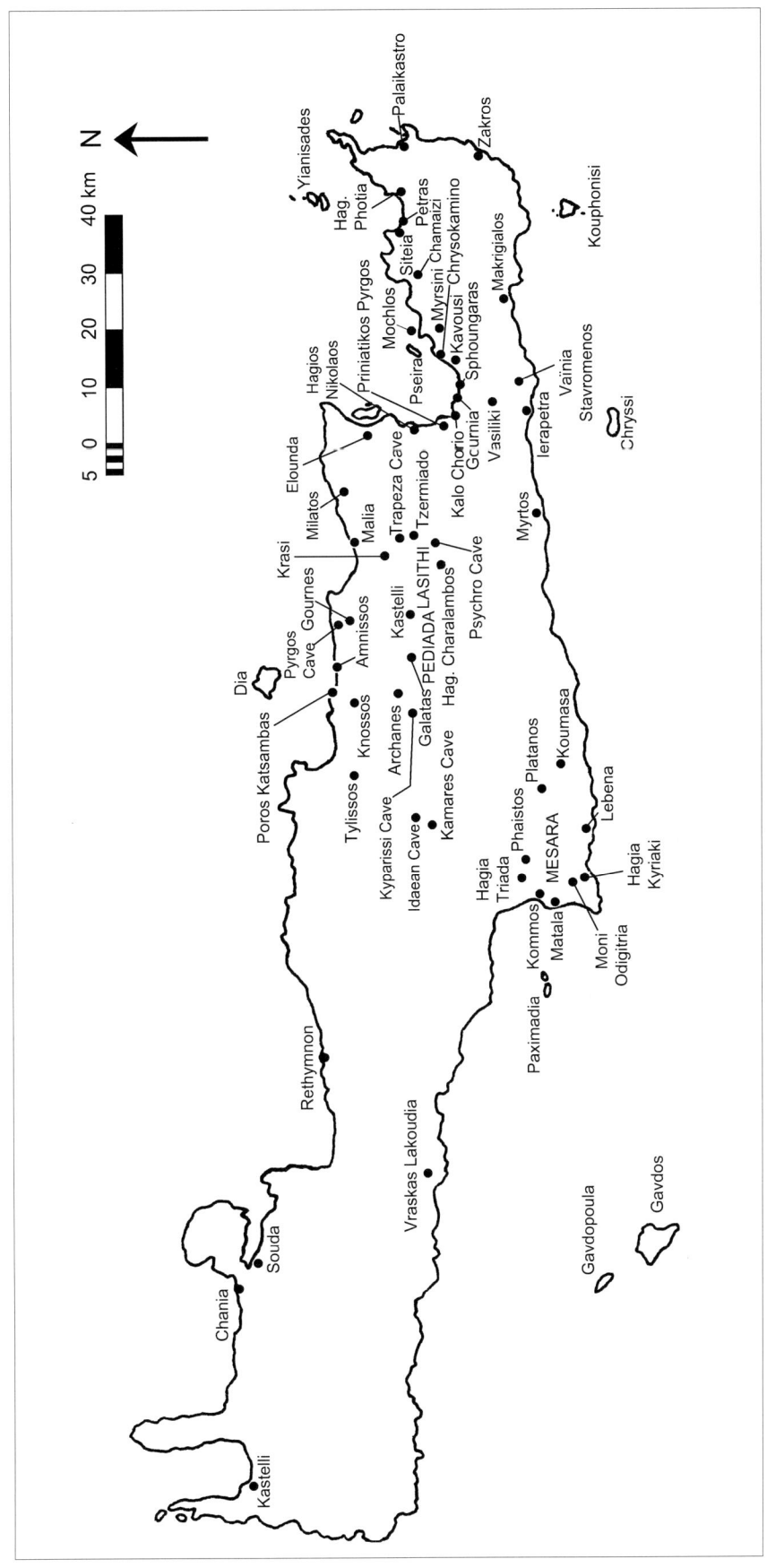

Figure 1. Map of Crete.

FIGURE 2

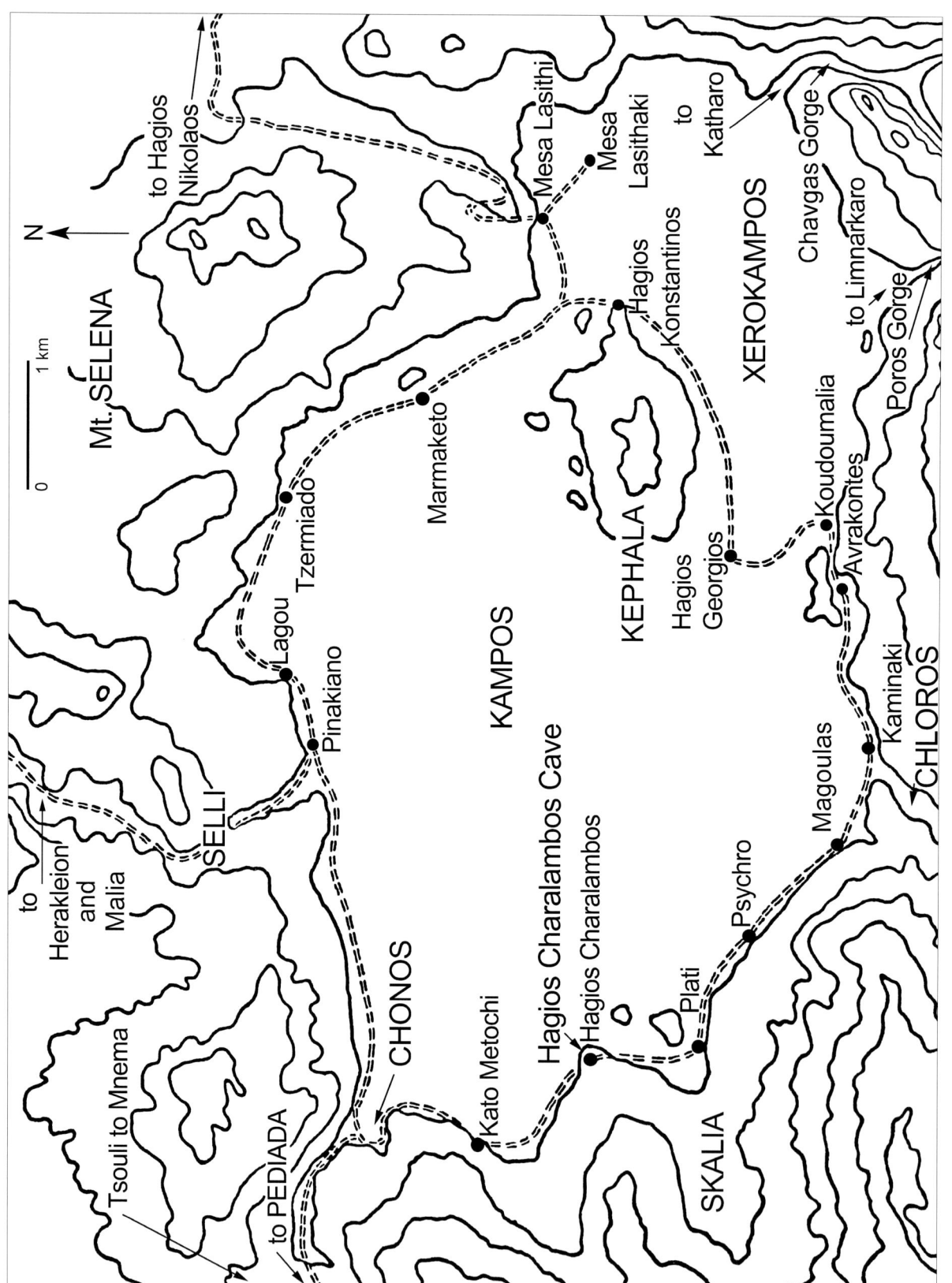

Figure 2. Map of the Lasithi Plain.

FIGURE 3

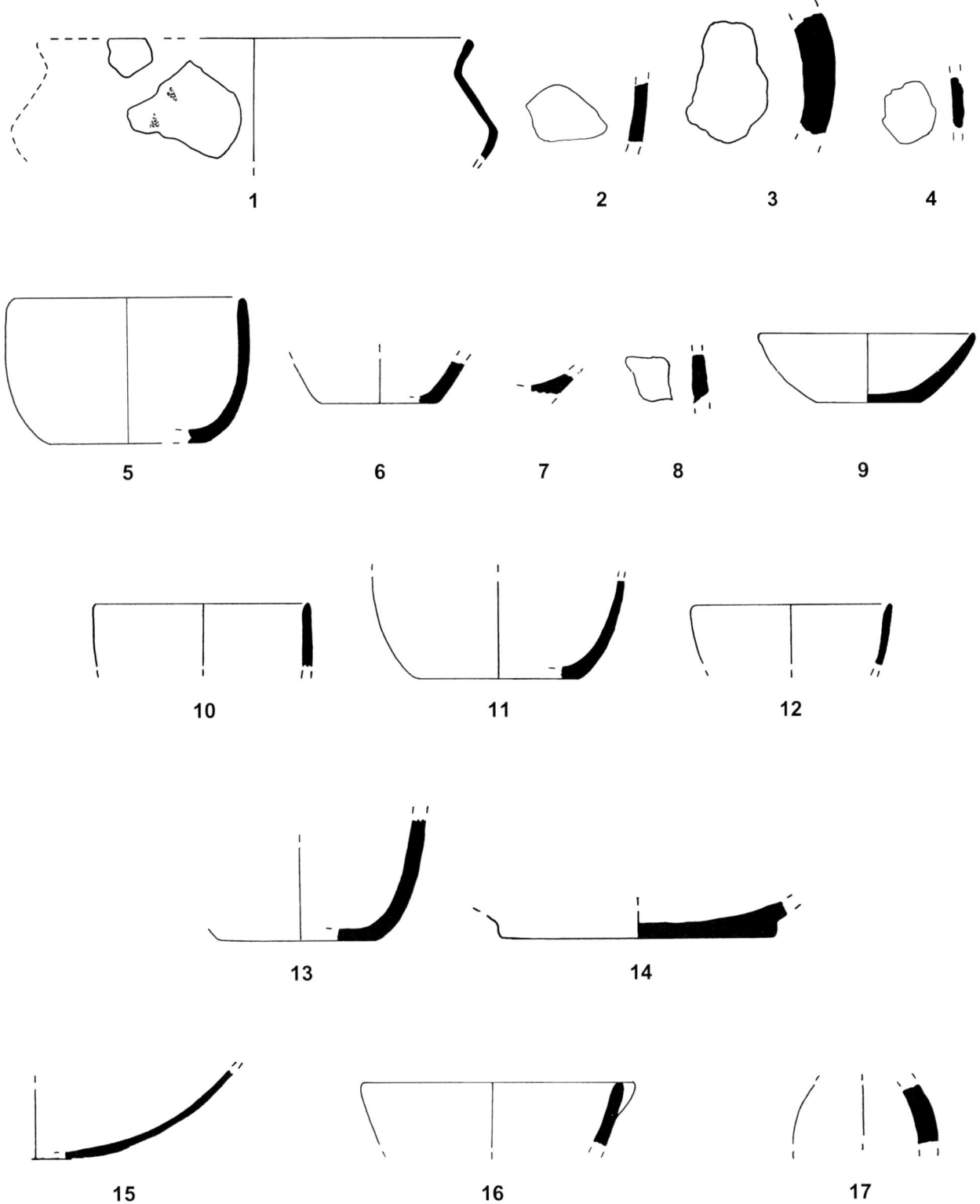

Figure 3. Neolithic to EM I coarse fabrics: carinated bowl with everted rim (**1**), cups and other open vessels (**2**, **4–15**), closed vessel with handle (**3**), cup with lug handle (**16**), and jar with strap handle (**17**). Scale 1:3.

FIGURE 4

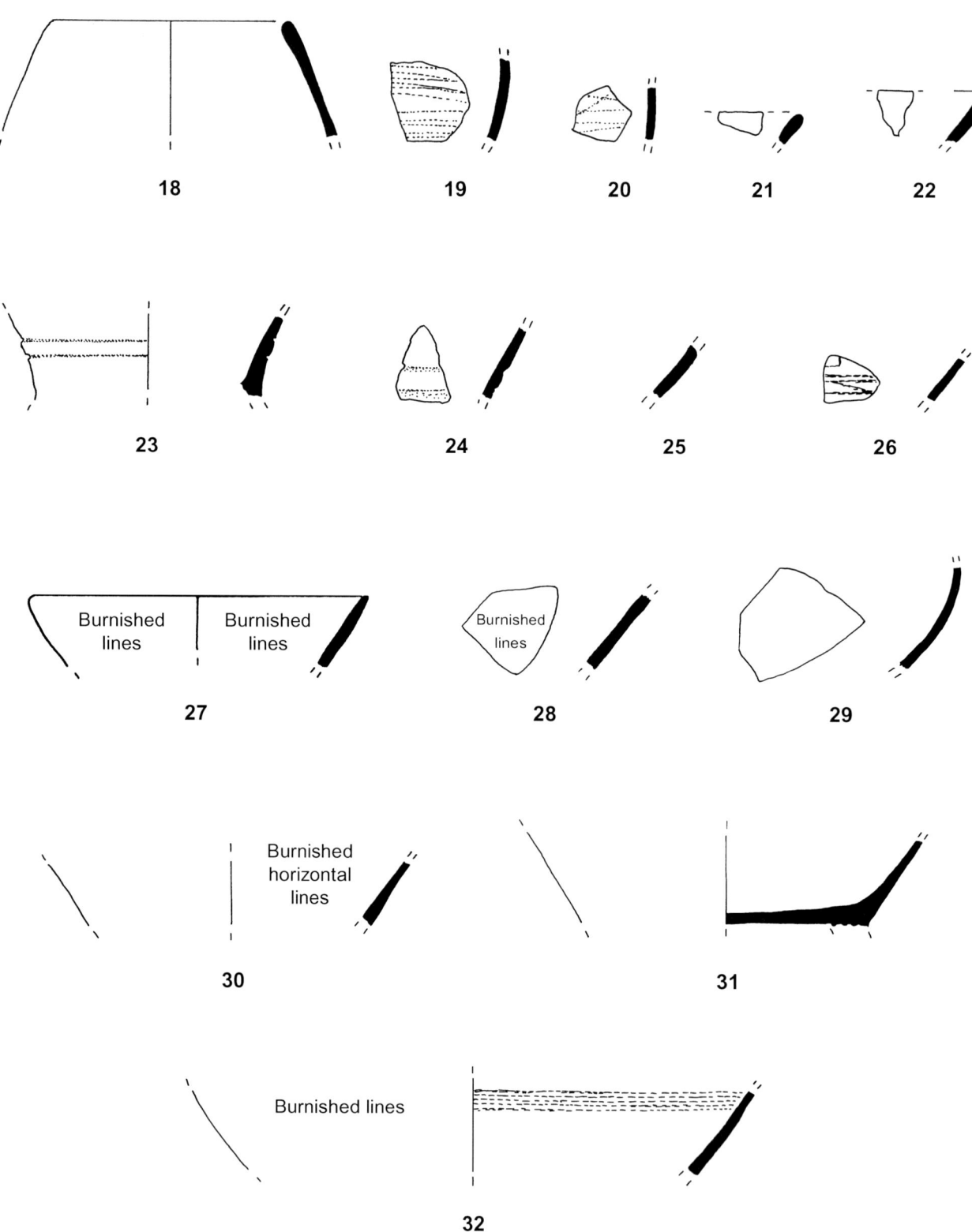

Figure 4. Neolithic to EM I coarse fabrics: jar (**18**), open vessels (**19–22, 28, 29**), Pyrgos Style chalices (**23–27**), and other chalices (**30–32**). Scale 1:3.

FIGURE 5

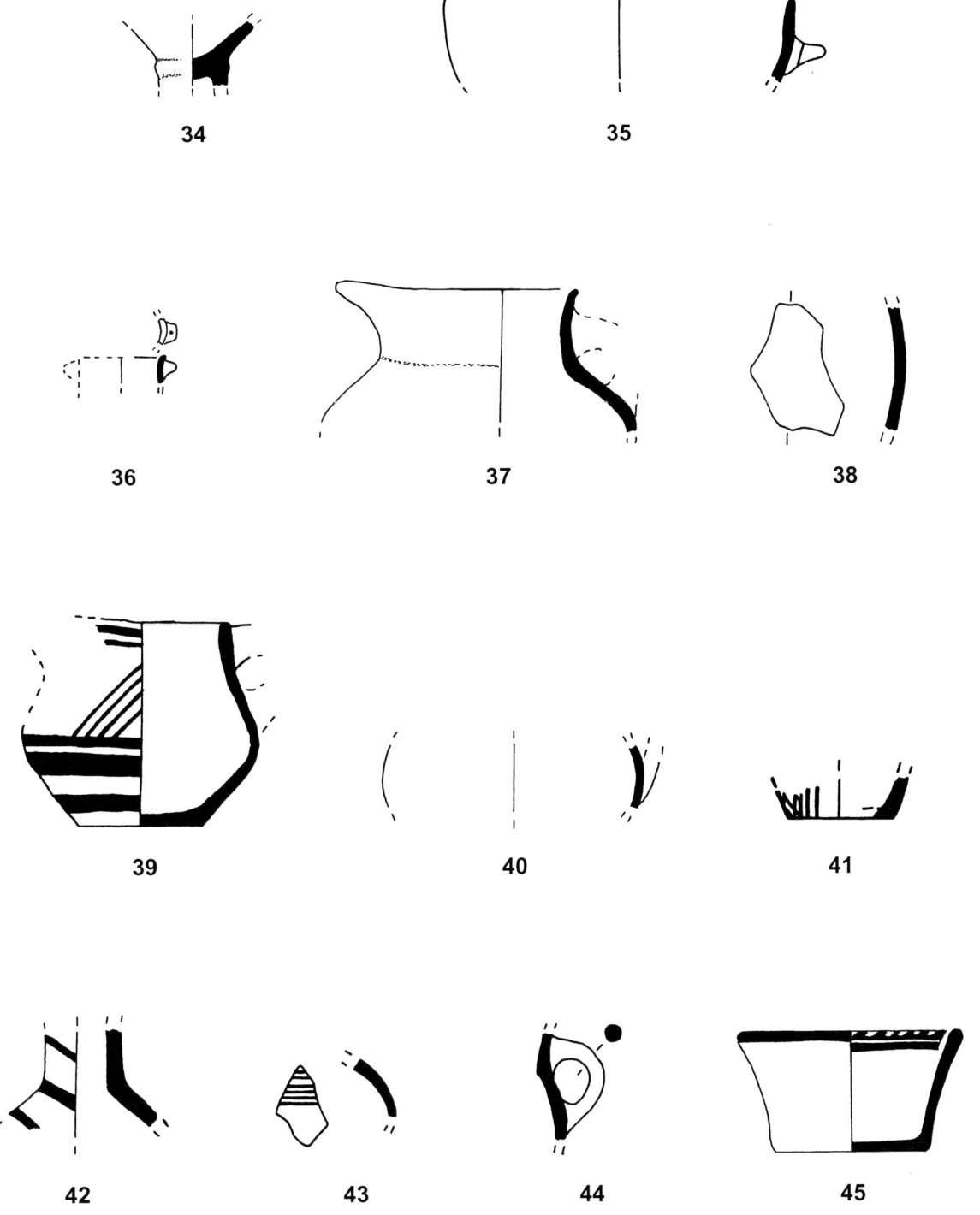

Figure 5. Neolithic to EM I coarse fabrics: goblet (**34**), open vessel with horizontal lug handle (**35**), miniature vessel with pierced lug handle (**36**), and large jug with pointed spout (**37**). EM I–IIA coarse and fine fabrics: open vessel (**38**), jugs (**39–44**), and Koumasa Style cup (**45**). Scale 1:3.

FIGURE 6

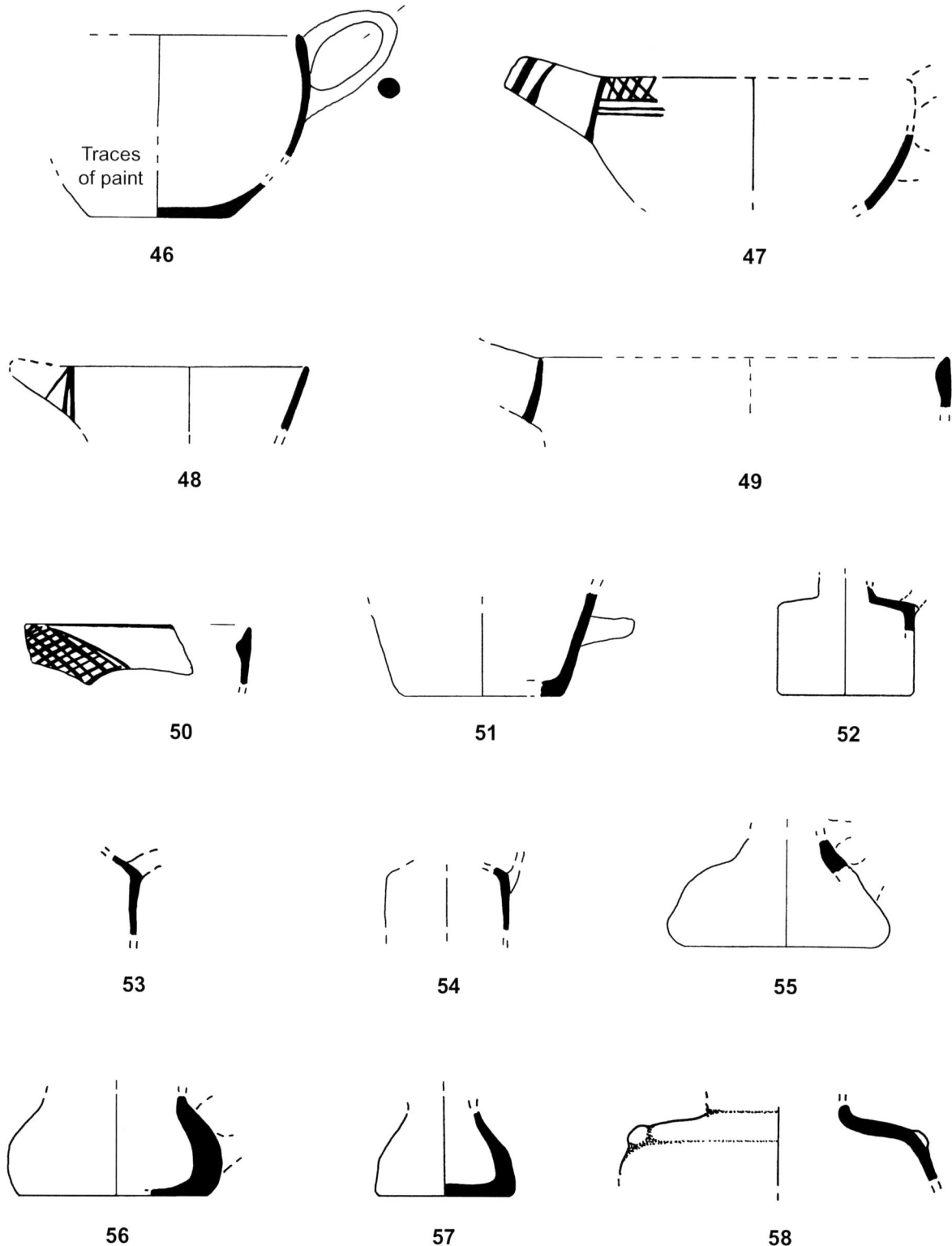

Figure 6. EM IIA Koumasa Style: cup (**46**) and wide-mouthed vessels with spout (**47–50**). EM IIA Monochrome Style: bowl with horizontal handle (**51**); cylindrical jugs (**52–54**); irregular conical jugs (**55–57**); and pyxis (**58**). Scale 1:3.

FIGURE 7

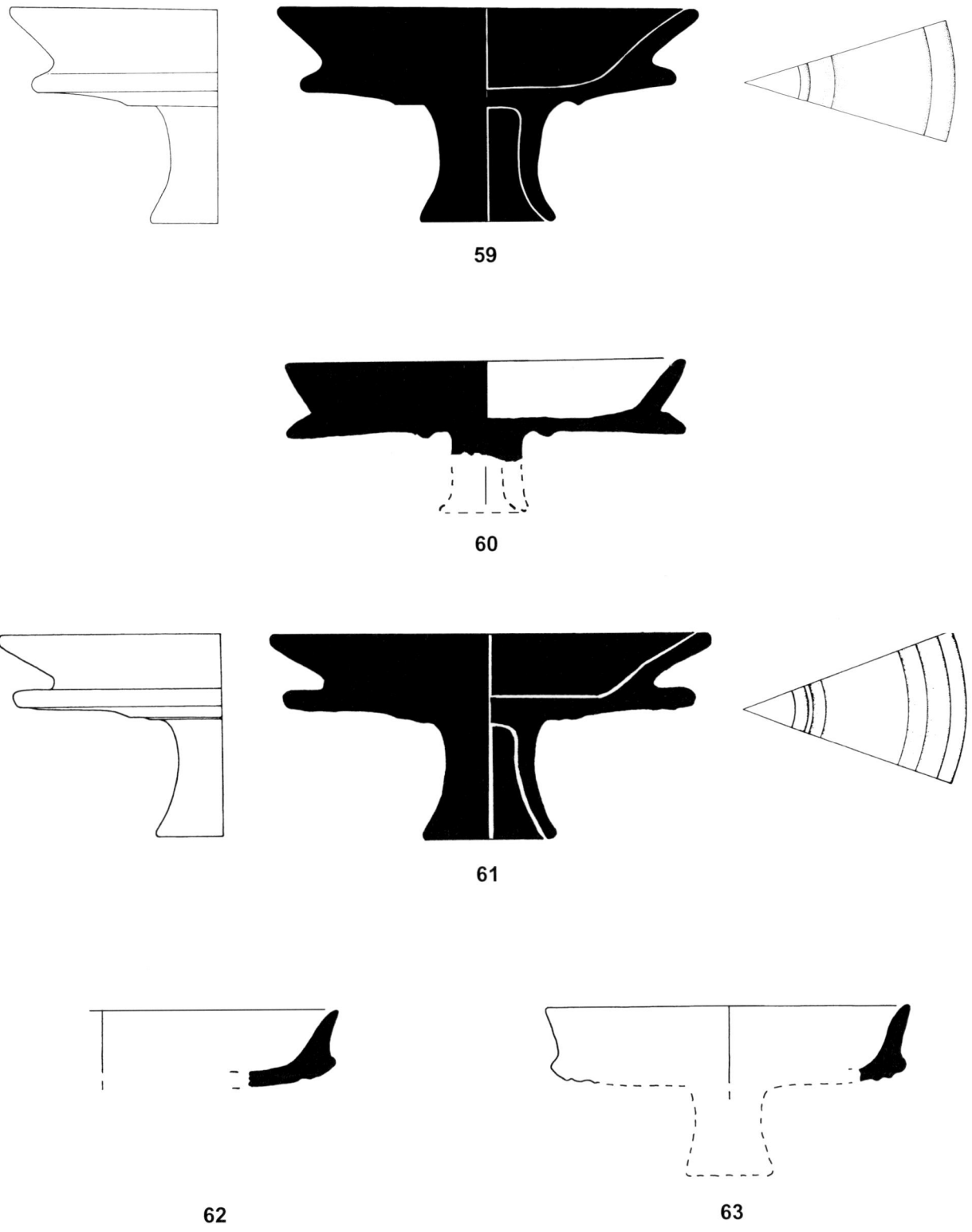

59

60

61

62

63

Figure 7. EM I–IIA dark burnished offering tables: Type B (**59**, **60**), Type D (**61**), and Type E (**62**, **63**). Scale 1:3.

FIGURE 8

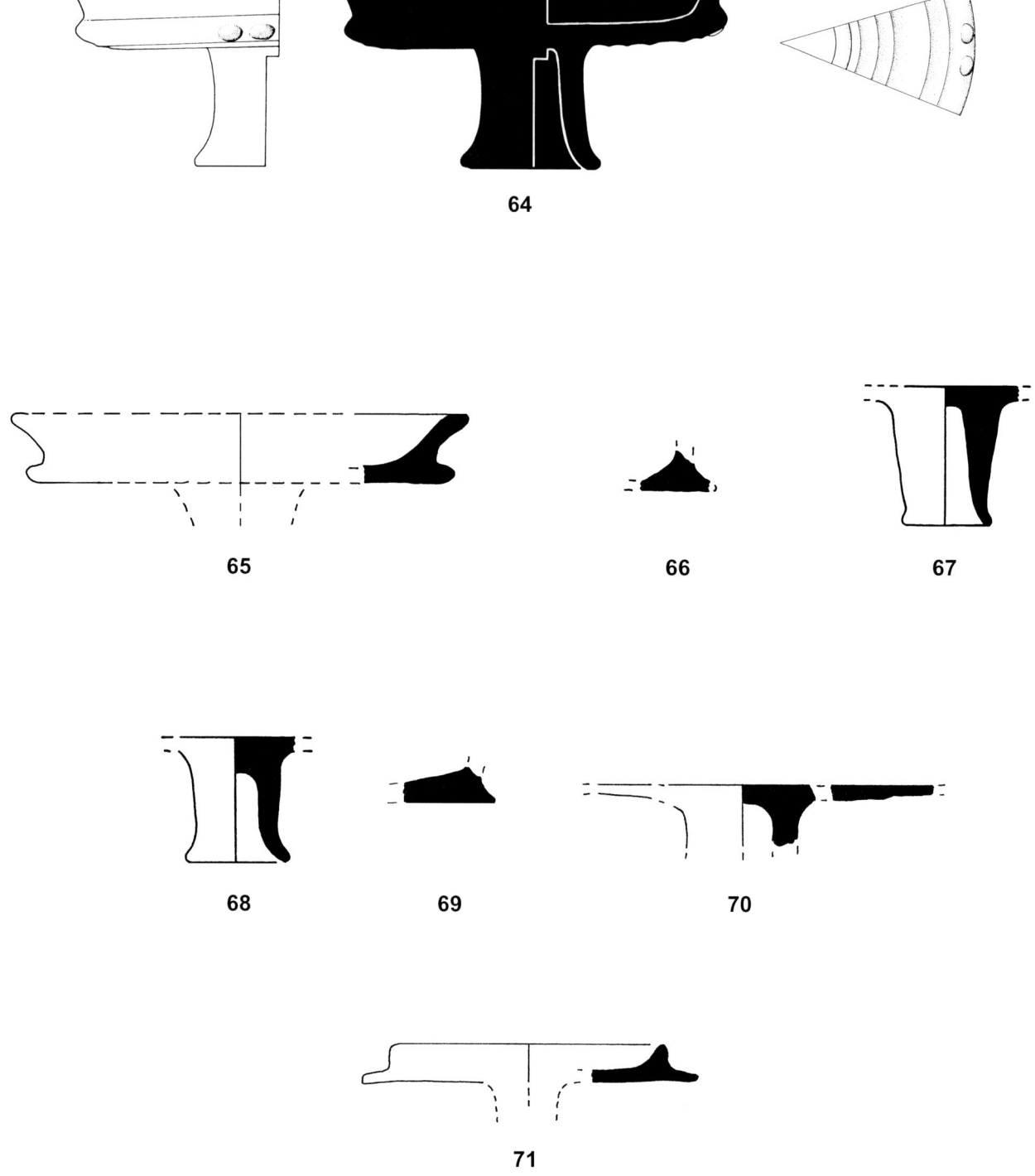

Figure 8. EM I–IIA dark burnished offering tables: Type E (**64**), Type I (**65**), and unknown plastic decoration (**66**–**71**). Scale 1:3.

FIGURE 9

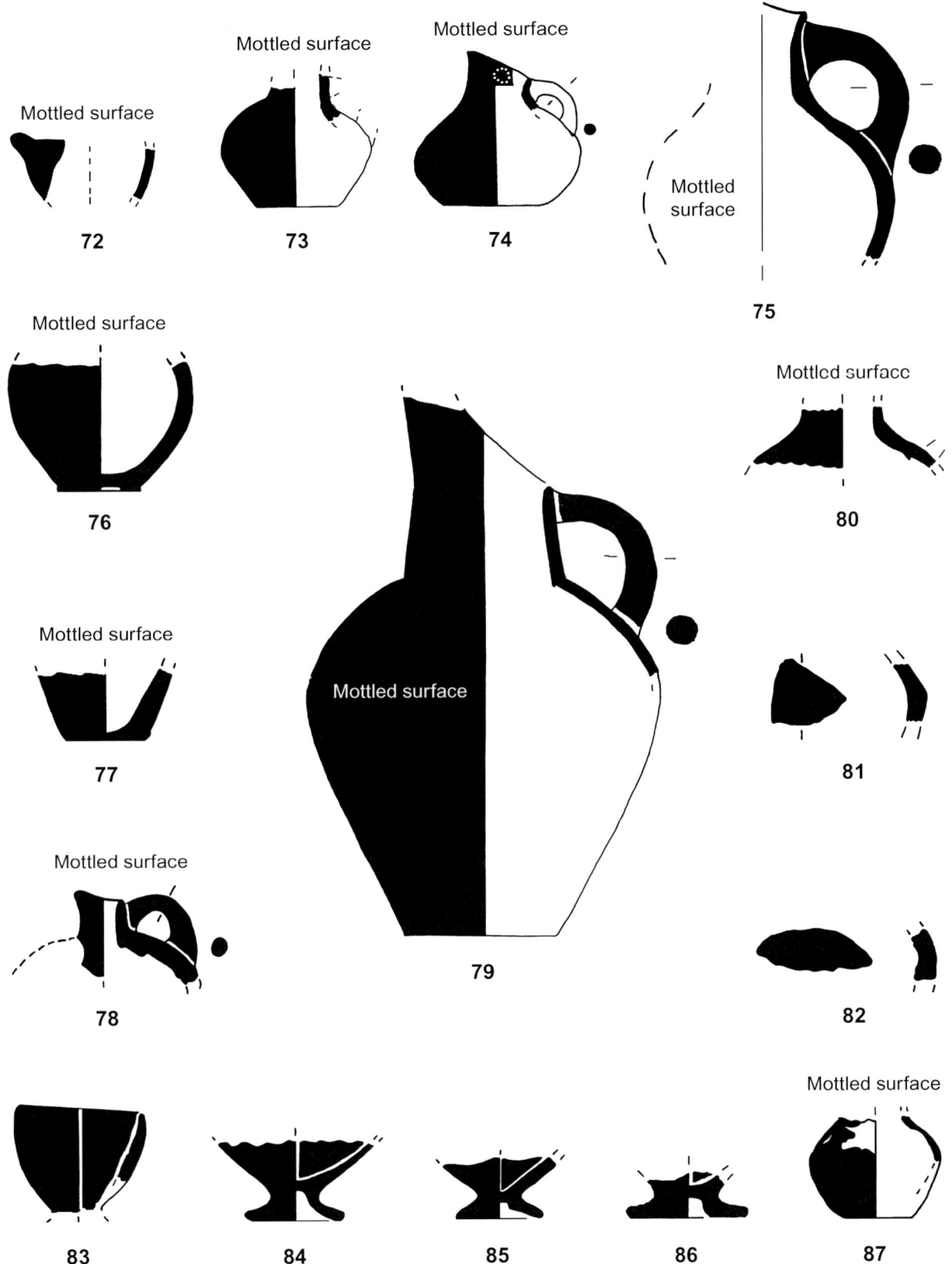

Figure 9. EM IIB Vasiliki Ware: spouted goblet (**72**), jugs (**73–82**). EM IIB imitations in pale fabrics of Vasiliki Style: goblet (**83–86**) and small jug (**87**). Scale 1:3.

FIGURE 10

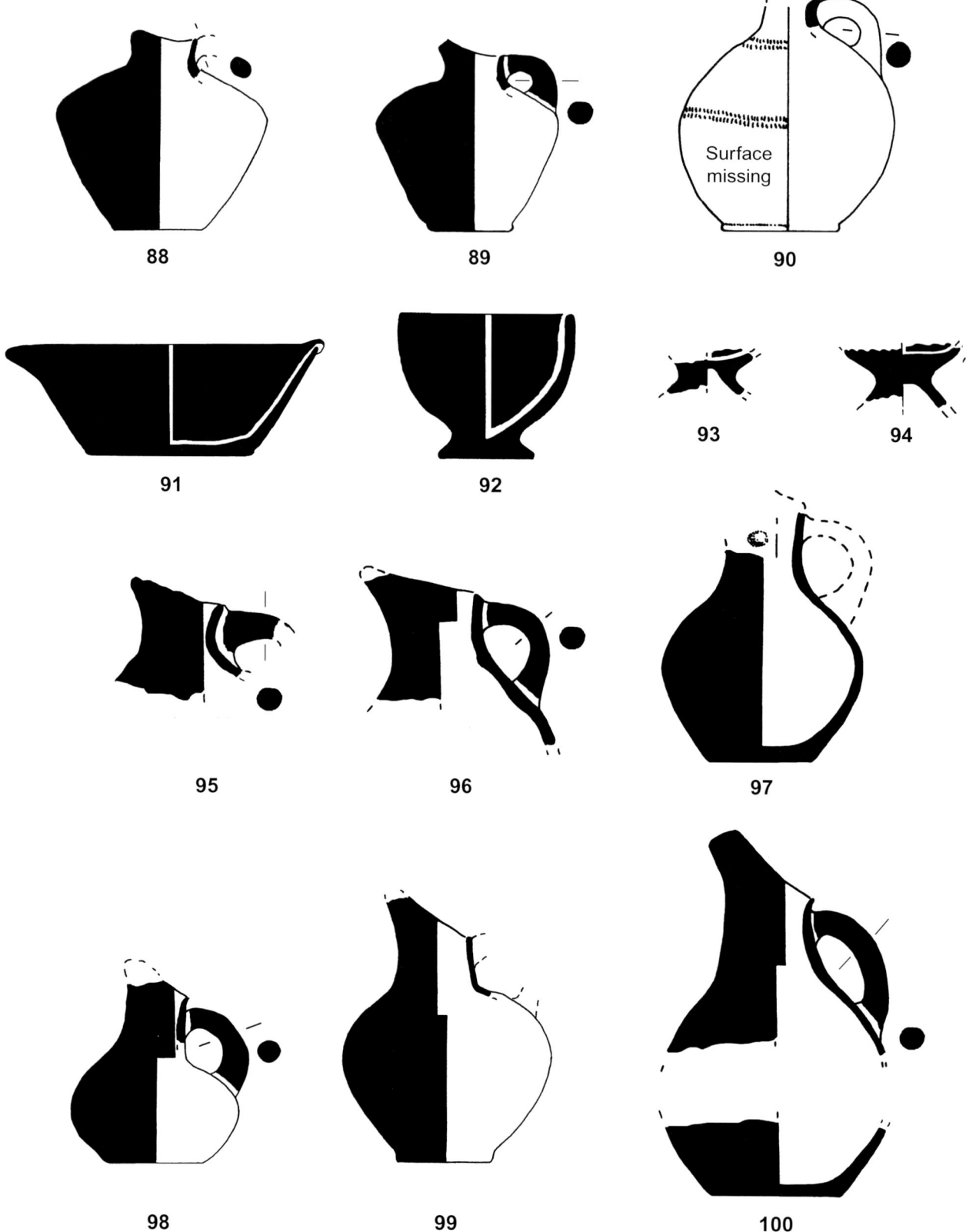

Figure 10. EM IIB Vasiliki Style imitations in pale fabrics: jugs (**88–90**). EM IIB Vasiliki Style imitations in the Lasithi Red Fabric Group: spouted conical bowl (**91**), goblets (**92–94**), and jugs (**95–100**). Scale 1:3.

FIGURE 11

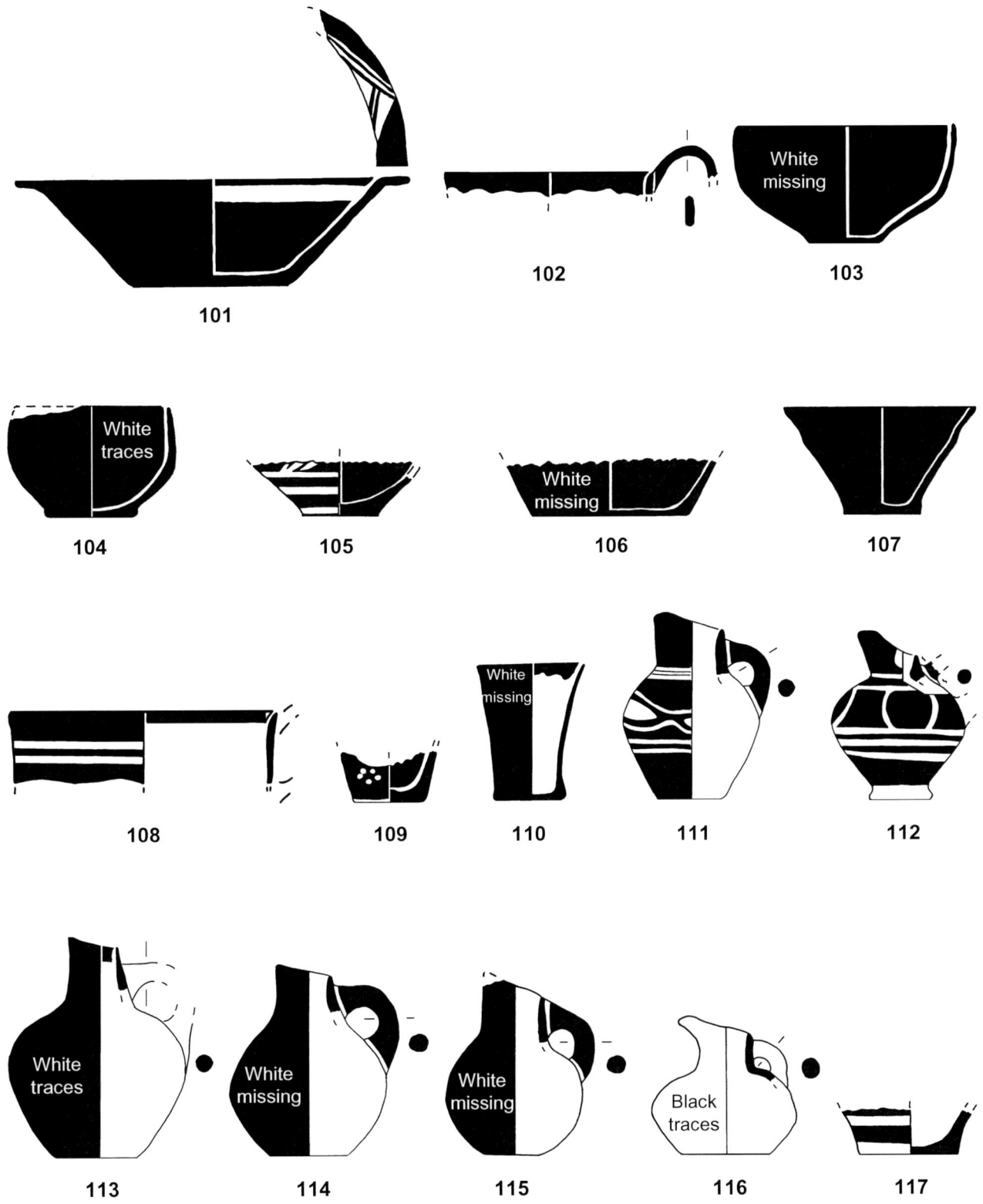

Figure 11. EM III–MM II vessels with pale fabrics, light-on-dark decoration: shallow bowl (**101**), cups (**102–108**), tumblers (**109–110**), and small sized jugs (**111–117**). Scale 1:3.

FIGURE 12

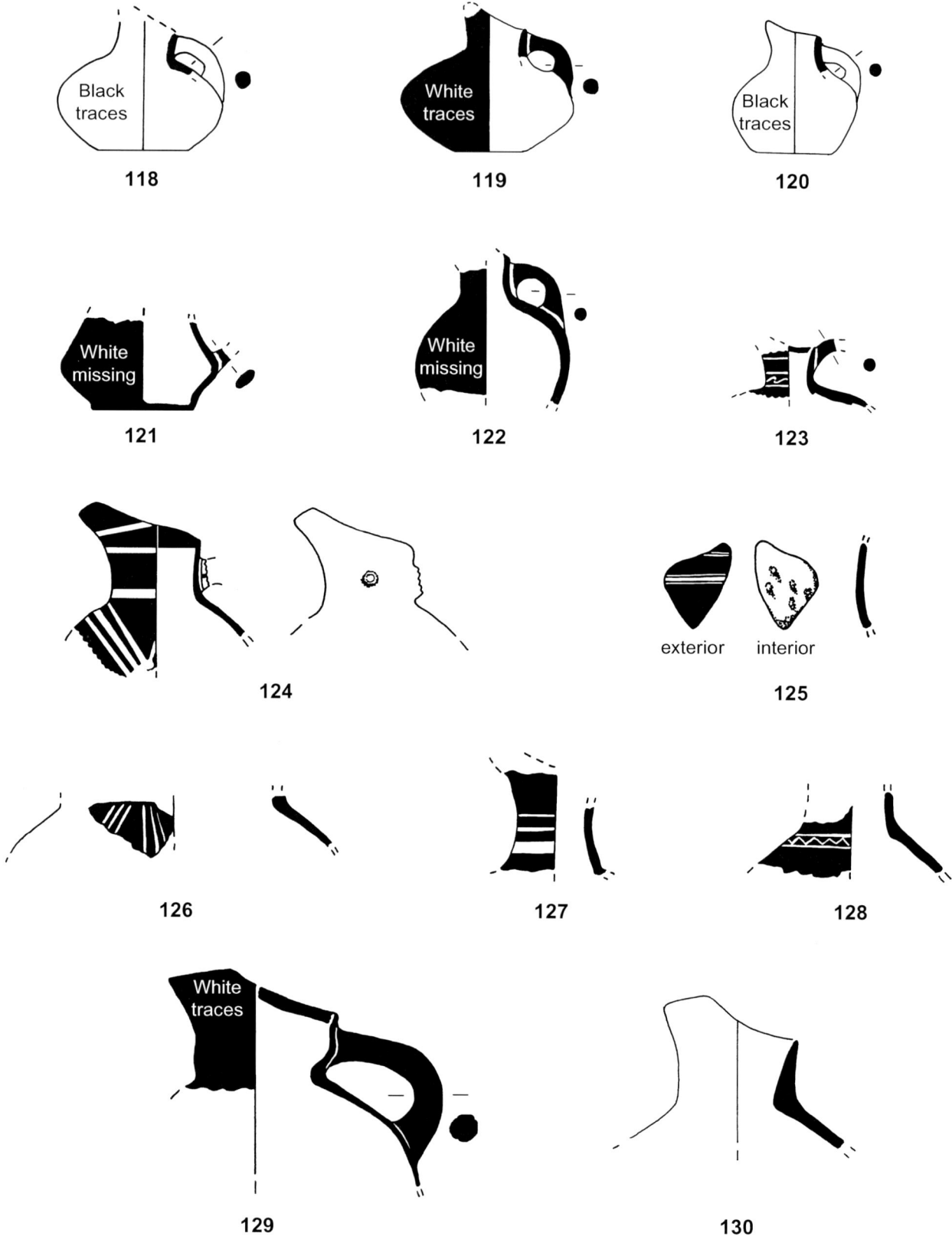

Figure 12. EM III–MM II vessels with pale fabrics and light-on-dark decoration: small jugs (**118–123**) and medium to large jugs (**124–130**). Scale 1:3.

FIGURE 13

Figure 13. EM III–MM II vessels with pale fabrics and light-on-dark decoration: medium to large jug (**131**), closed vessel (**132**), jar with pierced vertical handles (**133**), miscellaneous pouring vessels (**134–138**), jars (**139–141**), and pyxis (**142**). Scale 1:3.

FIGURE 14

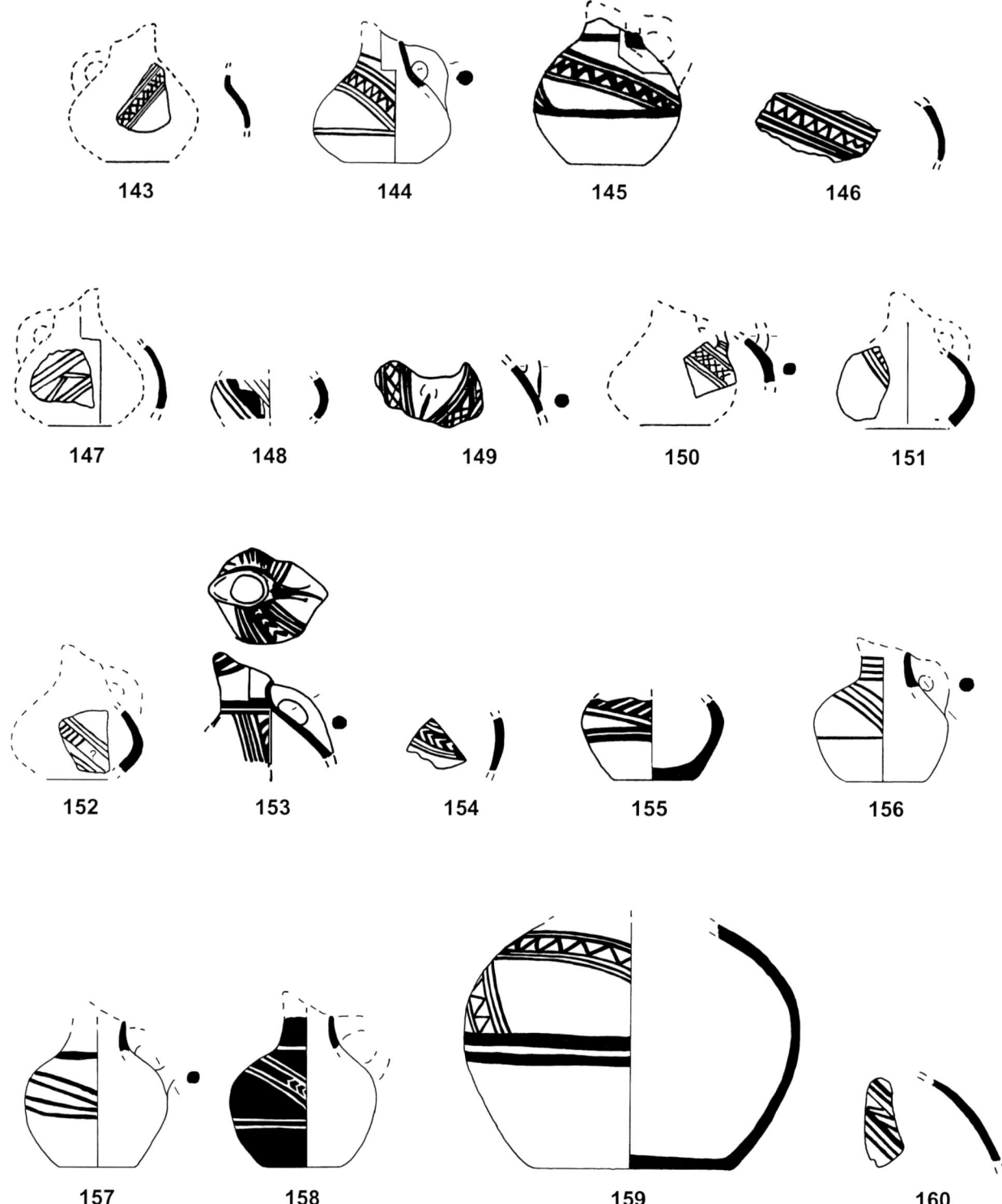

Figure 14. MM I Diagonal Line Style jugs (**143–160**). Scale 1:3.

FIGURE 15

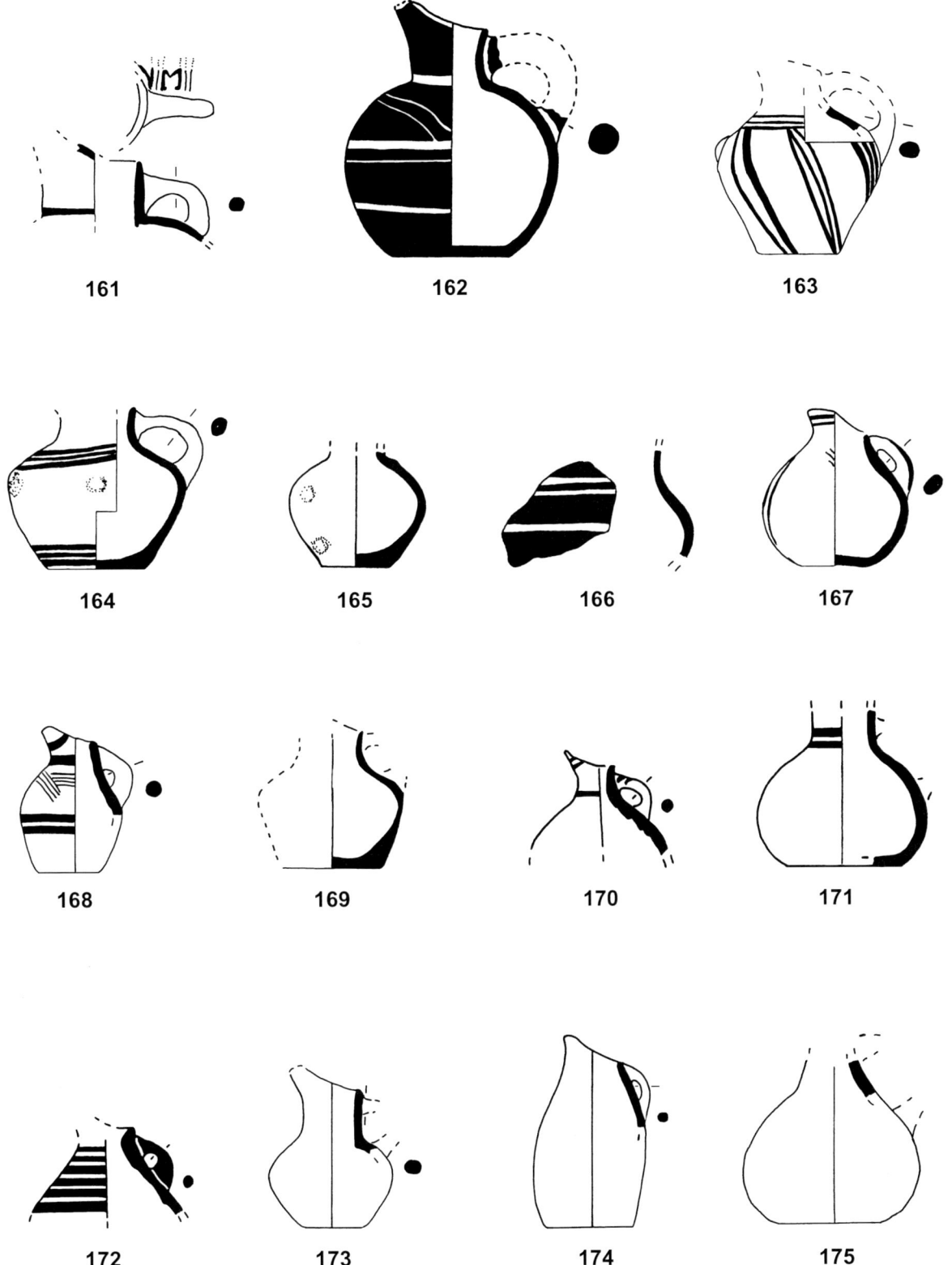

Figure 15. MM I: Diagonal Line Style jugs (**161**, **162**), jugs with bosses on body (**163–166**), and various classes of small jugs (**167–175**). Scale 1:3.

FIGURE 16

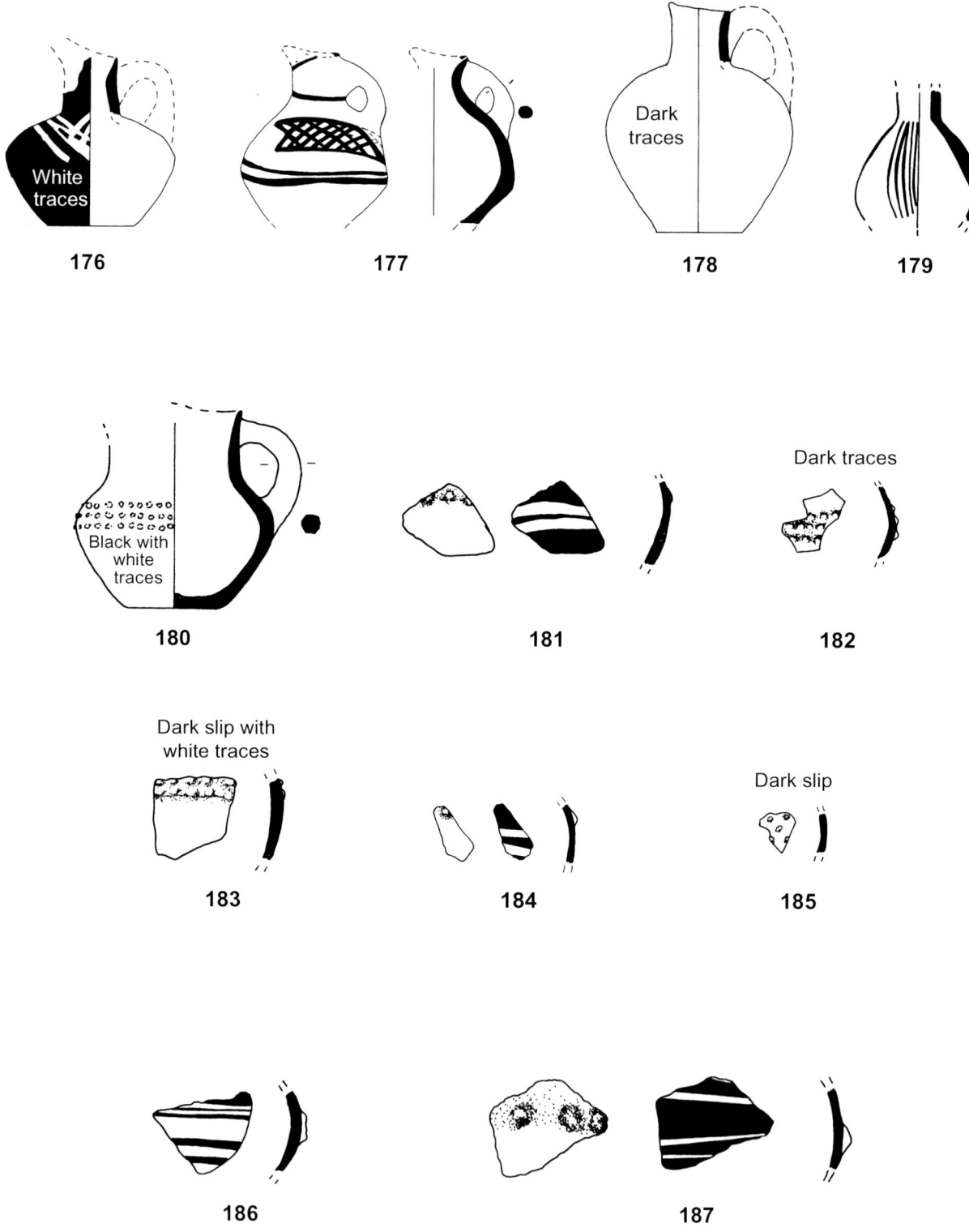

176   **177**   **178**   **179**

White traces

Dark traces

180   **181**   **182**

Black with white traces

Dark traces

Dark slip with white traces

183   **184**   **185**

Dark slip

186   **187**

Figure 16. MM I various classes of small jugs (**176–179**). MM I–II Barbotine Style jugs (**180–187**). Scale 1:3.

FIGURE 17

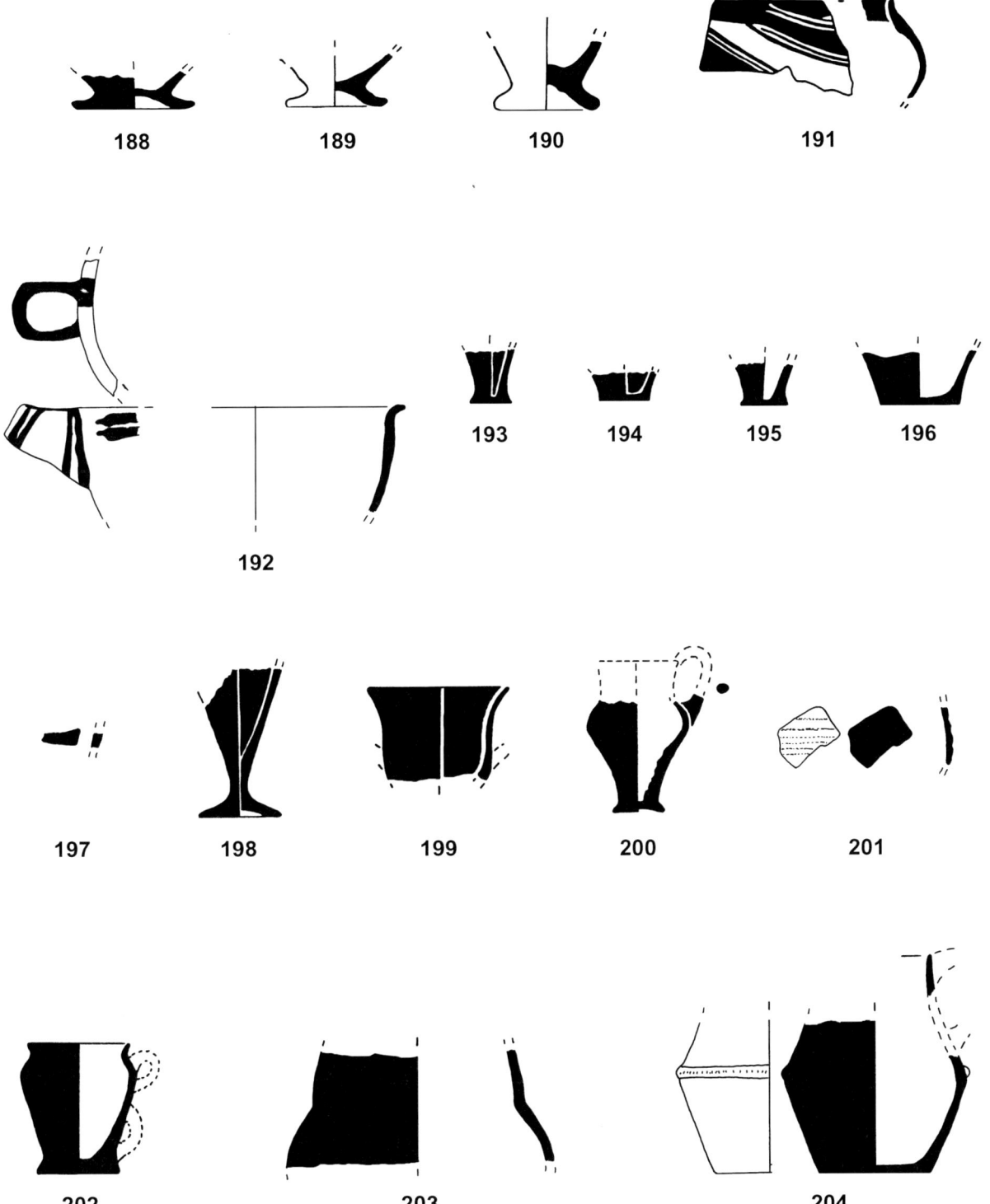

Figure 17. EM II–MM I Knossian pottery: goblets (**188–190**), spouted jar (**191**), and wide-mouthed vessel with a bridged-spout (**192**). Mostly MM IB fine red fabrics with dark red slip imported from sites in the Pediada: tumblers (**193–197**), goblet (**198**), kantharos (**199**), tall carinated cup (**200**), carinated cup with grooves (**201**), minitaure pithos (**202**), small jug(?) (**203**), and jug with incised band on body (**204**). Scale 1:3.

FIGURE 18

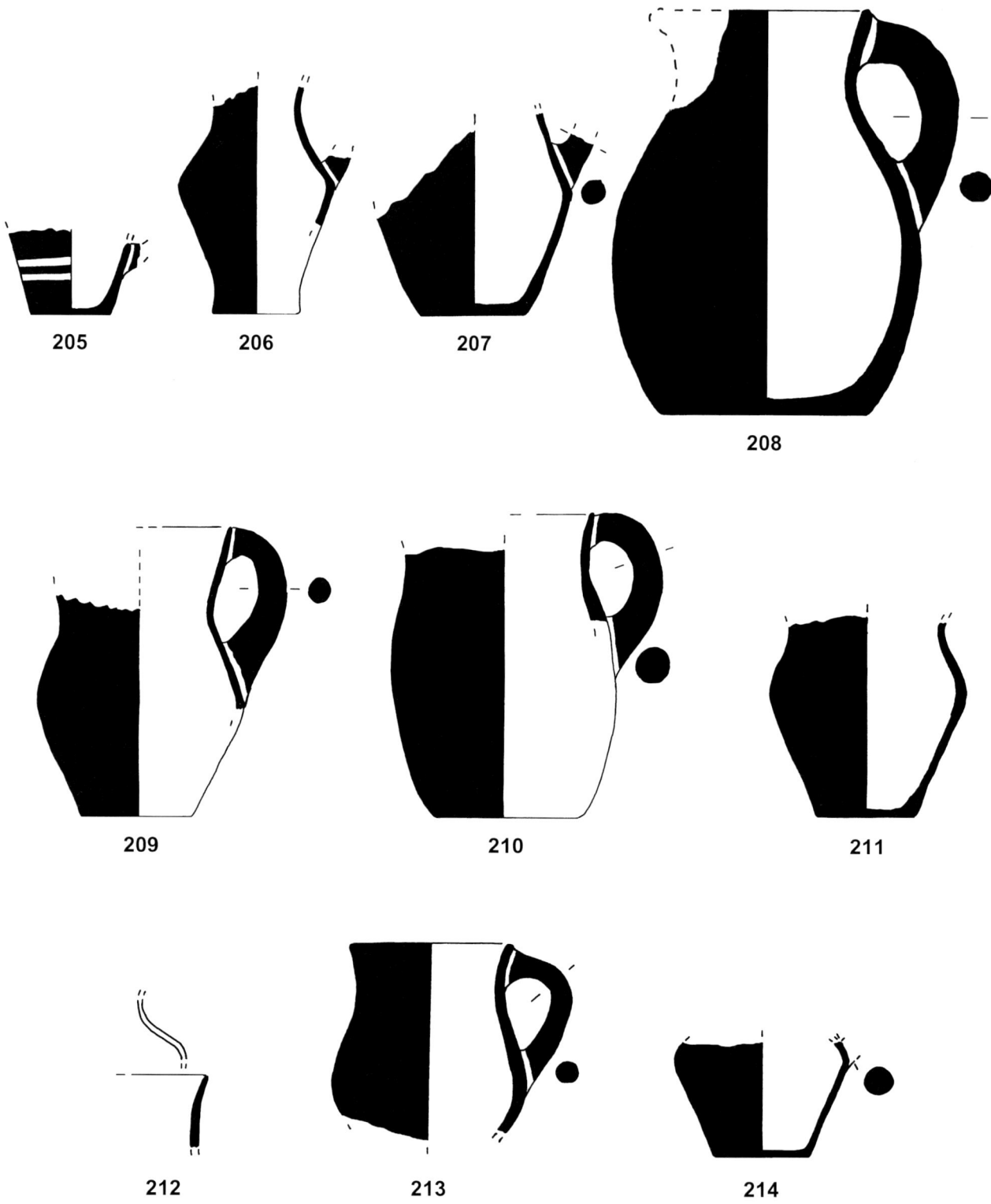

Figure 18. MM IB–IIA cup and trefoil-mouthed jugs with parallels from Galatas: straight-sided cup (**205**) and jugs (**206–214**). Scale 1:3.

FIGURE 19

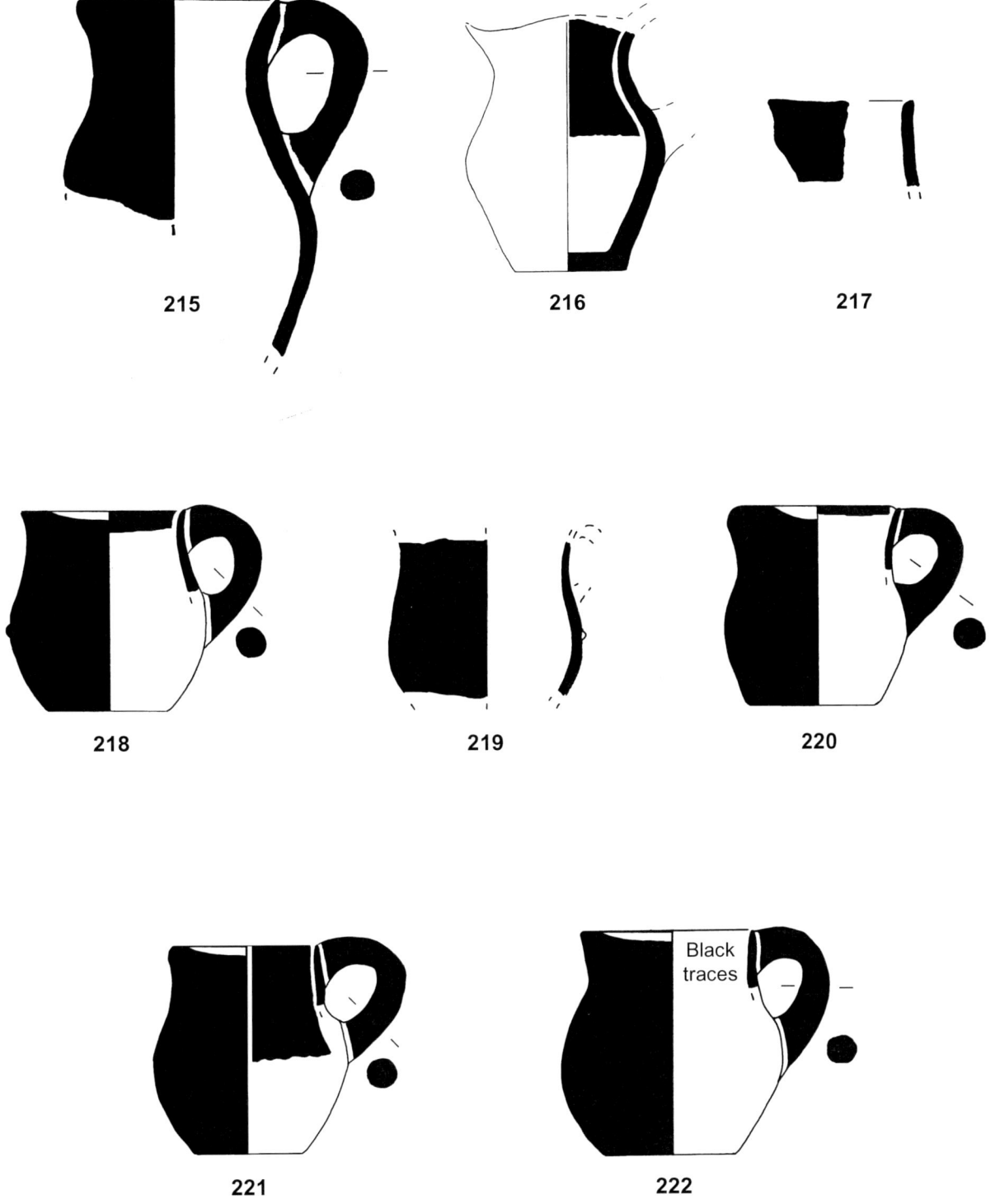

Figure 19. MM IB–IIA trefoil-mouthed jugs with parallels from Galatas (**215–217**) and side-spouted jugs or cups (**218–222**). Scale 1:3.

FIGURE 20

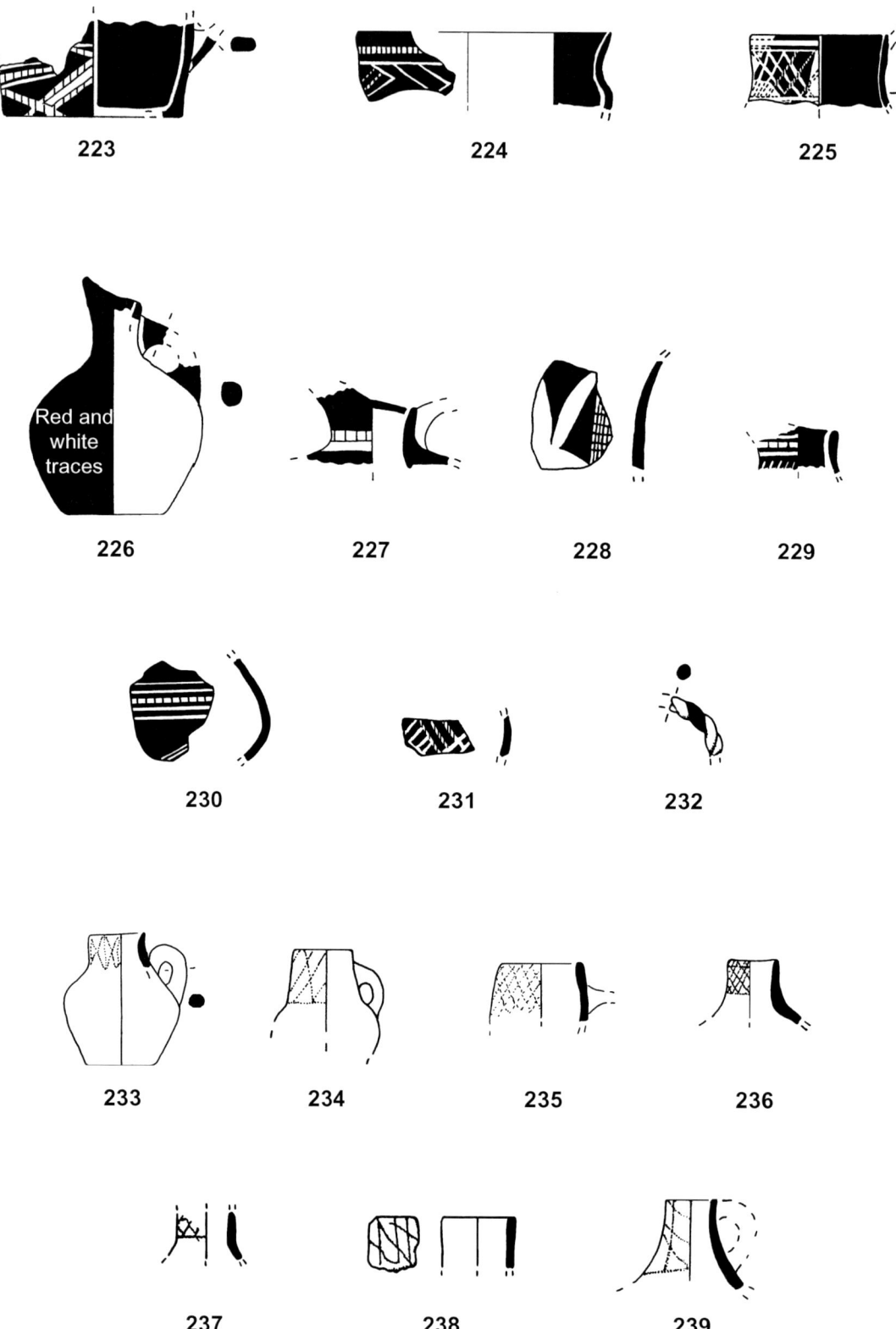

Figure 20. MM IB–IIB Kamares Style: straight-sided cup (**223**), semi-globular cup (**224**), carinated cup (**225**), jugs (**226–231**), and jar (**232**). MM IIB Chamaizi pots (**233–239**). Scale 1:3.

FIGURE 21

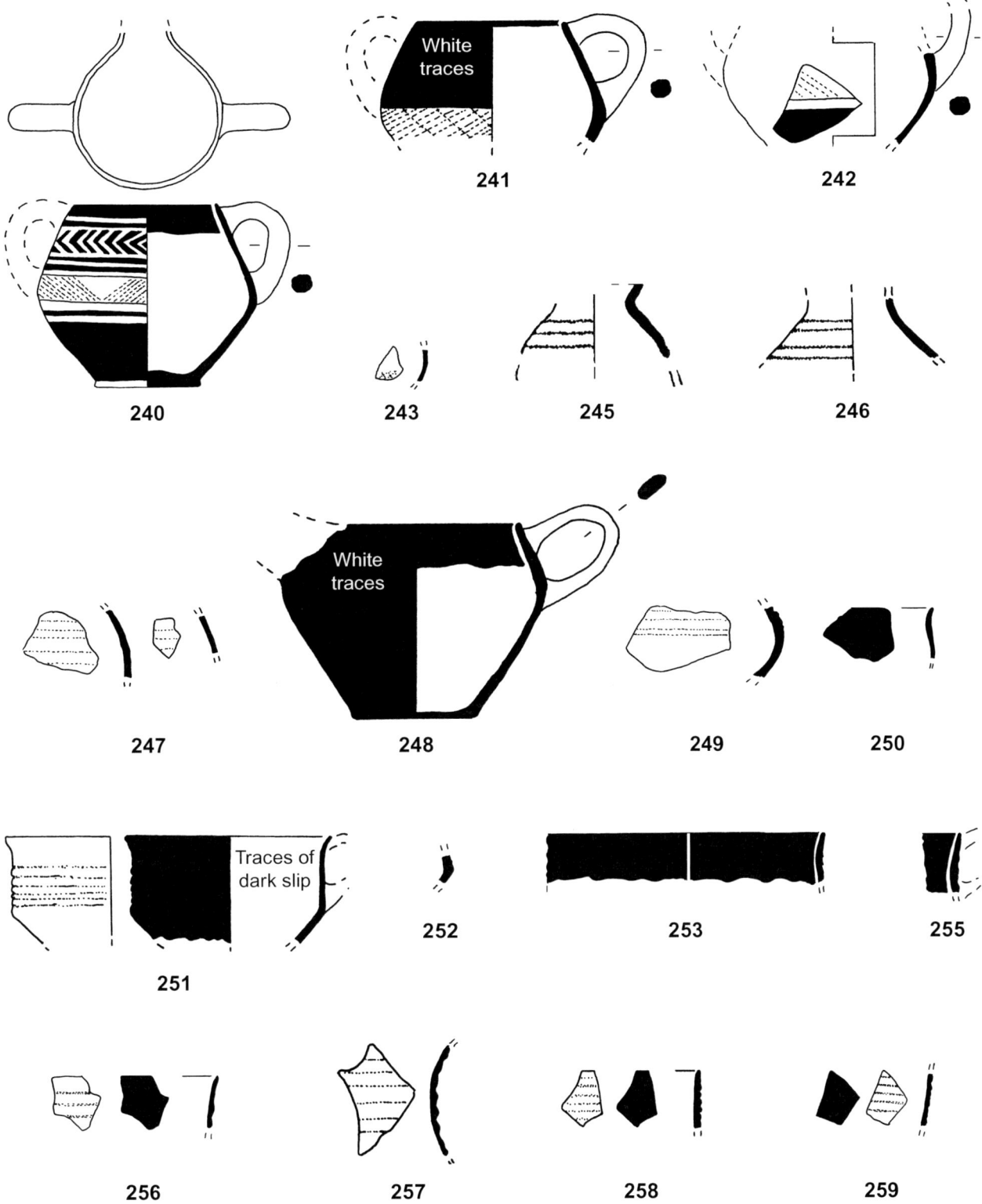

Figure 21. MM I–II imports to Lasithi with parallels from Malia: spouted two-handled cups (**240**–**243**), jugs (**245**–**248**), and carinated cups or kantharoi (**249**–**259**). Scale 1:3.

FIGURE 22

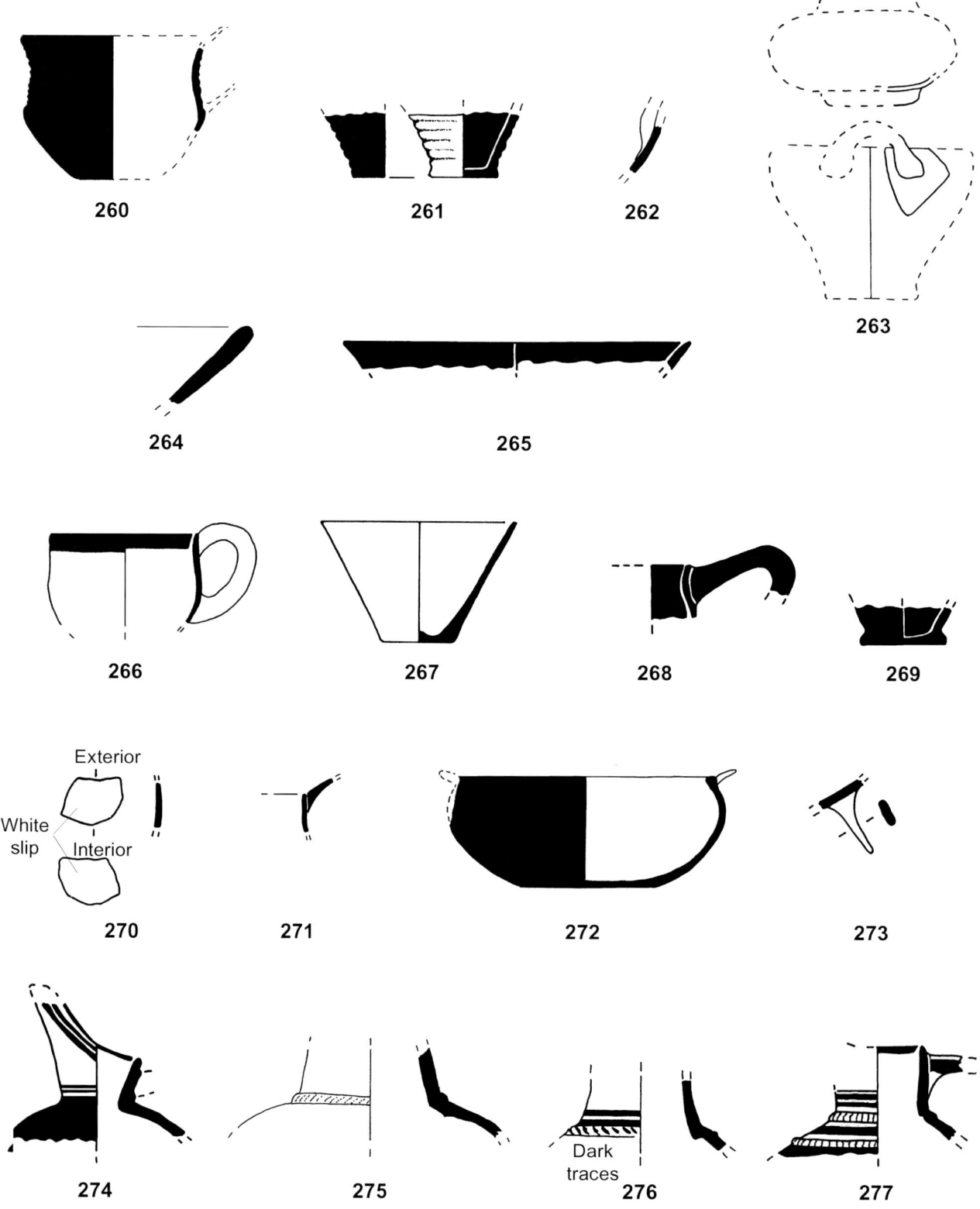

Figure 22. MM I–II imports to Lasithi with parallels from Malia: cups (**260, 261**), scoop (**262**), and basket-shaped vessel (**263**). EM III–MM IIB miscellaneous imported fabric groups: shallow bowls (**264, 265, 272**), cups (**266–271**), miniature tripod vessel (**273**), and jugs (**274–277**). Scale 1:3.

FIGURE 23

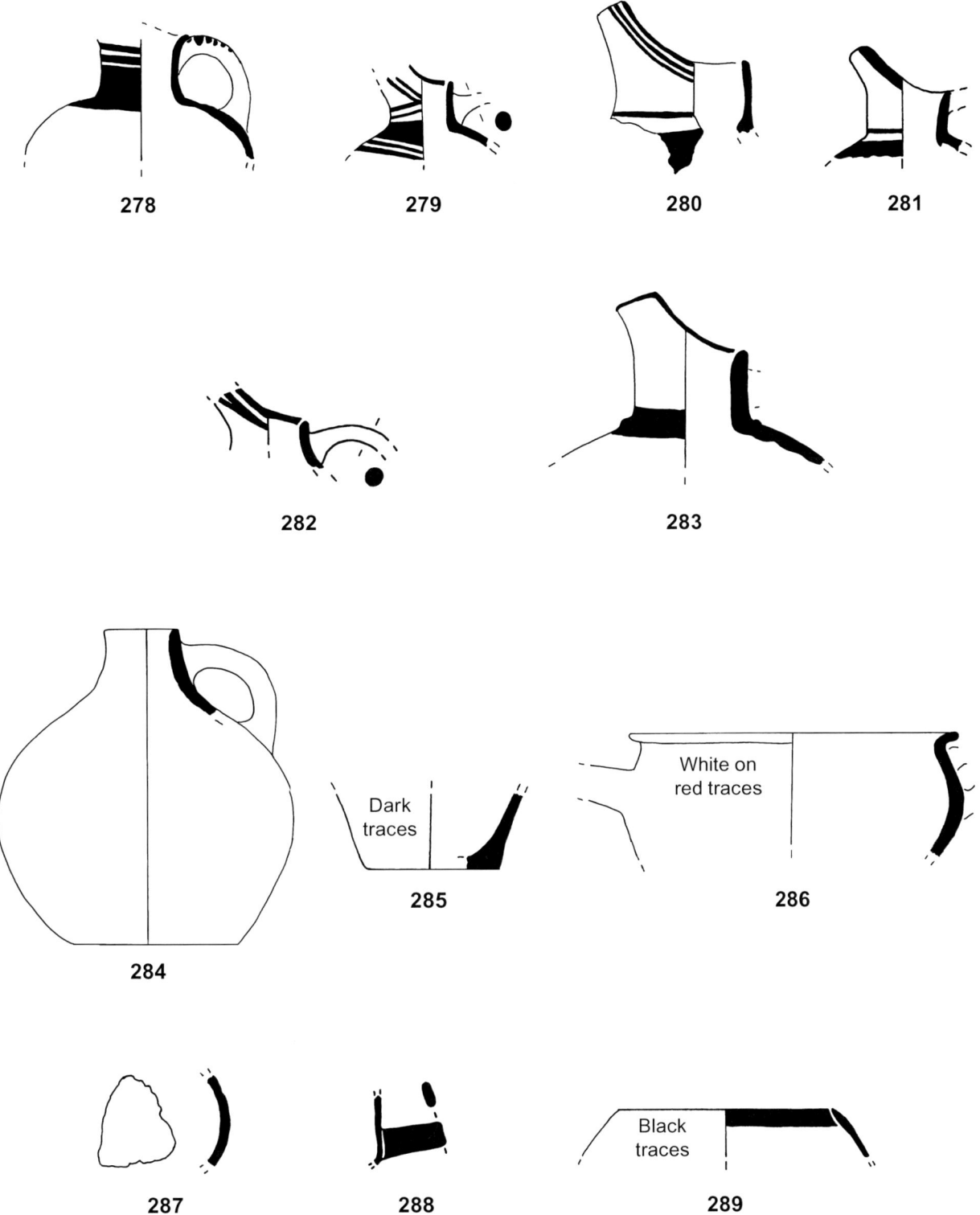

Figure 23. EM III–MM IIB miscellaneous imported fabric groups: jugs (**278–285**), wide-mouthed vessel (**286**), closed vessel (**287**), and bridge-spouted jars (**288, 289**). Scale 1:3.

FIGURE 24

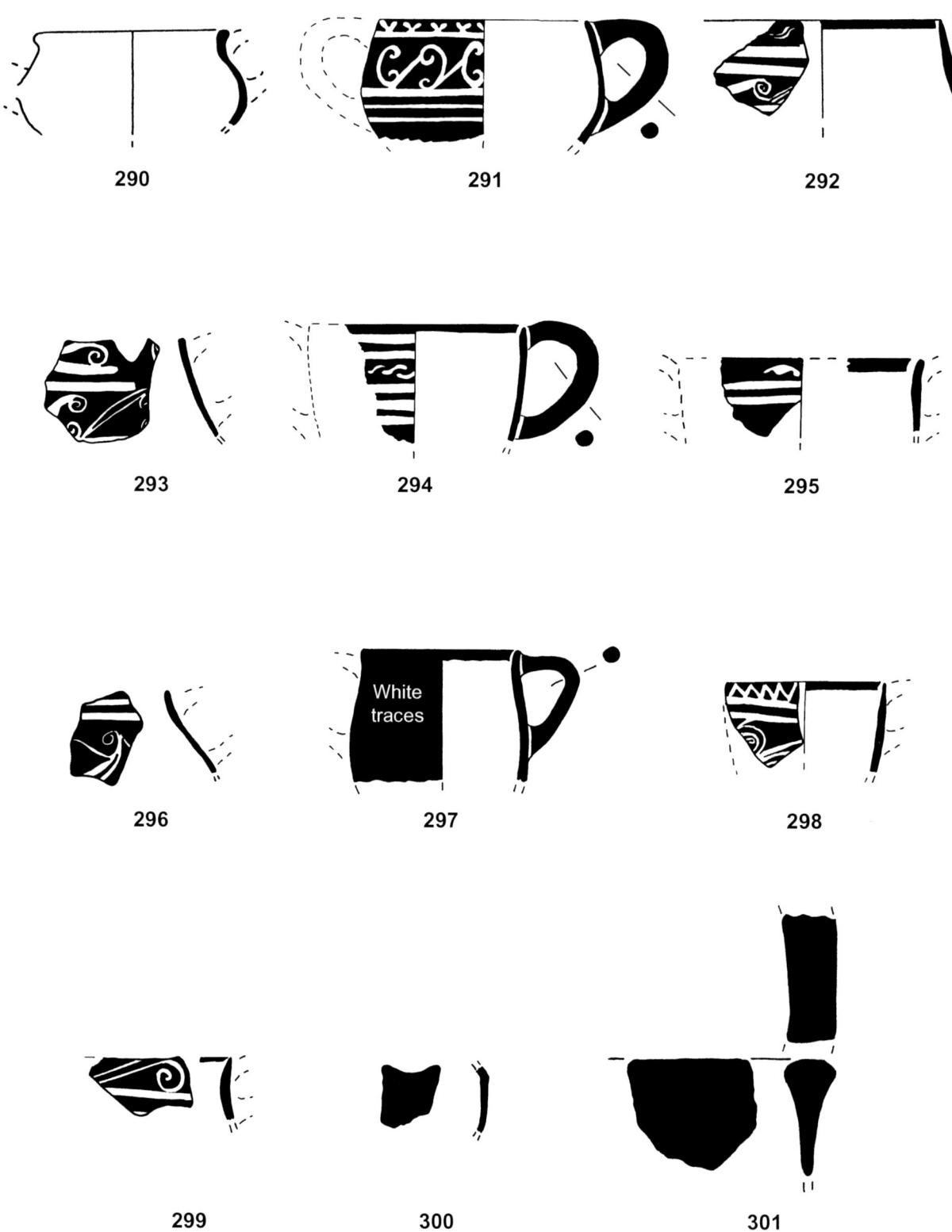

Figure 24. EM III–MM IIB miscellaneous imported fabric groups: side-spouted jug (**290**), spouted two-handled cups (**291–299**), small jar (**300**), and jar with thickened rim (**301**). Scale 1:3.

FIGURE 25

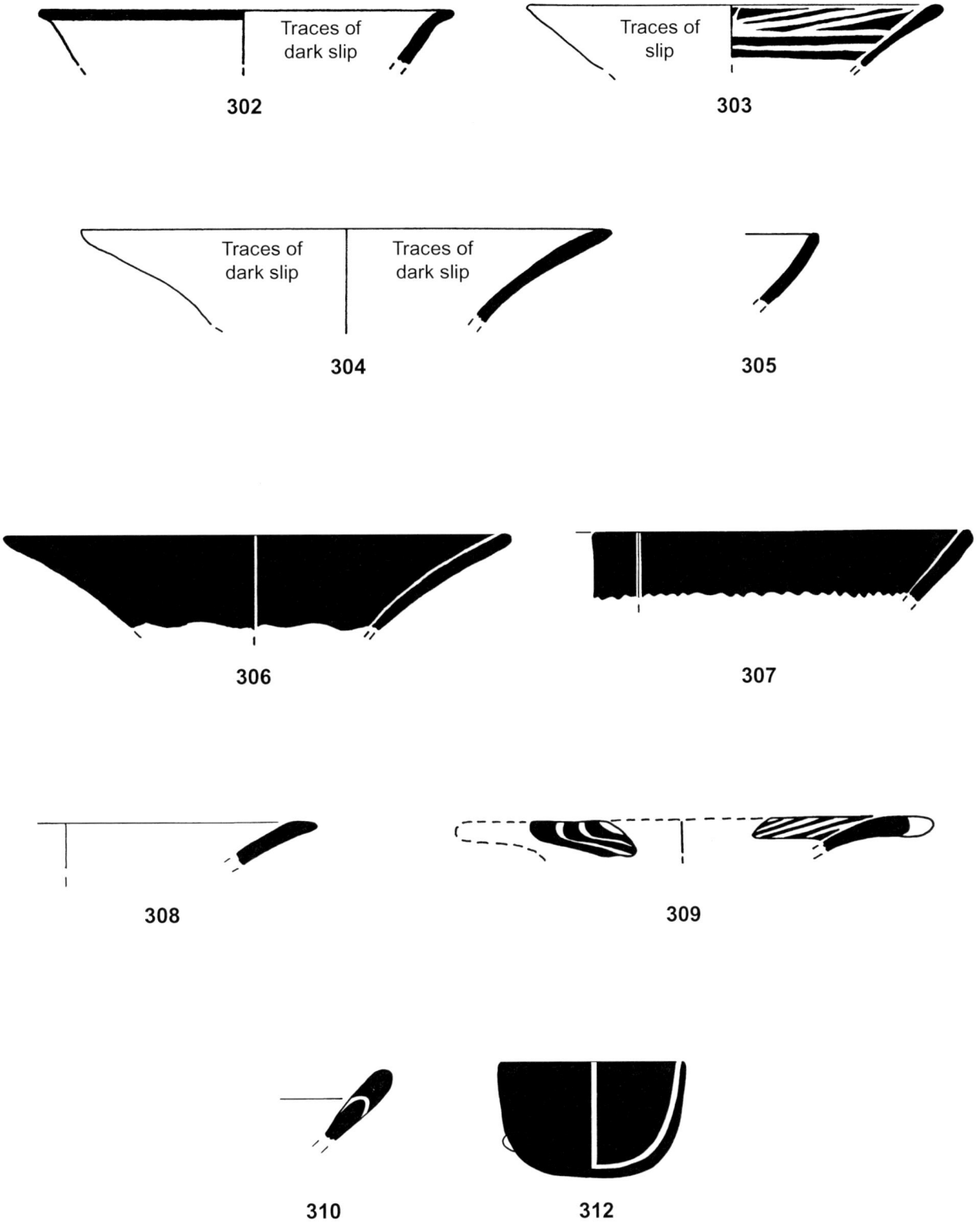

Figure 25. EM III–MM IA Lasithi Red Fabric Group: shallow conical bowls (**302–308**), shallow conical bowls with tab handles (**309, 310**), and rounded cup with lugs (**312**). Scale 1:3.

FIGURE 26

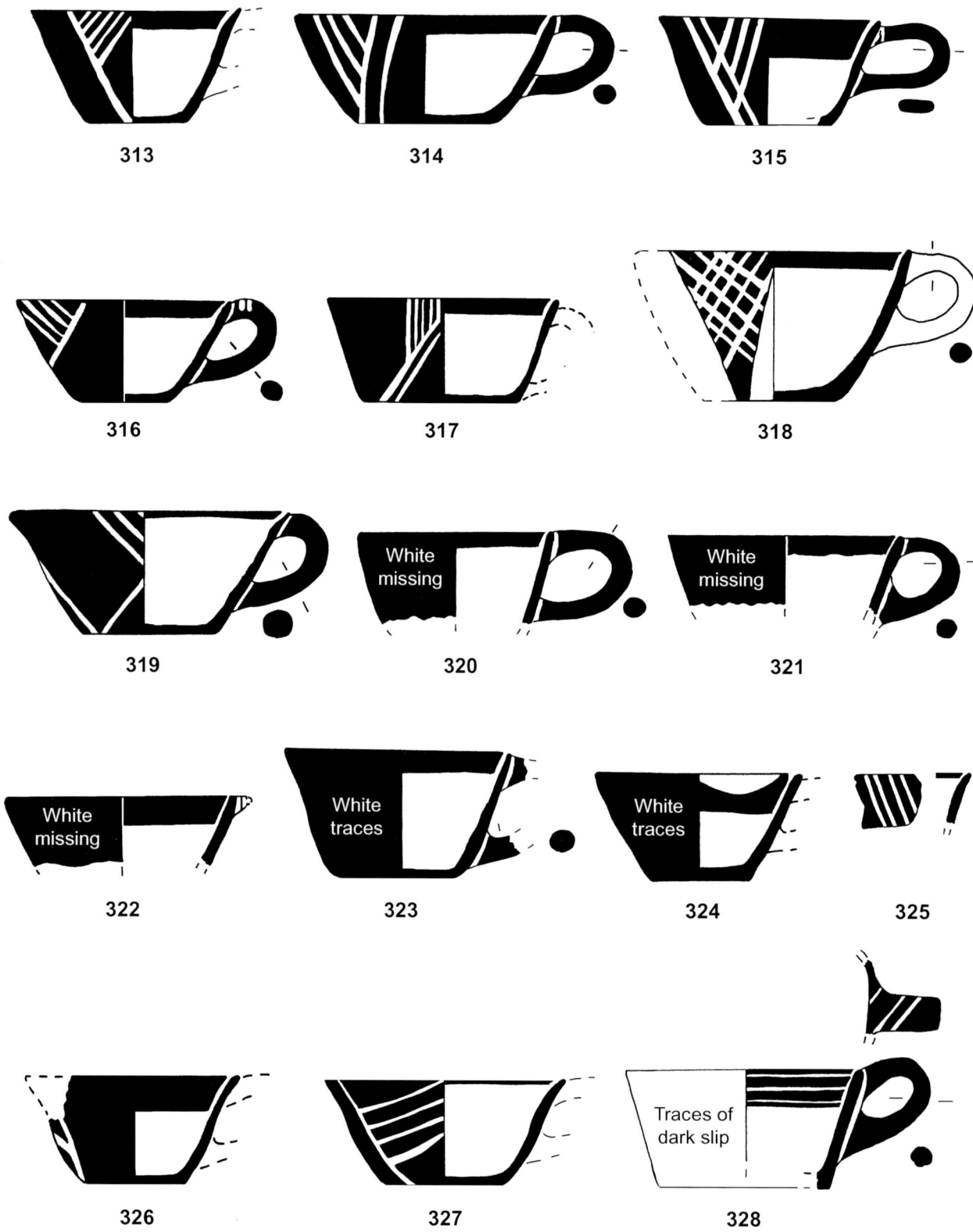

Figure 26. EM III–MM IA Lasithi Red Fabric Group: straight-sided cups with dark paint on interior of rim and exterior (**313**–**328**). Scale 1:3.

FIGURE 27

Figure 27. EM III–MM IA Lasithi Red Fabric Group: rounded cups with dark paint on interior of rim and exterior (**329–343**). Scale 1:3.

FIGURE 28

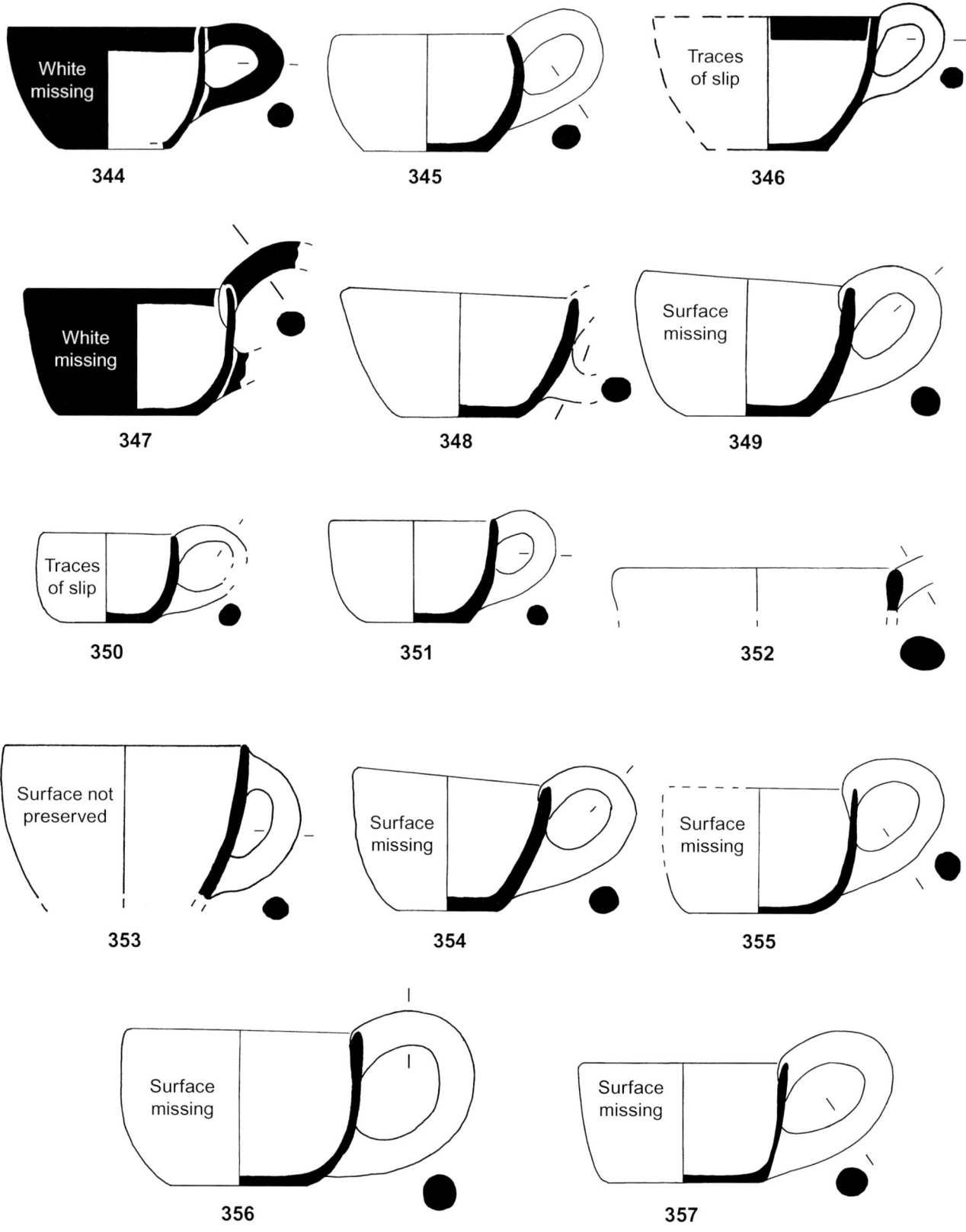

Figure 28. Lasithi Red Fabric Group: EM III–MM IA rounded cups with dark paint on interior of rim and exterior (**344–347**) and EM II–MM I plain rounded cups or cups with paint missing (**348–357**). Scale 1:3.

FIGURE 29

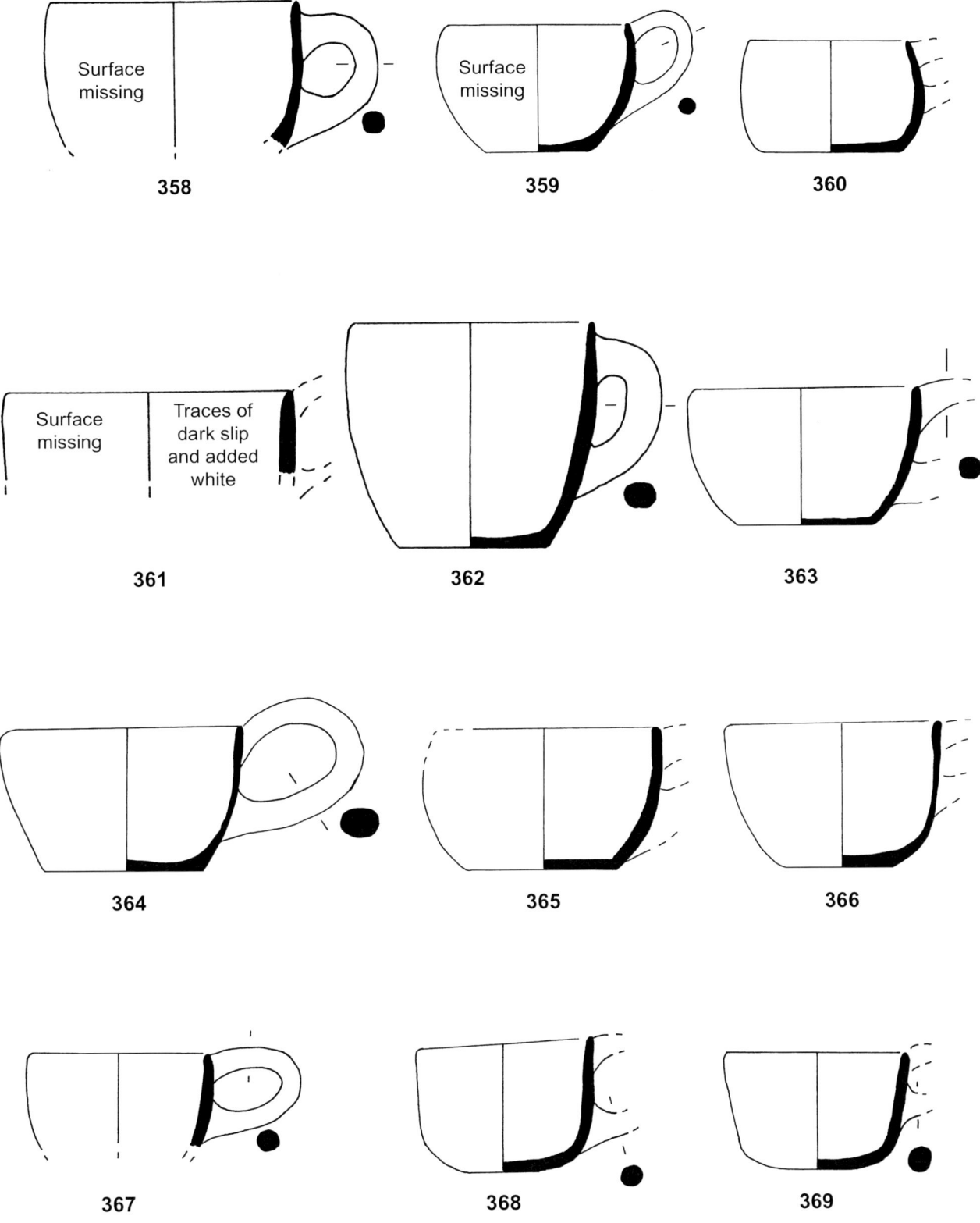

Figure 29. EM II–MM I Lasithi Red Fabric Group: rounded cups, plain or with paint missing (**358–369**). Scale 1:3.

FIGURE 30

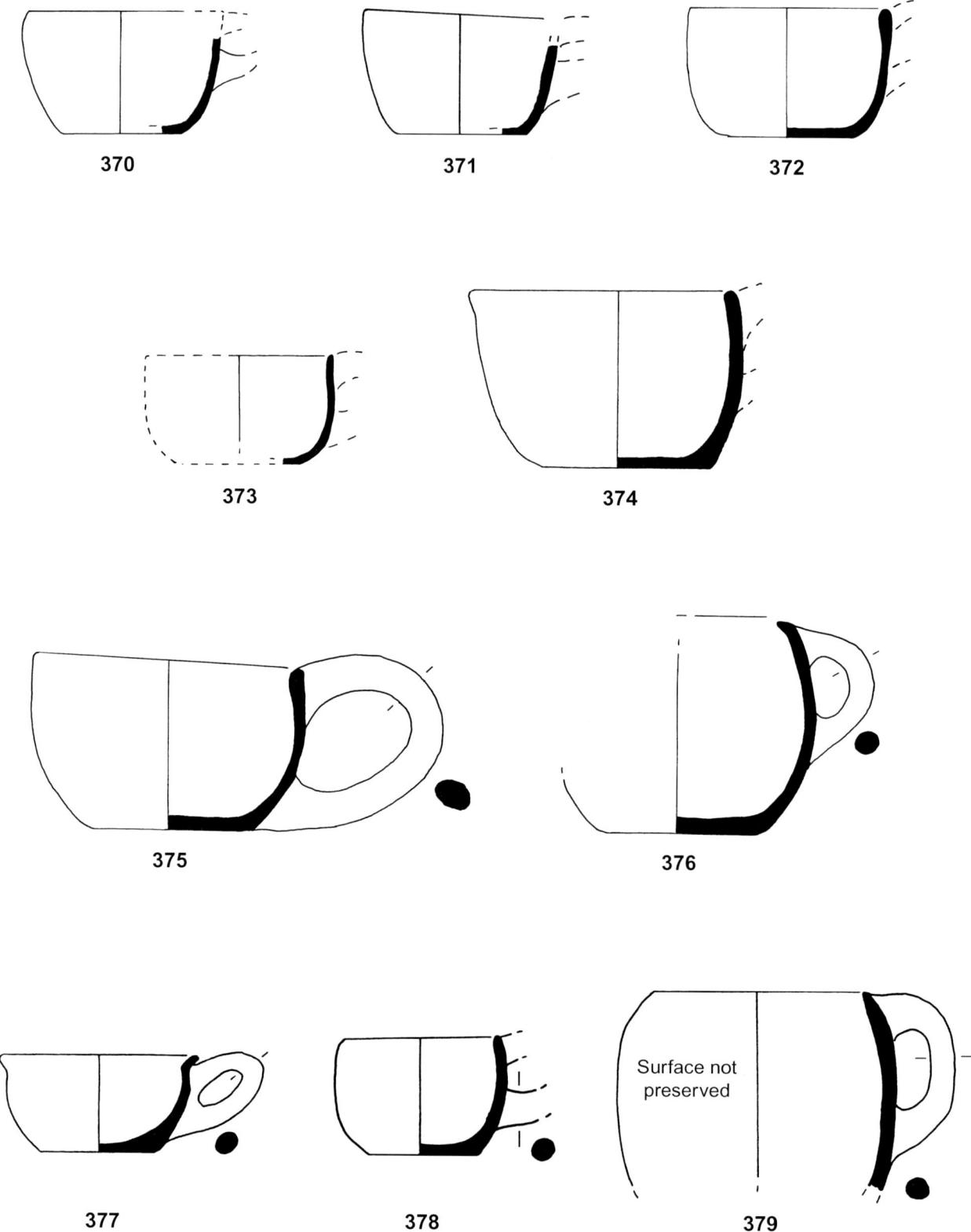

Figure 30. EM II–MM I Lasithi Red Fabric Group: rounded cups, plain or with paint missing (**370–379**). Scale 1:3.

FIGURE 31

Figure 31. Lasithi Red Fabric Group: EM II–MM I rounded cups, plain or with paint missing (**380–382**), MM I–II cylindrical cup with dark slip on interior and exterior (**383**), and rounded cups with dark slip on interior and exterior (**384–394**). Scale 1:3.

FIGURE 32

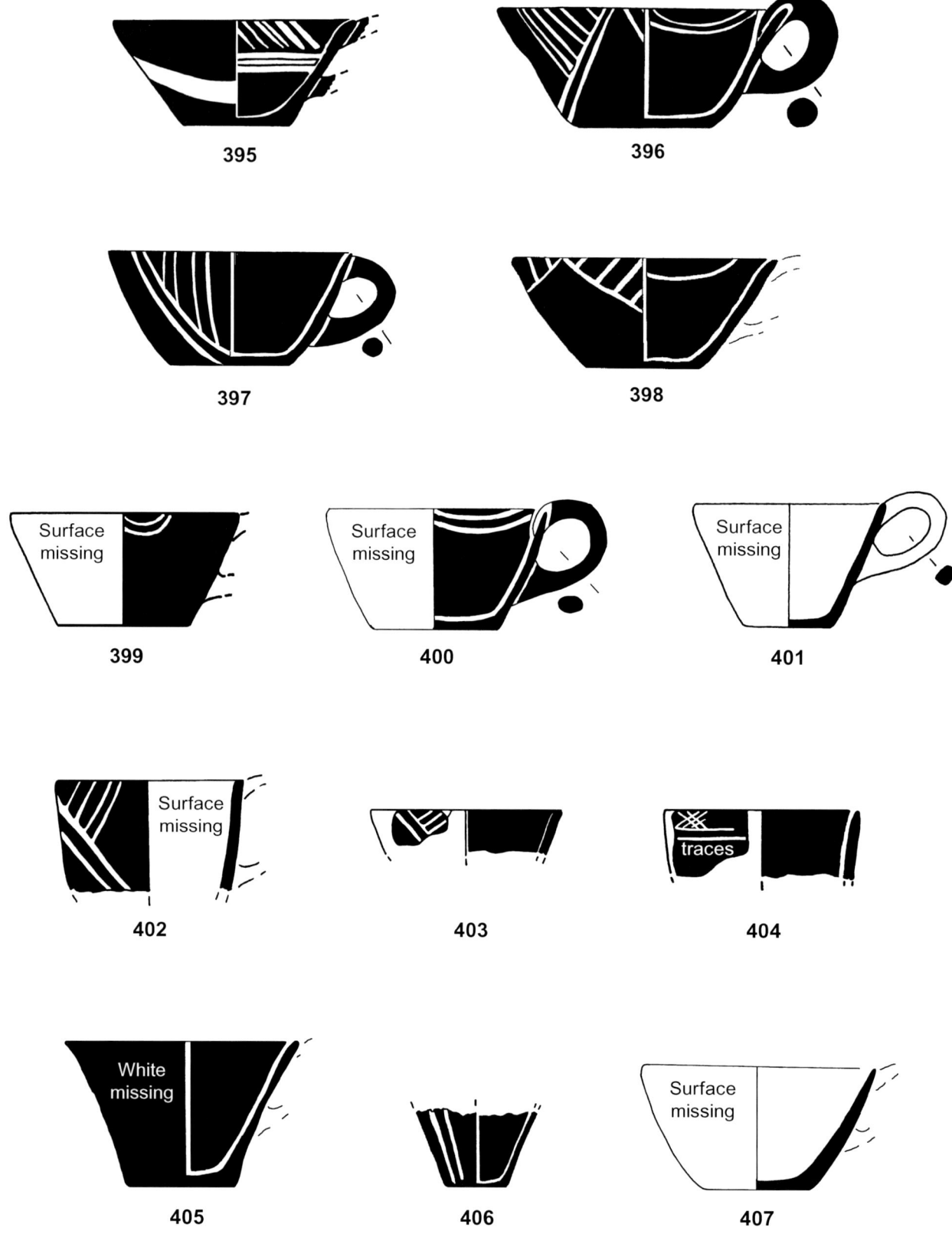

Figure 32. EM III–MM I Lasithi Red Fabric Group: straight-sided cups with conical shape (**395–407**). Scale 1:3.

FIGURE 33

Figure 33. EM III–MM II Lasithi Red Fabric Group: straight-sided cups with conical shape (**408–412**) and miscellaneous open shapes (**413–419**). Scale 1:3.

FIGURE 34

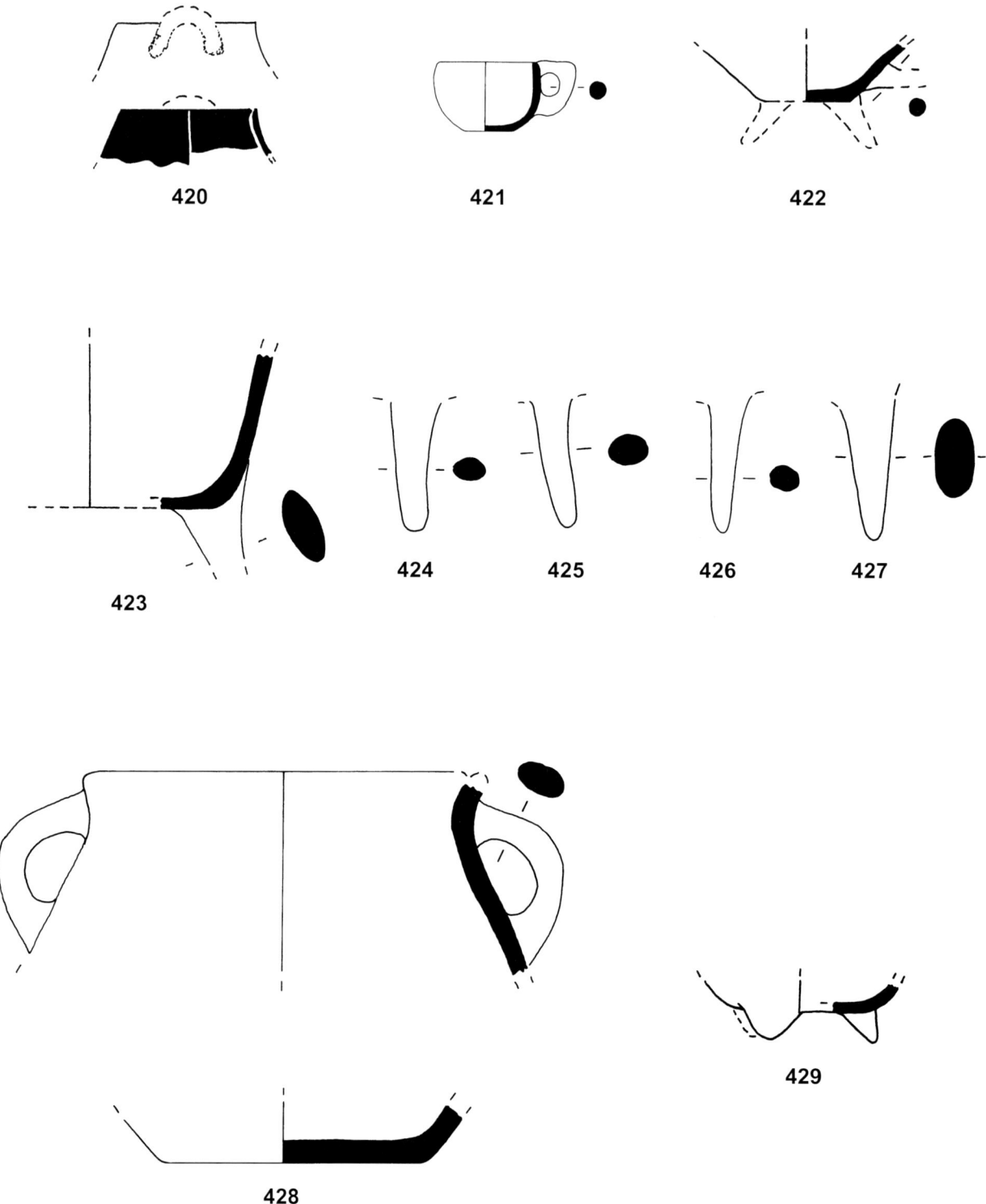

Figure 34. EM III–MM II Lasithi Red Fabric Group: miscellaneous open shapes (**420**, **421**), brazier (**422**), tripod vessels (**423–427**), cooking vessel without legs (**428**), and small tripod bowl (**429**). Scale 1:3.

FIGURE 35

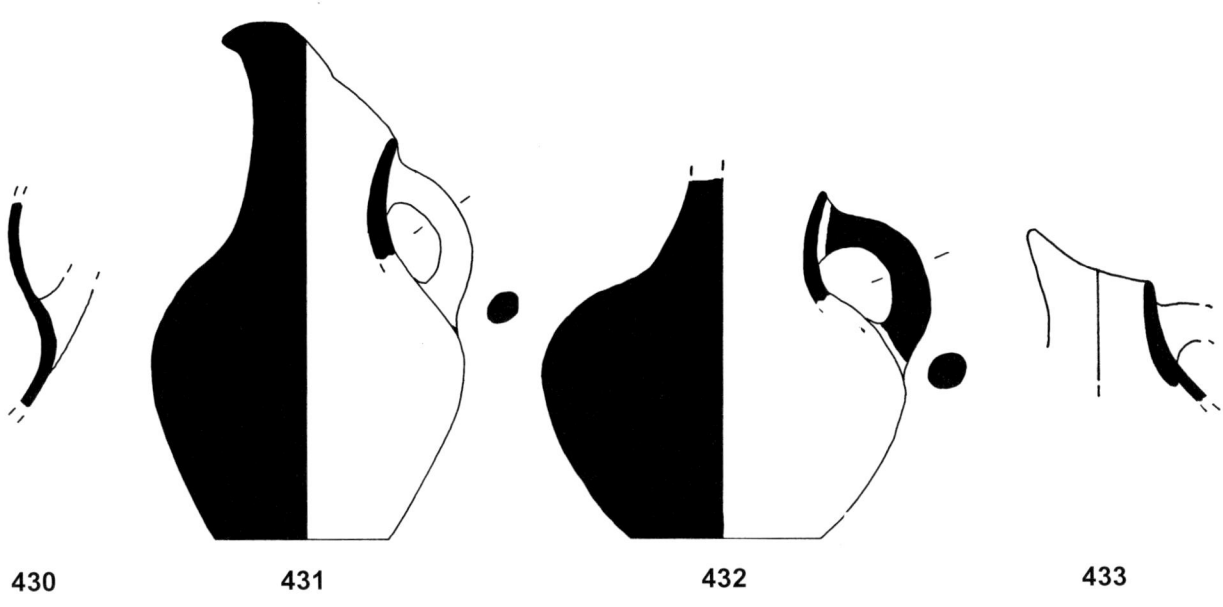

430   431   432   433

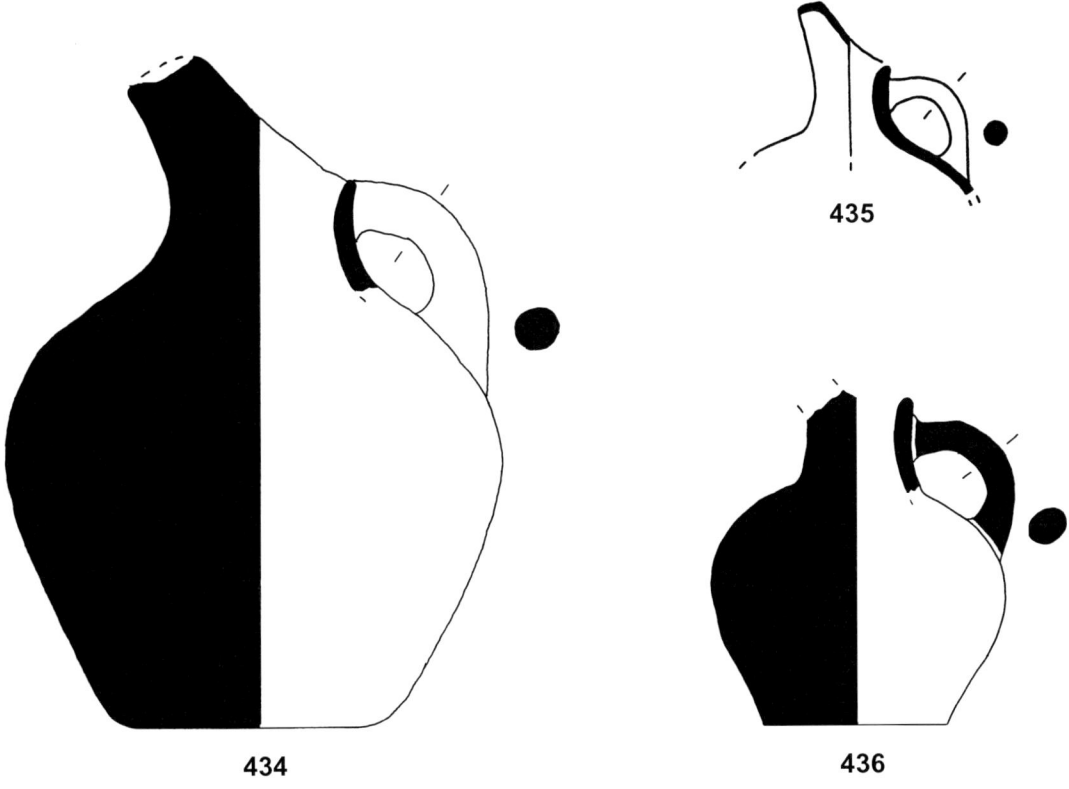

435

434   436

Figure 35. EM IIB–MM II Lasithi Red Fabric Group: jug or kantharos(?) (**430**) and jugs (**431–436**). Scale 1:3.

FIGURE 36

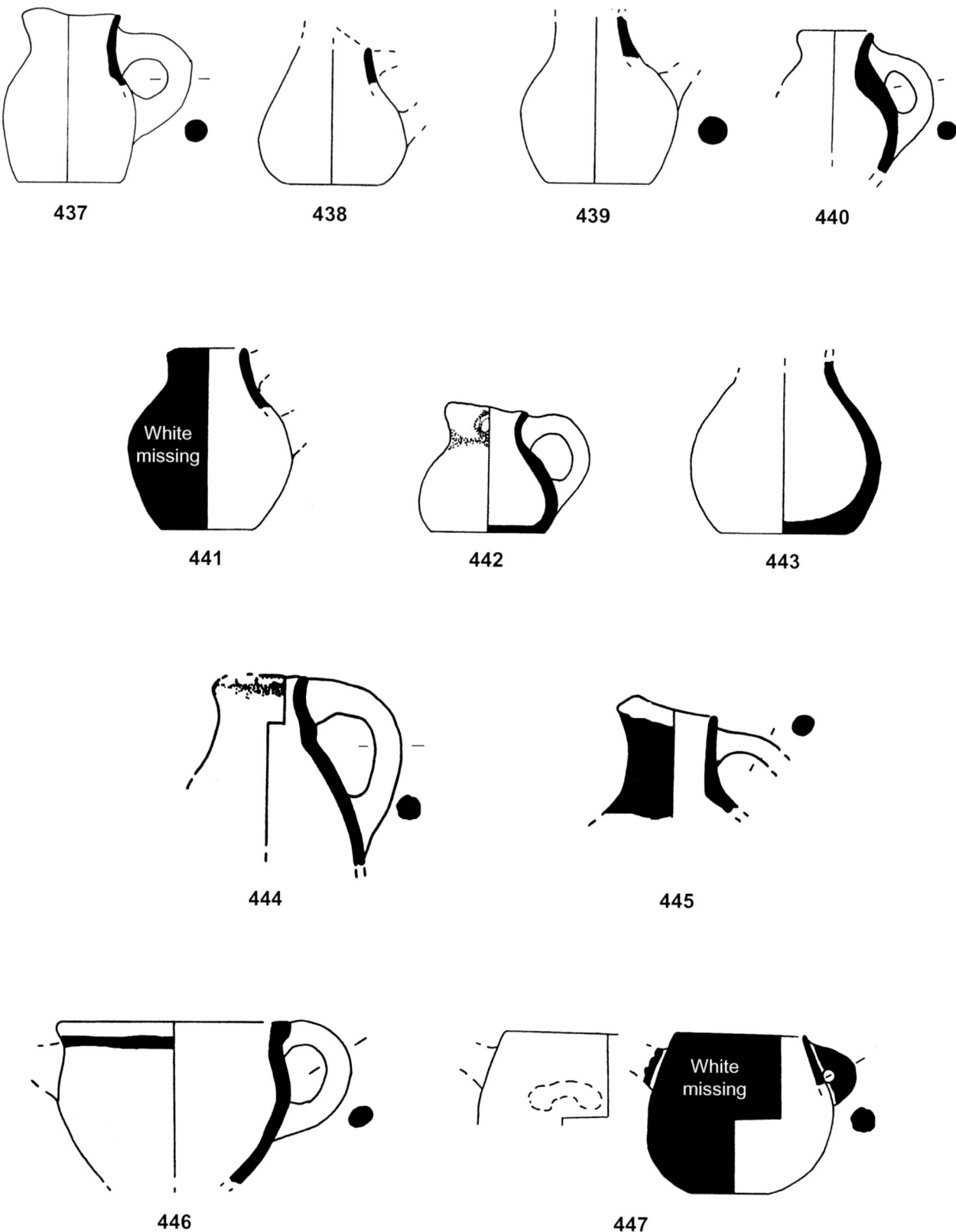

437

438

439

440

White missing

441

442

443

444

445

446

White missing

447

Figure 36. MM I–II Lasithi Red Fabric Group: jugs (**437–445**) and tube-spouted and bridge-spouted jugs (**446, 447**). Scale 1:3.

FIGURE 37

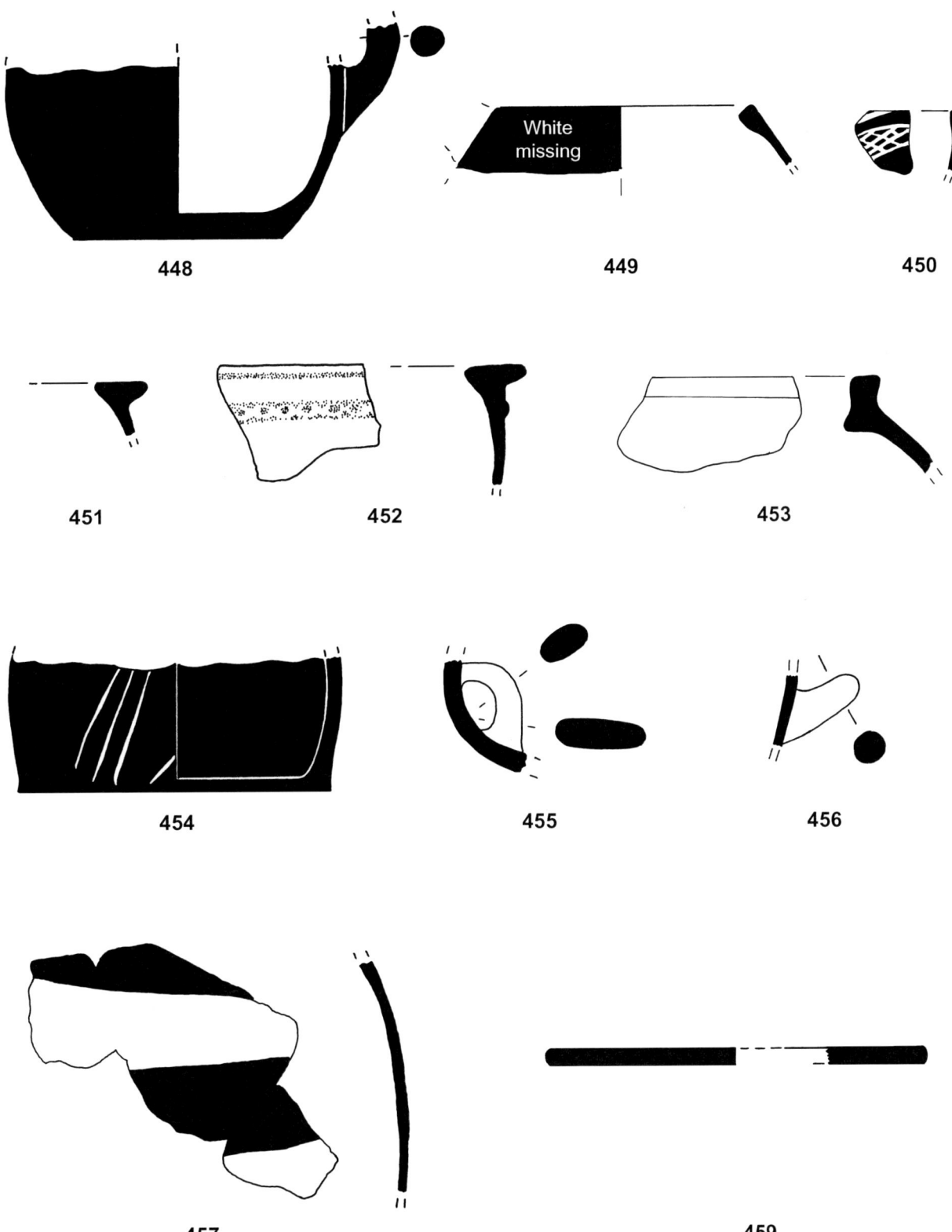

Figure 37. EM IIB–MM II Lasithi Red Fabric Group: tube-spouted and bridge-spouted jugs (**448, 449**), closed vessel (**450**), jars (**451–457**), and lid (**459**). Scale 1:3.

FIGURE 38

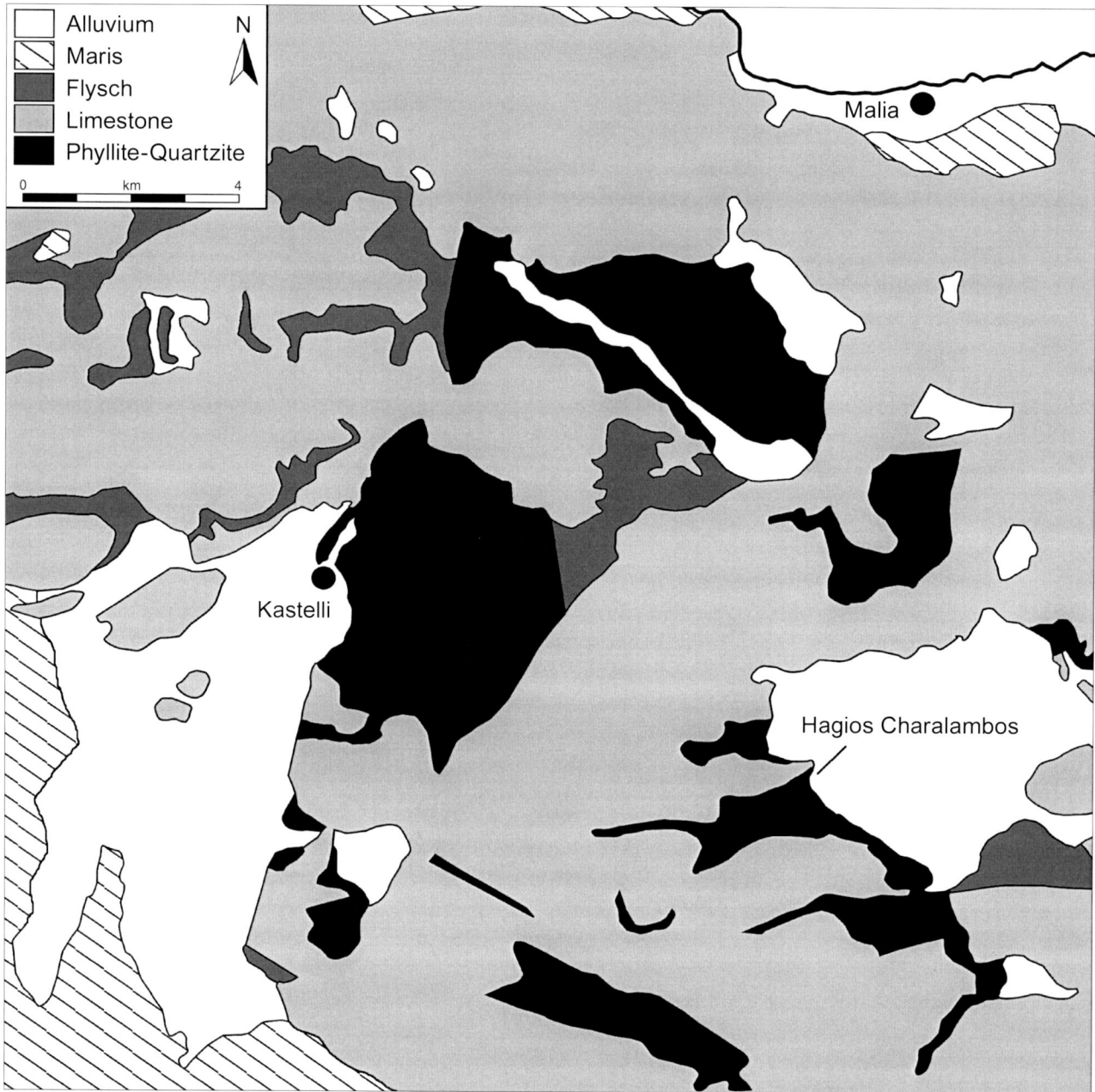

Figure 38. Geological map of the areas discussed in the text (adapted after IGME 1989).

Plates

PLATE 1

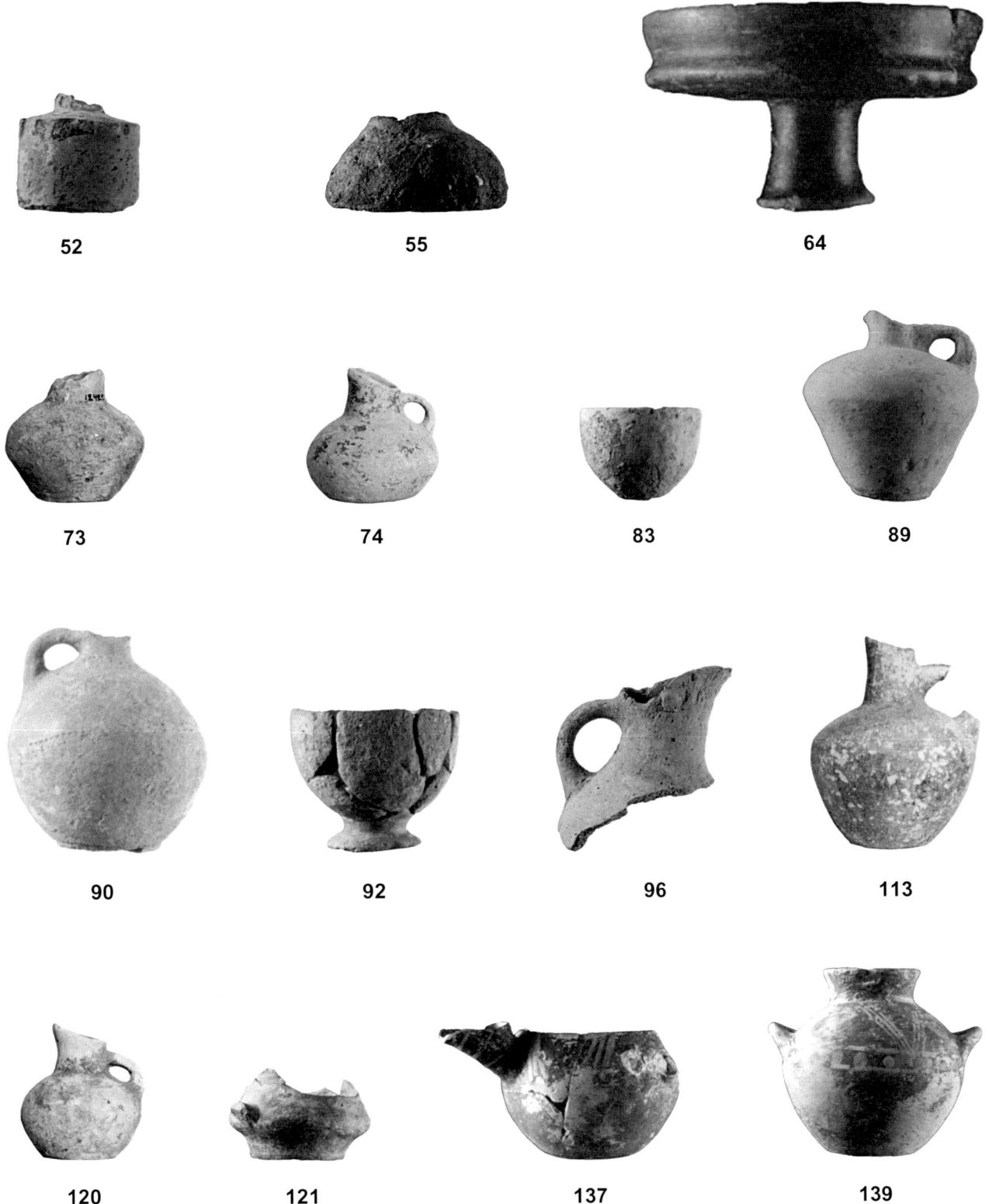

Plate 1. EM IIA Monochrome Style jugs (**52**, **55**). EM IIA dark burnished offering table (**64**). EM IIB Vasiliki Ware small jugs (**73**, **74**). EM IIB Vasiliki Style imitations in pale fabrics: goblet (**83**) and jugs (**89**, **90**). EM IIB Vasiliki Style imitations in the Lasithi Red Fabric Group: goblet (**92**) and jug (**96**). EM III–MM II vessels in pale fabrics with light-on-dark decoration: jugs (**113**, **120**, **121**), teapot (**137**), and jar (**139**). Scale 1:3.

PLATE 2

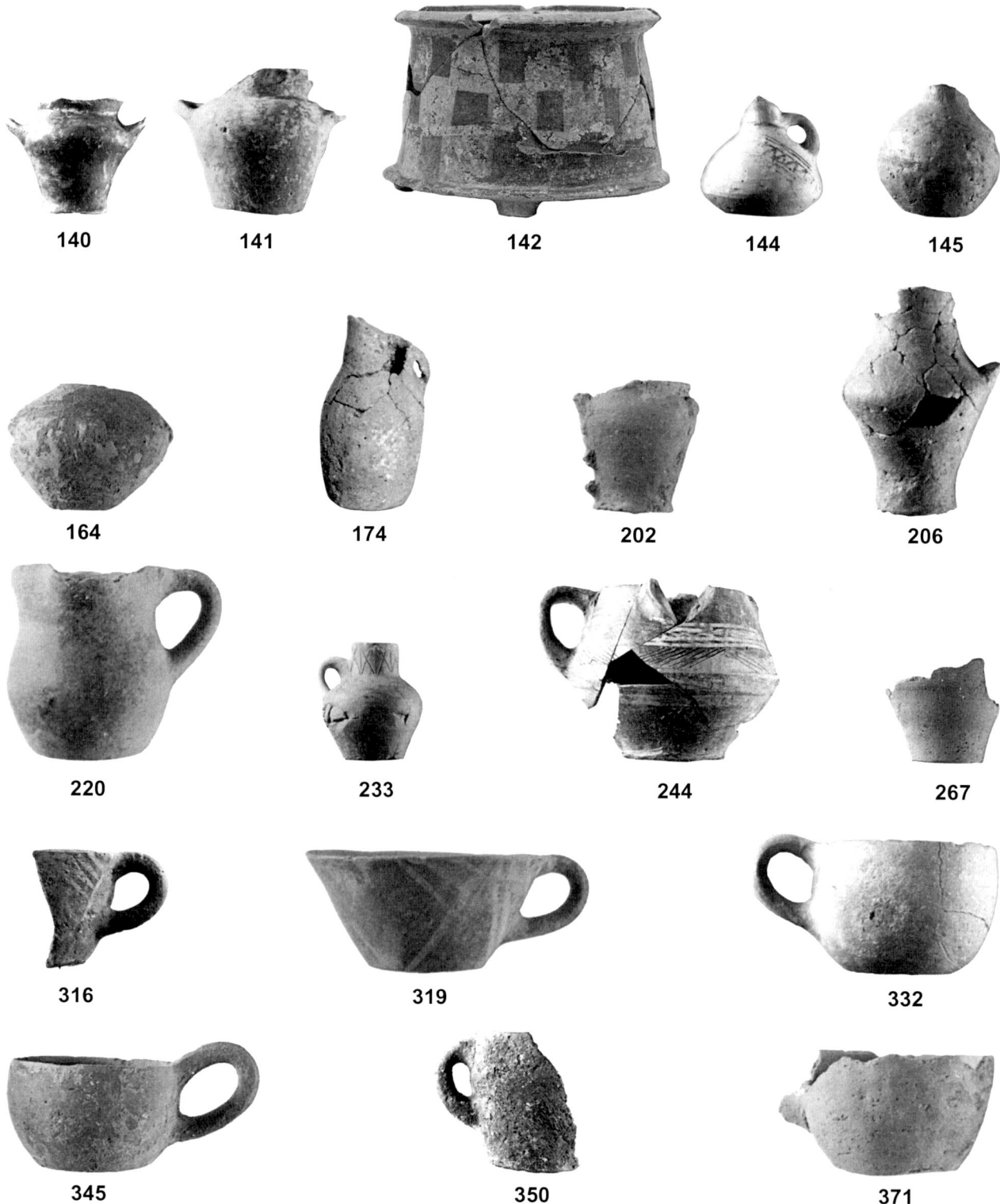

Plate 2. EM III–MM II vessels in pale fabrics with mostly light-on-dark decoration: jars (**140**, **141**) and pyxis (**142**). MM I Diagonal Line Style and other small jugs in pale fabrics (**144**, **145**, **164**, **174**). MM IB miniature pithos in fine red fabric with dark red slip (**202**). MM IB vessels with parallels from Galatas: jug (**206**) and side-spouted jug or cup (**220**). MM IIB Chamaizi pot (**233**). MM I–II imports into Lasithi: spouted two-handled cup (**244**) and conical cup (**267**). EM III–MM II Lasithi Red Fabric Group: straight-sided cups with conical shape (**316**, **319**) and rounded cups (**332**, **345**, **350**, **371**). Scale 1:3.

PLATE 3

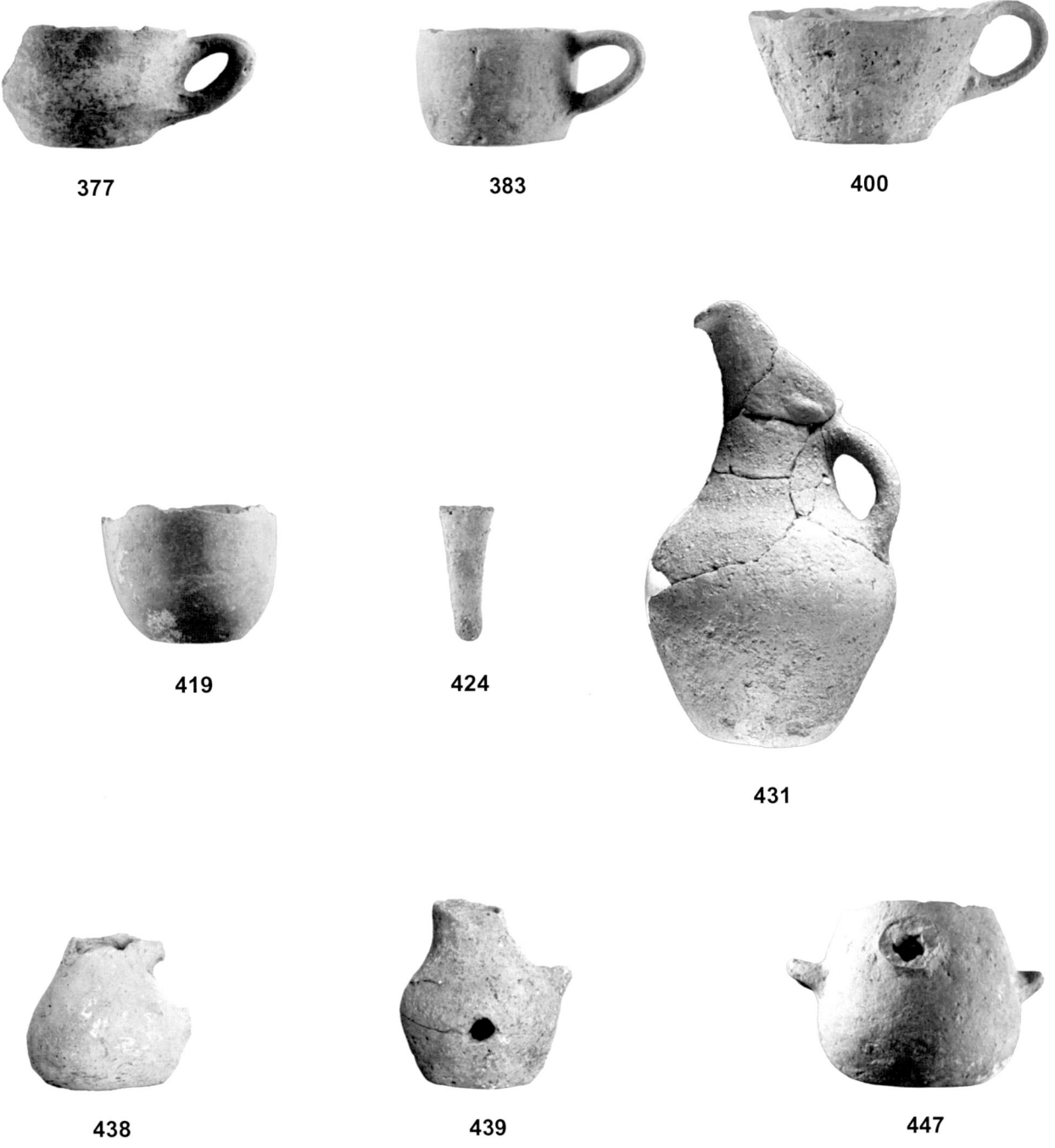

Plate 3. EM III–MM II Lasithi Red Fabric Group: rounded cup (**377**), cylindrical cup (**383**), straight-sided cup with conical shape (**400**), deep rounded cup (**419**), tripod vessel leg (**424**), jug with elevated spout (**431**), jugs (**438**, **439**), and side-spouted or tube-spouted jug (**447**). Scale 1:3.

PLATE 4

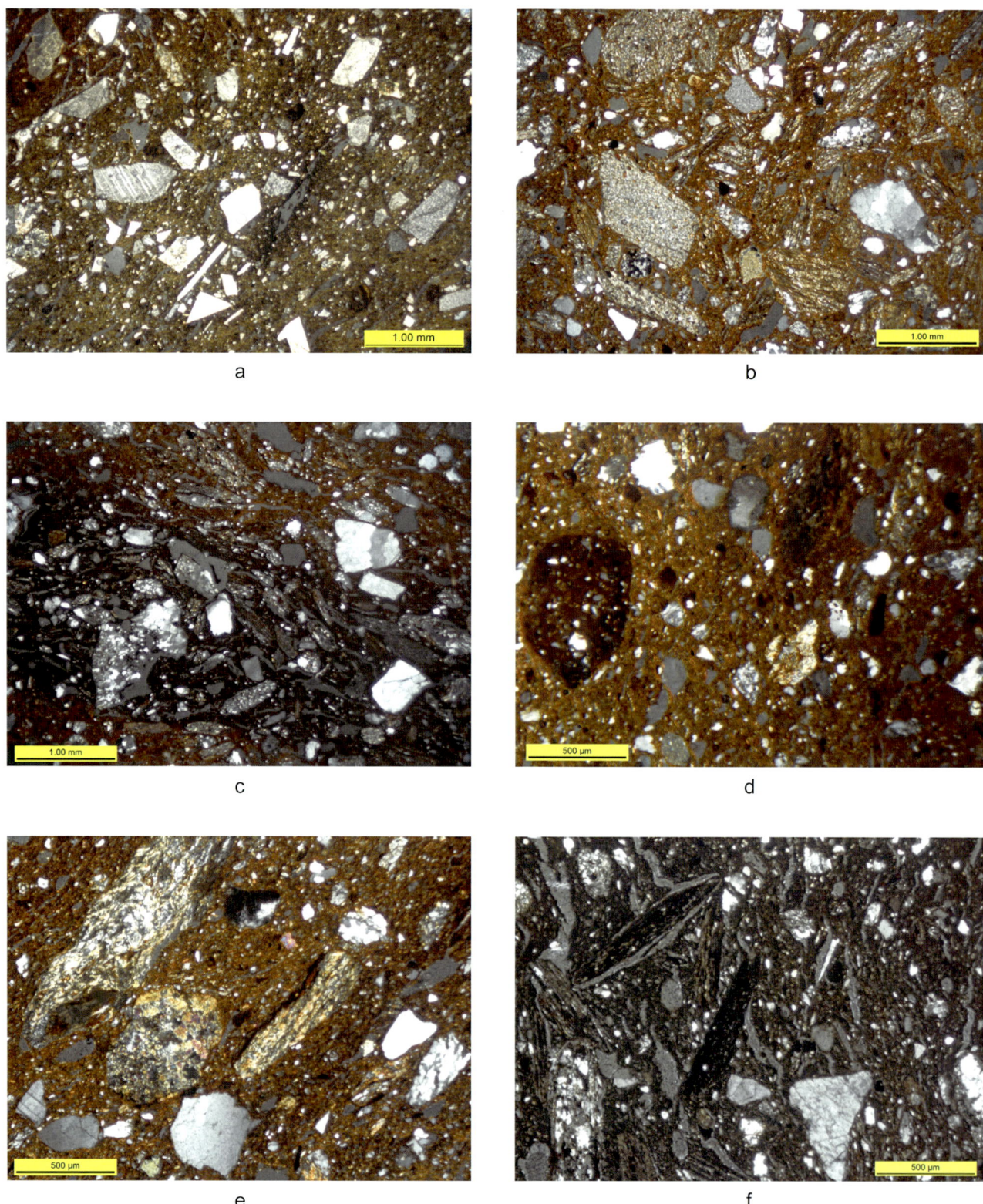

a

b

c

d

e

f

Plate 4. Fabric Group 1: calcite tempered (a). Fabric Group 2 (Lasithi Red Fabric Group): coarse phyllite fabric Group 2A (b), Group 2B (c). Small groups and loners with metamorphic rocks (d–f). All microphotographs with scale of 1.00 mm were taken under x25 magnification; all microphotographs with scale of 500 μm were taken under x50 magnification.

PLATE 5

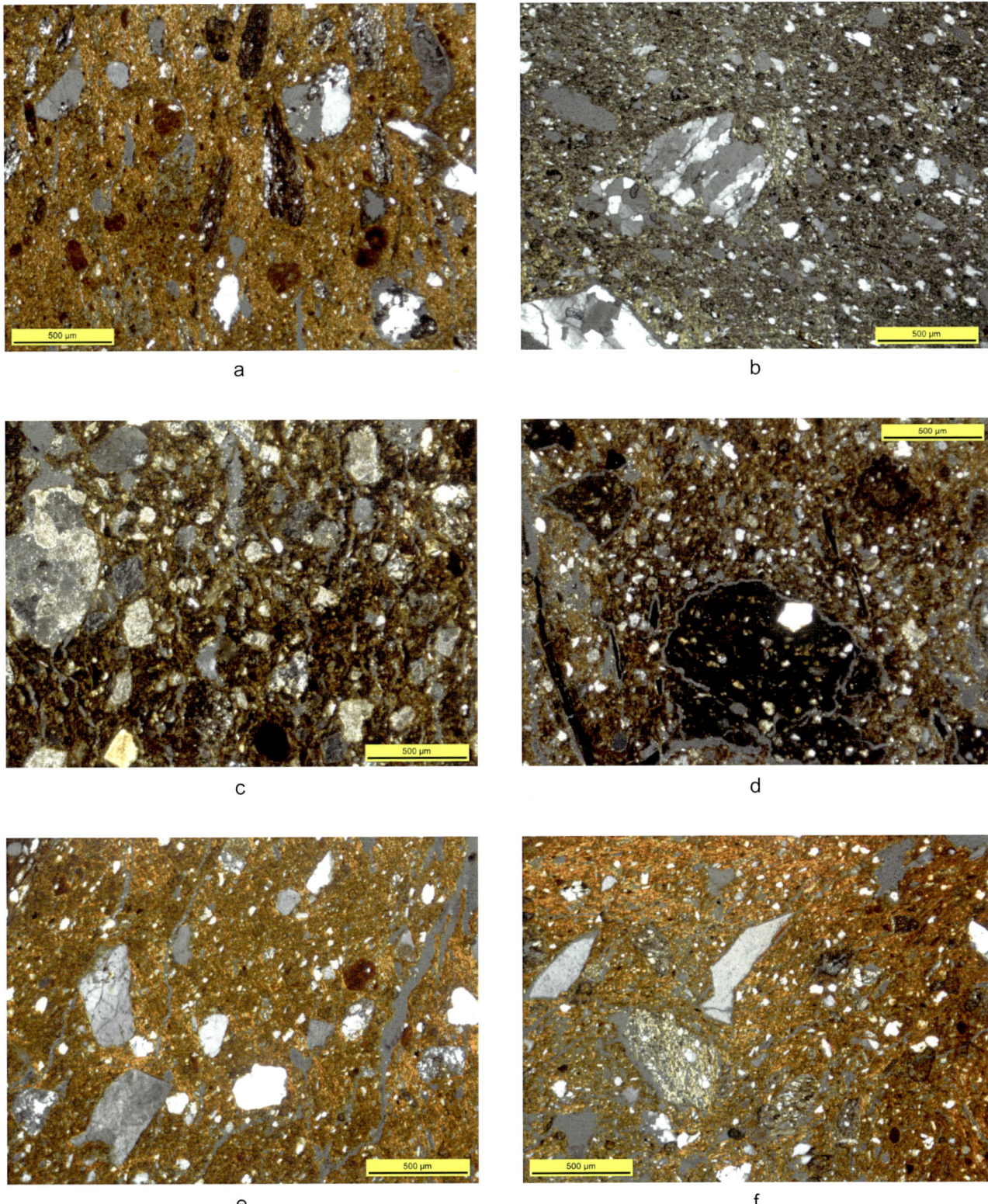

Plate 5. Small groups and loners with metamorphic rocks (a, b). Fabric Group 3: semicoarse with carbonate rock fragments (c). Fabric Group 4: calcareous grog-tempered (d); note elongate voids with remnants of burned organic material. Fabric Group 5: red fabric with quartz and clay pellets (e). Fabric Group 6: red fabric with quartz (f). Microphotographs with scale of 500 μm were taken under x50 magnification.

PLATE 6

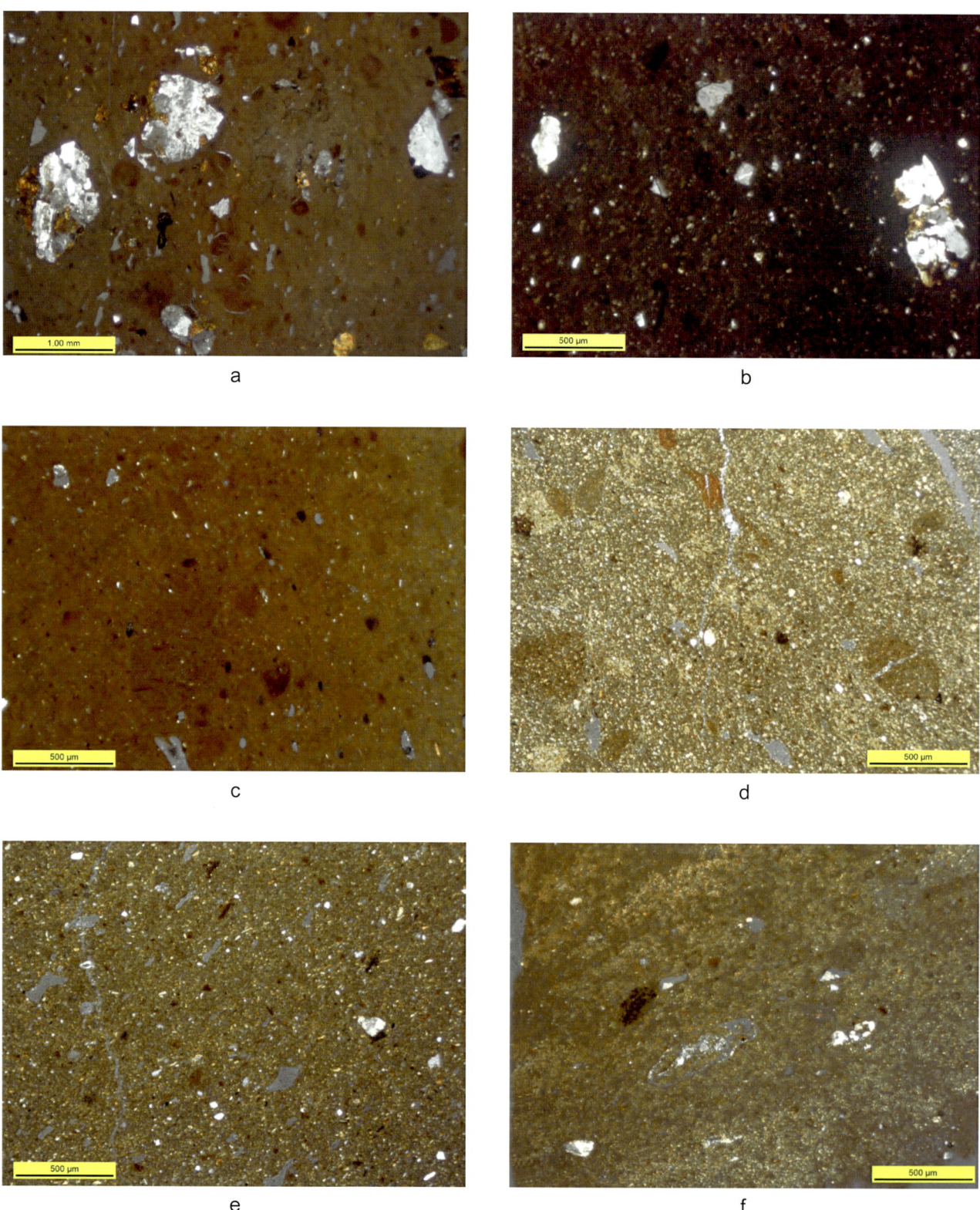

Plate 6. Fabric Group 7 (Mirabello Fabric): granodiorite fabric Group 7A (a), Group 7B (b). Fabric Group 8: very fine fabric with rare nonplastics (c). Fabric 9: fine calcareous with pellets (d). Fabric 10: fine calcareous with biotite mica (e). Fabric 11: fine fabric with rounded sandstones (f). All microphotographs with scale of 1.00 mm were taken under x25 magnification; all microphotographs with scale of 500 μm were taken under x50 magnification.

PLATE 7

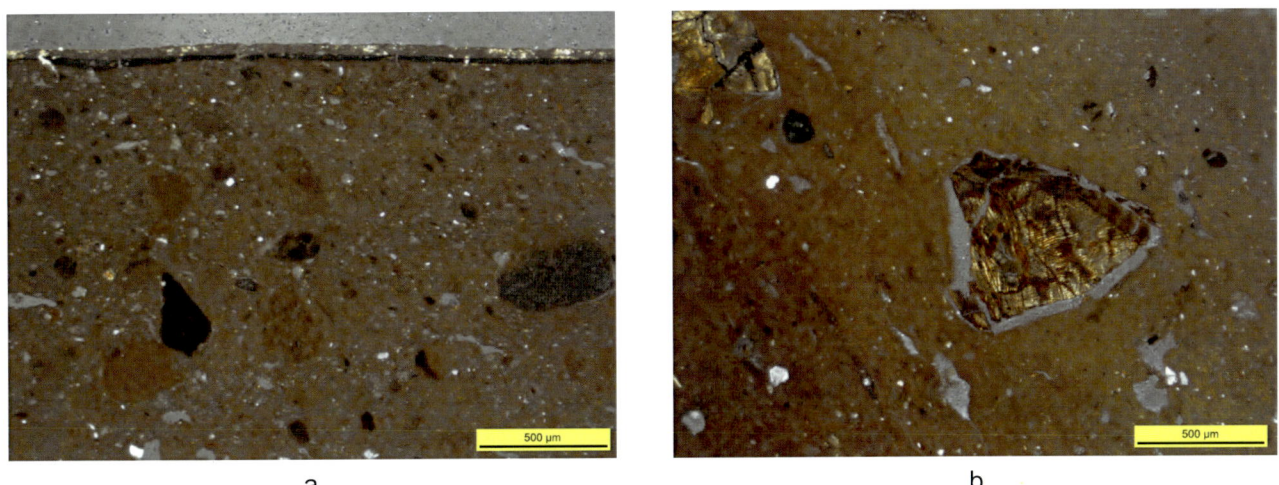

a

b

Plate 7. Fabric 12: very fine fabric with pellets (a); note layer of slip on the upper part of field. Fabric 13: very fine fabric with serpentinite (b). Microphotographs with scale of 500 μm were taken under x50 magnification.